Clean by Light Irradiation
Practical Applications of Supported TiO_2

Clean by Light Irradiation
Practical Applications of Supported TiO$_2$

Vincenzo Augugliaro, Vittorio Loddo, Giovanni Palmisano and Leonardo Palmisano
Dipartimento di Ingegneria Chimica dei Processi e dei Materiali, University of Palermo, Italy

Mario Pagliaro
Istituto per lo Studio dei Materiali Nanostrutturati, CNR, Palermo, Italy

ISBN: 978-1-84755-870-1

A catalogue record for this book is available from the British Library

© V. Augugliaro, V. Loddo, M. Pagliaro, G. Palmisano and L. Palmisano 2010

All rights reserved

Apart from fair dealing for the purposes of research for non-commercial purposes or for private study, criticism or review, as permitted under the Copyright, Designs and Patents Act 1988 and the Copyright and Related Rights Regulations 2003, this publication may not be reproduced, stored or transmitted, in any form or by any means, without the prior permission in writing of The Royal Society of Chemistry or the copyright owner, or in the case of reproduction in accordance with the terms of licences issued by the Copyright Licensing Agency in the UK, or in accordance with the terms of the licences issued by the appropriate Reproduction Rights Organization outside the UK. Enquiries concerning reproduction outside the terms stated here should be sent to The Royal Society of Chemistry at the address printed on this page.

The RSC is not responsible for individual opinions expressed in this work.

Published by The Royal Society of Chemistry,
Thomas Graham House, Science Park, Milton Road,
Cambridge CB4 0WF, UK

Registered Charity Number 207890

For further information see our web site at www.rsc.org

Preface

Chemical pollutants are harmful substances for human beings and for the environment. In water they are in the form of dissolved substances but in air they can be in the form of particulate, liquid droplets or gases. The contaminant species can be classified as primary or secondary pollutants; the first of these are substances emitted directly from a process, whereas the last are formed by reaction with primary pollutants. In the case of air pollution, we must distinguish between outdoor and indoor pollution. Figure P.1 shows the causes and effects of outdoor air pollution.

The levels of air pollution inside houses are often from two to five times higher than outdoor levels. There are many sources of indoor air pollution as household products, including paints, wood preservatives, aerosol sprays, cleansers and disinfectants, stored fuels and automotive products, hobby supplies and dry-cleaned clothing. Figure P.2 shows the air pollution sources in a house. Poor indoor air quality is associated with many health problems, including eye irritation, headaches, allergies and respiratory problems such as asthma.

Conventional methods of water and air decontamination, even if effective, are often chemically, energetically and operationally intensive and suitable only for large systems; moreover, the residuals coming from intensive chemical treatments can add to the problems of contamination. Advanced oxidation processes, developed in recent decades, may be used for degradation of pollutants and also for removing pathogens. These methods rely on the formation of highly reactive chemical species that degrade even the most recalcitrant molecules.

Among the advanced oxidation processes, photocatalysis in the presence of an irradiated semiconductor has proven to be very effective in the field of environment remediation. The use of irradiation to initiate chemical reactions is the principle on which heterogeneous photocatalysis is based; in fact, the irradiation of a semiconductor oxide with light of suitable energy determines the production of electron–hole pairs that eventually generate hydroxyl radicals on the catalyst surface. The radical mechanism of photocatalytic

---

Clean by Light Irradiation: Practical Applications of Supported TiO_2
By Vincenzo Augugliaro, Vittorio Loddo, Mario Pagliaro, Giovanni Palmisano and Leonardo Palmisano
© V. Augugliaro, V. Loddo, M. Pagliaro, G. Palmisano and L. Palmisano 2010
Published by the Royal Society of Chemistry, www.rsc.org

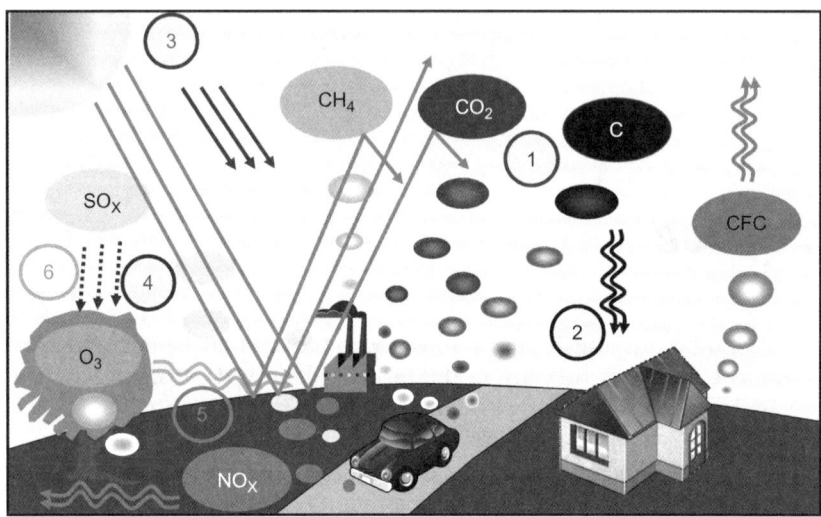

Figure P.1 Causes and effects of outdoor air pollution: (1) greenhouse effect, (2) particulate contamination, (3) increased UV radiation, (4) acid rain, (5) increased levels of nitrogen oxides and (6) increased ozone concentration.

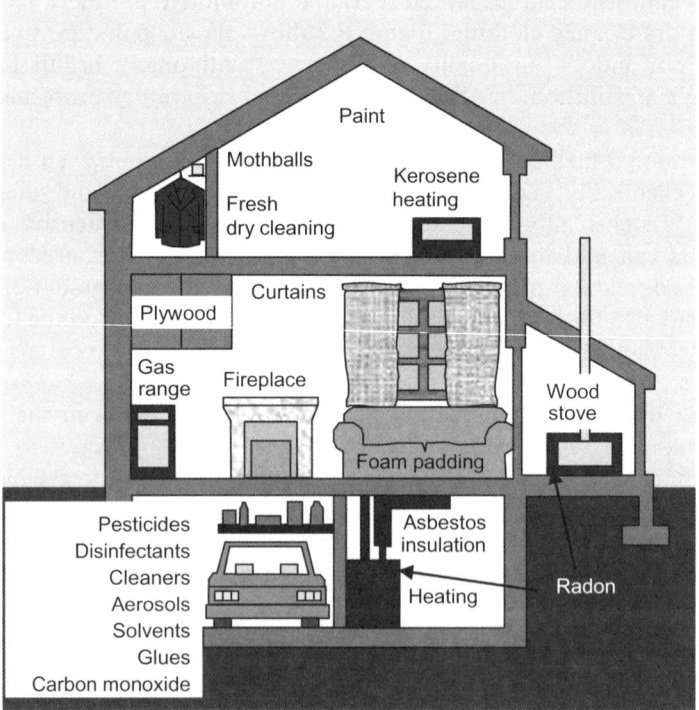

Figure P.2 Air pollution sources in a house.

Preface

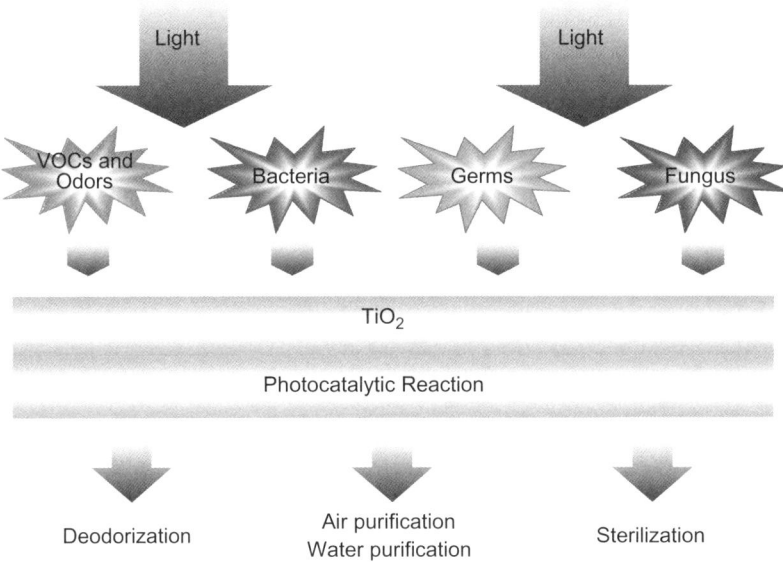

Figure P.3 Scheme of photocatalytic remediation.

reactions, which involve fast attack of strongly oxidant radicals, determines their unselective features. Figure P.3 shows a scheme of photocatalytic remediation.

The main advantage of heterogeneous photocatalysis, when compared with conventional methods, is that in most cases it is possible to obtain complete mineralization of the toxic substrate even in the absence of added reagents. The role of photocatalyst is therefore that of absorbing radiant energy and transforming it into highly reactive species. An ideal photocatalyst should exhibit the following features: (i) high reaction rate under band gap (or higher) irradiation; (ii) photostability; (iii) chemical and biological inactivity; and (iv) ready availability and low cost. Many semiconductors (such as TiO_2, ZnO, ZrO_2, CdS, MoS_2, Fe_2O_3, WO_3, *etc.*), both pure and/or doped, have been examined and used as photocatalysts for the degradation of environmental pollutants in air and water. TiO_2 has been widely used because it is inexpensive, it is harmless and its (photo)stability is very high. The use of different types of non-TiO_2-based materials, prepared by means of sophisticated methods, has been proposed by some authors, but it is hard to believe that such materials will find large-scale environmental applications in the near future, although the possibility of using such materials in a specific niche utilization, for instance in electronics or in very expensive device, cannot be excluded. Despite the great research activity developed in last two decades in search for a photocatalyst with optimal features, titania remains a benchmark against which any alternative photocatalyst must be compared.

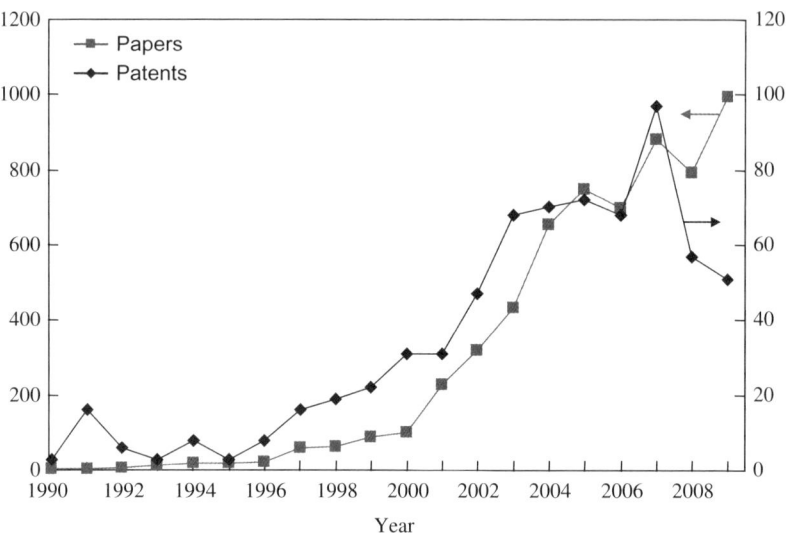

Figure P.4 Number of papers published (■) and patents (♦) related to heterogeneous photocatalysis by TiO$_2$ since 1990. (Source: Scopus.)

Scientific and industrial interest worldwide in heterogeneous photocatalysis by TiO$_2$ has increased greatly in recent decades. A clear indicator of this interest is that the number of papers and patents published in the last two decades has increased exponentially, as it may be noted from Figure P.4, which reports the annual number of publications and patents over the last two decades.

TiO$_2$ samples prepared both as powders and films have been characterized and used for many practical applications. Attention has focused mainly on the anatase and rutile polymorphs, although some papers reporting the photoactivity of the brookite phase can be found in the literature. One serious problem is the difficulty, sometimes, in obtaining pure phases, and the properties attributed to one phase are due in many cases to a mixture of two or three phases.

On these grounds this book is focused only on the applications of pure TiO$_2$ supported on various materials. The most popular techniques used to prepare TiO$_2$-supported materials are also presented together with some specific preparations useful to obtain various TiO$_2$ colloidal dispersions or powders. The presence of metal and non-metal dopants is neglected, with a few exceptions, although the authors are well aware of the importance of the presence of foreign species in shifting light absorption towards visible region or in slowing down the recombination rate of the photoproduced electron–hole pairs. Nevertheless, the use of dopant species is not always beneficial for the photoactivity; products commercially proposed and/or sold do not contain doped-TiO$_2$ but, instead, only pure TiO$_2$, generally in the anatase phase.

To satisfactorily understand the working principles of photocatalytic materials, artifacts and devices presented in this book (which is also aimed at

Preface

relatively inexperienced readers), it is important to have knowledge of some of the fundamentals of heterogeneous photocatalysis. Therefore, Chapter 1 introduces the working principles of the thermodynamic and kinetic of photocatalytic processes; in this chapter some properties and definitions of semiconductors related mainly to TiO_2 are reported, together with the peculiarity of other conductor, insulating and semiconductor materials. Particular attention is devoted to the surface processes that are important for the occurrence of all types of kinetic mechanisms. The latter can be explained according to the Langmuir–Hinshelwood or Eley–Rideal models. Moreover, some studies on the influence of light intensity on photodegradation rate of the molecules under investigation are highlighted.

Most of the practical applications of TiO_2 imply its deposition in various ways on different materials. Despite this, it was considered useful to report in Chapter 2 not only the preparation, characterization and testing of TiO_2 films, mainly on bench scale, but also of powdered crystalline samples with different particle sizes. The obtainment of TiO_2 coated surfaces, in fact, cannot ignore in many cases the preparation methods of the bare TiO_2. Some of the most popular methods reported are the sol-gel, hydrothermal and solvothermal methods. As far as films are concerned, few preparation methods can be used for large-scale film deposition; one method is the magnetron sputtering technique, due to its low-cost and easy controllability.

Some properties of TiO_2 such as its high optical index, refraction and transparency in the visible wavelength range, and superhydrophilicity allow its practical application and are reported in Chapter 3. The most important among them, *i.e.*, the superhydrophilicity exhibited under illumination by suitable TiO_2 thin films supported glasses, for instance, can be invoked to explain why droplets of water spread completely across the surface during the contemporaneous occurrence of a self-cleaning effect. Figure P.5 illustrates these phenomena.

Chapter 4 deals with the improved performances of various types of TiO_2-supported glasses and their practical uses. Buildings covered with self-cleaning glasses now exist.

When used in a blend with a concrete structure, TiO_2 can decompose organic and inorganic fouling species, including not only oil, particulates, soot and grime but also biological organisms and airborne pollutants such as the most common and harmful volatile organic compounds (VOCs). Chapter 5 describes the use of cementitious materials, ceramic tiles and various kinds of pavers, including asphalts TiO_2 treated for outdoor and indoor applications.

The preparation of photocatalytic polymers is useful for a wide range of applications in household appliances, automobile industry, soundproof road barriers and tents for outdoors applications. Several methods are used to prepare these materials. Chapter 6 is concerned with TiO_2 on plastic, textile, metal and paper. Indeed, organic matter deposited on textiles coated with a thin transparent layer of TiO_2 can be decomposed, together with smoke smell and pathogens. Such self-cleaning textiles could be used for sportswear, military uniforms and carpets. The field of photocatalytic paper is pioneering and it

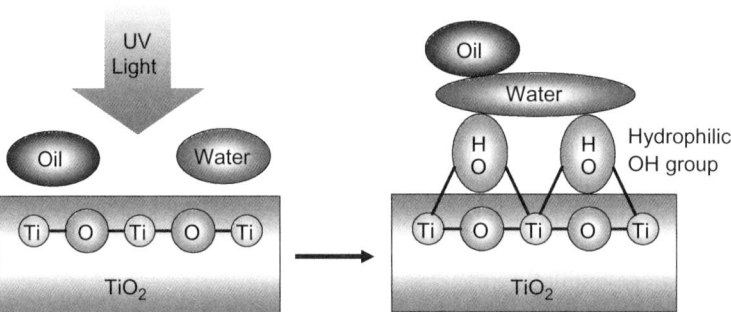

Figure P.5 Superhydrophilic effect.

is worth noting that the presence of TiO_2 also improves the opacity and the whiteness.

Metals such as stainless steel, aluminium alloy, *etc.* can be functionalized by means of TiO_2 to gain photocatalytic activity, surface superhydrophilicity and antibacterial activity.

Heterogeneous photocatalysis is one of the most efficient and cost-effective air purification technologies and scientists estimate that it could be the main method of air purification in the near future. This method has attracted the attention of industry devoted to producing apparatuses for air conditioning, with the main objective of purifying the indoor air of houses, hospitals, buildings, farms, *etc.* Chapter 7 describes not only some devices for air purification but also some field applications of photocatalytic degradation of pollutants in water.

Finally, Chapter 8 deals with the most relevant standardization methods reported in the literature for determining figures of merit in heterogeneous photocatalysis.

Contents

Chapter 1	**Fundamentals**		**1**
	1.1 Working Principles and Thermodynamics of Heterogeneous Photocatalysis		1
		1.1.1 Conductors, Insulators and Semiconductors	1
		1.1.2 Properties of Semiconductor Materials	2
		1.1.3 Photocatalytic Processes	11
	1.2 Kinetics of Photocatalytic Processes		16
		1.2.1 Langmuir Isotherm	23
		1.2.2 Freundlich Isotherm	25
		1.2.3 Redlich–Peterson Isotherm	26
		1.2.4 Light Intensity Dependence	28
	1.3 Radiation Sources		30
		1.3.1 Arc Lamps	31
		1.3.2 Fluorescent Lamps	31
		1.3.3 Incandescent Lamps	32
		1.3.4 Lasers	32
		1.3.5 Light Emitting Diodes (LEDs)	33
		1.3.6 Solar Resources	34
	References		35
Chapter 2	**Powders *versus* Thin Film Preparation**		**41**
	2.1 Introduction		41
	2.2 Structures and some Properties of the various TiO_2 Polymorphs		43
	2.3 Preparation Methods of Powdered TiO_2		43
		2.3.1 Sulfate Process	45
		2.3.2 Chloride Process	45
		2.3.3 Flame Pyrolysis Process	46

Clean by Light Irradiation: Practical Applications of Supported TiO_2
By Vincenzo Augugliaro, Vittorio Loddo, Mario Pagliaro, Giovanni Palmisano and Leonardo Palmisano
© V. Augugliaro, V. Loddo, M. Pagliaro, G. Palmisano and L. Palmisano 2010
Published by the Royal Society of Chemistry, www.rsc.org

		2.3.4	Preparation of Powdered and Nanocrystalline TiO_2 by the Sol-gel Method	46
		2.3.5	Preparation of Nanoparticles, Nanorods, Nanotubes and Nanowires of TiO_2 by the Hydrothermal Method	49
		2.3.6	Solvothermal Method	57
		2.3.7	Sol Method	57
		2.3.8	Laser Pyrolysis Method	58
		2.3.9	Microwave Method	58
	2.4	Preparation of Films		59
		2.4.1	Wet Coating Technologies to Prepare TiO_2 Films	59
		2.4.2	Physical Vapor Deposition (PVD) Techniques	70
		2.4.3	Chemical Vapor Deposition (CVD)	78
		2.4.4	Chemical Bath Deposition	80
		2.4.5	Thermal Oxidation and Anodic Oxidation	81
		2.4.6	Electrophoretic Deposition	83
	References			84

Chapter 3	Unique Properties of Supported TiO_2		98
	3.1	Superhydrophilic *versus* Photocatalytic Character	98
	3.2	Antireflection	102
	3.3	Photo-protection and Anticorrosive Effects	105
	3.4	Bactericidal Properties	107
	References		114

Chapter 4	Photocatalytic Glass		116
	4.1	Improving Glass Performance by Functionalization with TiO_2	116
	4.2	TiO_2 on Glass: More Tasks or Benefits?	119
	4.3	Antireflection and Composite Multilayer Films for Advanced Applications	127
	4.4	Industrial Overview and Commercial Products	131
	References		142

Chapter 5	TiO_2-modified Cement and Ceramics		144
	5.1	Keeping Structures and Air Clean Indoors and Outdoors	144
	5.2	Merging TiO_2 and Cementitious Materials	146
	5.3	Photocatalytic Ceramic Tiles	156
	5.4	New Concepts	163
	References		166

Contents

Chapter 6 TiO$_2$ on Plastic, Textile, Metal and Paper — **168**

    6.1 TiO$_2$ Supported on Plastic Materials — 168
    6.2 Photocatalytic Textiles — 178
    6.3 Photocatalytic Paper — 184
    6.4 TiO$_2$ on Metals — 187
    6.5 Practical Applications — 192
    References — 196

Chapter 7 Devices for Water and Air Purification — **199**

    7.1 Devices for Water Purification — 199
        7.1.1 Pesticide Degradation in a Solar Photoreactor — 199
        7.1.2 Cyanide Degradation in a Pilot Plant Photoreactor — 200
        7.1.3 Photo-CREC-Water Reactors — 207
        7.1.4 UBE Photocatalytic Fiber Reactor — 209
    7.2 Devices for Air Purification — 210
        7.2.1 Photo-CREC-Air Reactor — 213
        7.2.2 AirSteril Purifier — 214
        7.2.3 Airlife Purifier — 214
        7.2.4 Daikin Purifier — 215
        7.2.5 Genesis Air Purifier — 217
        7.2.6 Airwise® Purifiers — 217
        7.2.7 "Luch" Series Cleaners — 220
        7.2.8 Aero Super Element Cleaners — 221
        7.2.9 Zand-Air Cleaners — 222
        7.2.10 Comefresh Electronic Industry Cleaner — 224
        7.2.11 Airpura Purifiers — 226
        7.2.12 Air Oasis™ Purifier — 229
        7.2.13 Air Sterilizer "Medicare" — 231
        7.2.14 Photocatalytic Cold Fluorescent Lamp — 231
    References — 232

Chapter 8 Standardization — **235**

    8.1 Introduction — 235
    8.2 Efficiency Parameters of Photocatalytic Systems — 236
        8.2.1 Quantum Yield — 236
        8.2.2 Experimental Method for the Determination of Absorbed Photons — 238
        8.2.3 Photochemical Thermodynamic Efficiency Factor — 240
        8.2.4 Technological Parameters — 241

	8.3	Experimental Comparison of Photocatalytic Systems		243
		8.3.1	Photocatalytic Films	244
		8.3.2	Cementitious Building Materials	249
		8.3.3	Paving Blocks	255
	References			258

Subject Index   **262**

CHAPTER 1
Fundamentals

1.1 Working Principles and Thermodynamics of Heterogeneous Photocatalysis

1.1.1 Conductors, Insulators and Semiconductors

The valence bond theory is useful in explaining the structure and the geometry of the molecules but it does not provide direct information on bond energies and fails to explain the magnetic properties of certain substances. The molecular orbitals (MO) theory solves these drawbacks. It is based on the assumption that the electrons of a molecule can be represented by wave functions, ψ, called molecular orbitals, characterized by suitable quantic numbers that determine their form and energy. The combination of two atomic orbitals gives rise to two molecular orbitals, indicated as ψ_+ and ψ_-. If an electron occupies the ψ_+ molecular orbital, a stable bond is formed between two nuclei, and so this is called *bonding orbital*. Conversely, when an electron occupies the ψ_- orbital it is called *anti-bonding* orbital and the presence of an electron promotes the dissociation of the molecule. Molecular orbitals of a solid consisting of n equal atoms are obtained by means of a linear combination of atomic orbitals. The number of molecular orbitals formed is equal to that of the atomic ones. By increasing the number of atoms the difference between the energetic levels decreases and a continuous band of energy is formed for high values of n.

The width of the various bands and the separation among them depends on the internuclear equilibrium distance between adjacent atoms. If the energetic levels of isolated atoms are not so different, progressive enlargement of the bands may lead to their overlapping by decreasing the internuclear distance. The most external energetic band full of electrons is called the *valence band* (*VB*).

The energy band model for electrons can be applied to all crystalline solids and allows one to establish if a substance is a conductive or an insulating material. Indeed, the properties of a solid are determined by the difference in energy between the different bands and the distribution of the electrons contained within each band.

Clean by Light Irradiation: Practical Applications of Supported TiO_2
By Vincenzo Augugliaro, Vittorio Loddo, Mario Pagliaro, Giovanni Palmisano and Leonardo Palmisano
© V. Augugliaro, V. Loddo, M. Pagliaro, G. Palmisano and L. Palmisano 2010
Published by the Royal Society of Chemistry, www.rsc.org

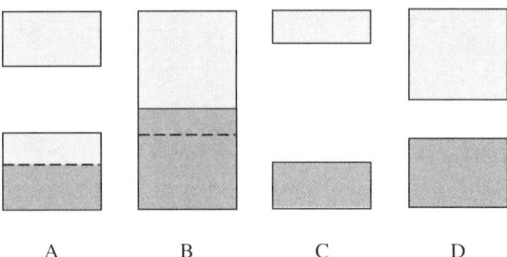

Figure 1.1 Schematic of the energetic bands for conductors, (a) and (b); insulators (c); and semiconductors (d).

If the valence band is partially filled or it is full and overlapped with a band of higher energy, electrons can move, allowing conduction (*conductors*), as in the case of metals that have relatively few valence electrons that occupy the lowest levels of the most external band.

In contrast, the valence band is completely filled in ionic or covalent solids but it is separated by a high energy gap from the subsequent empty band. In this situation no electrons can move even if high electric fields are applied and the solid is an *insulator*. If the forbidden energy gap is not very high, some electrons could pass into the energetic empty band by means of thermal excitation and the material behaves as a weak conductor, *i.e.*, as a *semiconductor*. The empty band, which allows the movement of the electrons, is called the *conduction band* (*CB*).

The energy difference between the lowest conduction band edge and the highest valence band edge is called the band gap (E_G). A material is generally considered a semiconductor when $E_G \leq 3$ eV, whereas it is considered a wide band gap semiconductor when its band gap value ranges between 3 and 4 eV.

Figure 1.1 shows the position of the energy bands of different types of materials.

1.1.2 Properties of Semiconductor Materials

The de Broglie relation[1] associates a wavelength with the electron as:

$$\lambda = \frac{h}{p} = \frac{h}{m \cdot v} \quad (1.1)$$

where p is the momentum and h is the Planck constant. It shows that the wavelength is inversely proportional to the momentum of a particle and that the frequency is directly proportional to the particle's kinetic energy:

$$f = \frac{E}{h} \quad (1.2)$$

Fundamentals

The wave number corresponds to the number of repeating units of a propagating wave per unit of space. It is defined as:

$$\bar{v} = \frac{1}{\lambda} \quad (1.3)$$

In the case of non-dispersive waves the wave number is proportional to the frequency, f:

$$\bar{v} = \frac{f}{v} \quad (1.4)$$

where v is the propagation velocity of the wave. For electromagnetic waves propagating in vacuum the following relation is obtained:

$$\bar{v} = \frac{f}{c} \quad (1.5)$$

where c is the velocity of light.

The wave vector is a vector related to a wave and its amplitude is equal to the wave number while its direction is that of the propagation of the wave:

$$\bar{k} = \frac{2\pi}{\lambda} \quad (1.6)$$

Therefore, the electron momentum can be expressed as:

$$\bar{p} = \frac{\hbar}{\lambda} = \hbar \bar{k} \quad (1.7)$$

where $\hbar$ is the reduced Plank constant also known as the Dirac's constant ($\hbar = h/2\pi$).

The electron energy is therefore:

$$E = \frac{p^2}{2m} = \frac{\hbar^2 \bar{k}^2}{2m} \quad (1.8)$$

where the coefficient m is the inertial mass of the wave-particle. As m varies with the wave vector, it is called effective mass, m^*, defined as:

$$m^* = \frac{\hbar^2}{\partial^2 E/\partial k^2} \quad (1.9)$$

A semiconductor is called a *direct band gap semiconductor* if the energy of the top of the valence band lies below the minimum energy of the conduction band without a change in momentum, whereas it is called an *indirect band gap*

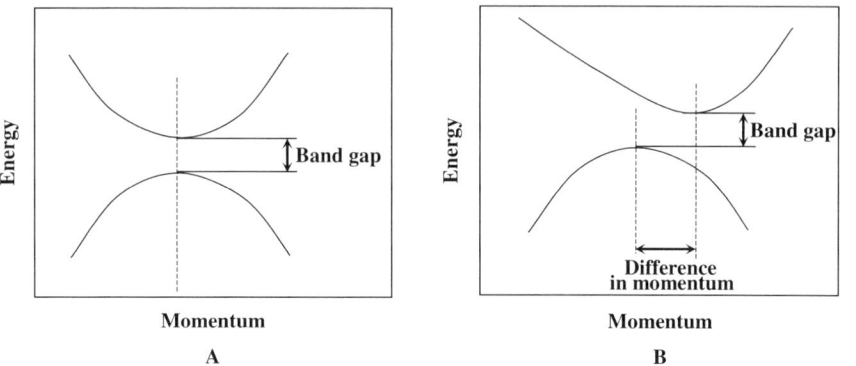

Figure 1.2 Energy *versus* momentum for (a) a direct band gap semiconductor and (b) an indirect band gap semiconductor.

semiconductor if the minimum energy in the conduction band is shifted by a difference in momentum (Figure 1.2).

The probability $f(E)$ that an energetic level of a solid is occupied by electrons can be determined by the *Fermi–Dirac distribution function*.[2] It applies to fermions (particles with half-integer spin, including electrons, photons and neutrons, which must obey the Pauli exclusion principle) and states that a given allowed level of energy E is function of temperature and of the Fermi level, E_F^0, according to the equation:

$$f(E) = \frac{1}{1 + \exp\left(\frac{E - E_F^0}{k_B T}\right)} \tag{1.10}$$

where k_B is the Boltzmann constant. The level E_F^0 represents the probability of 50% of finding an electron in it. For intrinsic semiconductors and for insulating materials, E_F^0 falls inside the energetic gap and its value depends on the effective mass of electrons present at the end of the conduction band (m_e^*), on the effective mass of electrons at the beginning of the valence band (m_h^*), and on the amplitude of the band gap (E_G) according to the following equation:

$$E_F^0 = \tfrac{1}{2}E_G + \tfrac{3}{4}kT \ln \frac{m_h^*}{m_e^*} \tag{1.11}$$

The value of E_F^0 is equivalent to the electrochemical potential of the electron, *i.e.*, it can be considered as the work necessary to transport an electron from an infinite distance to the semiconductor.

Figure 1.3 shows that the Fermi–Dirac distribution is a step function at $T = 0$ K. It has the value of 1 and 0 for energies below and above the Fermi level, respectively. As the temperature increases above $T = 0$ K, the distribution of electrons in a material changes. At $T > 0$ K the probability that energy levels

Fundamentals

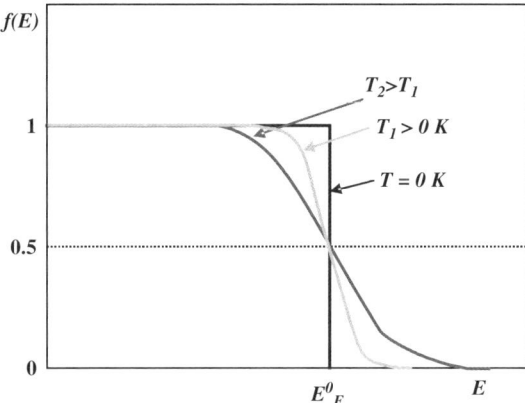

Figure 1.3 Fermi–Dirac distribution for several temperatures.

above E_F^0 are occupied, and similarly energy levels below E_F^0 are empty, is not zero. Moreover, the probability that an energy level above E_F^0 is occupied increases with temperature (distribution sigmoidal in shape) because some electrons begin to be thermally excited to energy levels above the chemical potential, $\mu(E_F^0)$. To understand the meaning of the chemical potential some considerations are presented here.

In determining the surface composition it is important to consider the thermodynamics of surfaces. The surface contribution to the free energy, G, of a solid is always positive. To create a new surface, it is necessary to perform a work on the system, which is employed to break bonds between the surface atoms. The work required to increase the surface area by dA, by means of a reversible path, at constant temperature and pressure, depends on the surface tension γ according to the following relation:

$$dW_{P,T} = \gamma \cdot dA \tag{1.12}$$

For a single-component system, it is possible to express γ as a function of the surface free energy per unit area, G^S, as:

$$\gamma = G^S + A \left(\frac{\partial G^S}{\partial A} \right)_{P,T} \tag{1.13}$$

where G^S represents the work (per unit surface) done to form a new surface dA and the second term of the right-hand side is the work done in stretching a pre-existent surface to increase it by dA. In this last case the number of surface atoms is fixed, whereas the state of strain of the surface changes. If the surface is unstrained, the second term of the right-hand side of Equation (1.13) is zero and γ is equal to the surface free energy:

$$\gamma = G^S = \left(\frac{\partial G}{\partial A} \right)_{P,T} \tag{1.14}$$

For a multicomponent system the Gibbs equation can be written as:

$$d\gamma = -S^S dT - \sum_i \Gamma_i d\mu_i \qquad (1.15)$$

were S^S is the surface entropy, μ_i the chemical potential of species i, and Γ_i is the number of moles of component i per unit surface (n_i^S/A) in excess with respect to those that would be present if the system was homogeneous from bulk up to the surface.

The chemical potential can be expressed as:

$$\mu_i = \mu_i^0 + RT \ln x_i \qquad (1.16)$$

where μ_i^0 is the standard chemical potential of the pure component i.

By substituting Equation (1.16) in Equation (1.15), at constant temperature, the following relation is obtained:

$$\Gamma_i = -\frac{1}{RT}\left(\frac{\partial \gamma}{\partial \ln x_i}\right)_{P,T,\mu_{j \neq i}} \qquad (1.17)$$

From Equation (1.17) it is possible to determine the excess concentration of a component by taking the derivative of the surface tension with respect to its concentration in the bulk.

Equation (1.17) states that if the surface tension decreases by increasing the bulk concentration of a component, the component will have a greater surface concentration ($\Gamma_i > 0$). Moreover, the component with the lowest surface tension will form a surface layer (if the temperature is high enough to allow its diffusion).

Owing to the difficulty in measuring the surface tension of the solids, the above equations allow us to determine the surface composition only qualitatively; to obtain more precise data it is necessary to apply statistic thermodynamics and experimental techniques.

Some types of impurities and imperfections may drastically affect the electric properties of a semiconductor. In fact the conductivity of a semiconductor can be significantly increased by adding foreign atoms in the lattice (*doping*) that make electrons available in the conduction band and holes available in the valence band. For example, silicon has a crystal structure similar to that of diamond (Figure 1.4) and each silicon atom forms four covalent bonds with four nearest atoms, corresponding to a chemical valence equal to four. The addition of atoms, for instance arsenic, phosphorous or antimony, having one valence electron more will lead to an excess of positive charge (Figure 1.5a), due to the transfer of an electron from the foreign atom to the conduction band (*donor doping*).

If the foreign atom, for instance boron, gallium, indium, has one valence electron less it can accept one electron from the valence band (*acceptor doping*) (Figure 1.5b). In the first case (Figure 1.5a) an energetic level close to the

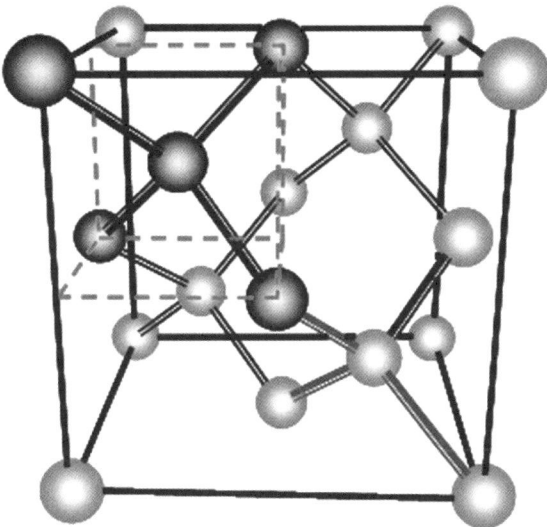

Figure 1.4 Crystal structure of silicon.

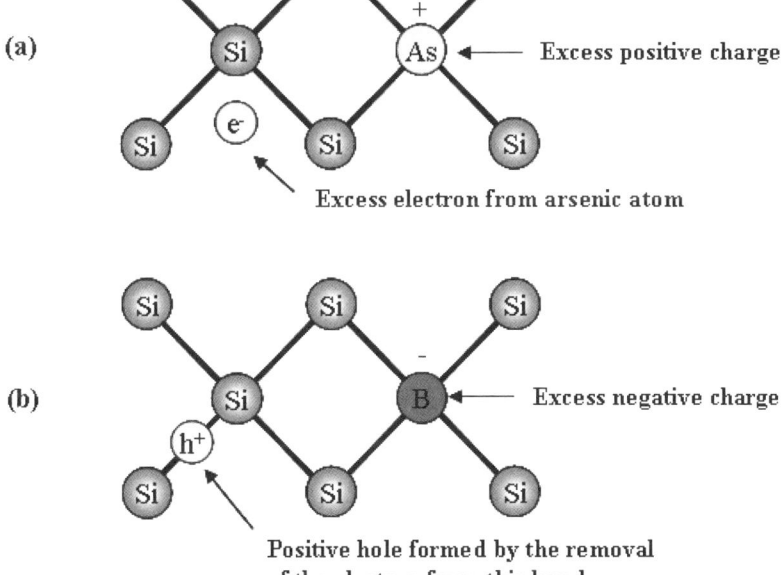

Figure 1.5 (a) With an arsenic impurity, silicon has an electron available for conduction; (b) with a boron impurity, silicon has a positive hole available.

conduction band is introduced; consequently, electrons can pass more easily in it. In this case the solid is called an "n" type semiconductor and the Fermi level will be close to the conduction band (Figure 1.6b). In the second case an energetic level close to the valence band is formed, in which electrons can be promoted with the formation of holes. Here, the semiconductor is of "p" type and its Fermi level will be close to the valence band (Figure 1.6c).

The notion of energetic levels of electrons in solids can be extended to the case of an electrolytic solution containing a redox system.[3] The occupied electronic levels correspond to the energetic states of the reduced species whereas the unoccupied ones correspond to the energetic states of the oxidized species. The Fermi level of the redox couple, $E_{F,redox}$, corresponds to the electrochemical potential of electrons in the redox system and it is equivalent to the reduction potential, V_0. To correlate the energetic levels of a semiconductor to those of a redox couple in an electrolyte, two different scales can be used. The first is expressed in eV, and the other in V (Figure 1.7a). The two scales differ because in solid-state physics zero is the level of an electron in a vacuum, whereas in electrochemistry the reference is the potential of the normal hydrogen electrode

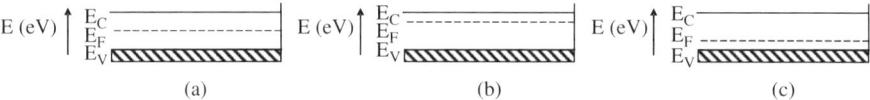

Figure 1.6 Energetic levels of a semiconductor: (a) intrinsic semiconductor; (b) "n" type semiconductor; (c) "p" type semiconductor.

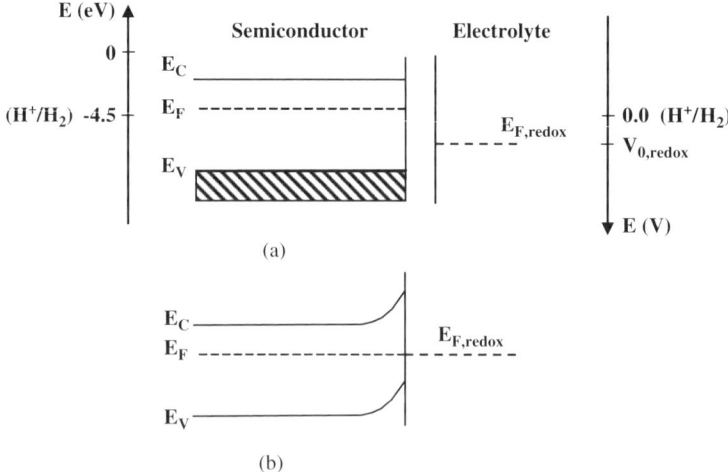

Figure 1.7 Formation of a junction between an "n" type semiconductor and a solution: (a) before contact and (b) at equilibrium.

Fundamentals

(NHE). The two scales can be correlated using the potential of NHE, which is equal to $-4.5\,\text{eV}$ when it is referred to that of the electron in vacuum.[4]

If a semiconductor is placed in contact with a solution containing a redox couple, equilibrium is reached when the Fermi levels of both phases become equal. This occurs by means of an electron exchange from solid and electrolyte, which leads to the generation of a charge inside the semiconductor. This charge is distributed in a spatial charge region near the surface, in which the values of holes and electrons concentrations also differ considerably from those inside the semiconductor. Figure 1.7(a) shows schematically the energetic levels of a "n" type semiconductor and a redox electrolyte before contact. In particular, as the energy of the Fermi level is higher than that of the electrolyte, equilibrium is reached by electron transfer from the semiconductor to the solution. The electric field produced by this transfer is represented by the upward band bending (Figure 1.7b). Owing to the presence of the field, holes in excess generated in the spatial charge region move toward the semiconductor surface, whereas electrons in excess migrate from the surface to the bulk of the solid. Figure 1.8 shows the contact between a redox electrolyte and a "p" type semiconductor. In this case transfer of electrons occurs from the electrolyte to the semiconductor and the band bending is downward.

If the potential of the electrode changes due to an anodic or cathodic polarization, a shift of the Fermi level of the semiconductor with respect to that of the solution occurs with an opposite curvature of the bands (Figure 1.9).

For a particular value of the electrode potential, the charge excess disappears and the bands become flat from the bulk to the surface of the semiconductor. The corresponding potential is called the *flat band potential*, V_{FB}, and the determination of this potential allows us to calculate the values of the energy of the conduction and the valence bands.[5]

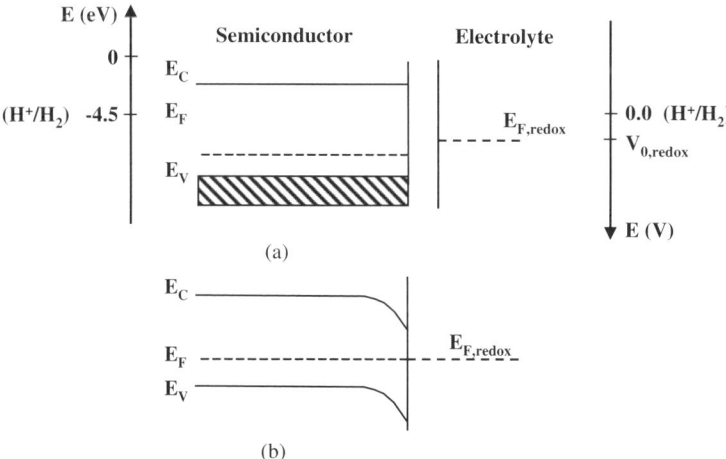

Figure 1.8 Formation of a junction between a "p" type semiconductor and a solution: (a) before contact and (b) at equilibrium.

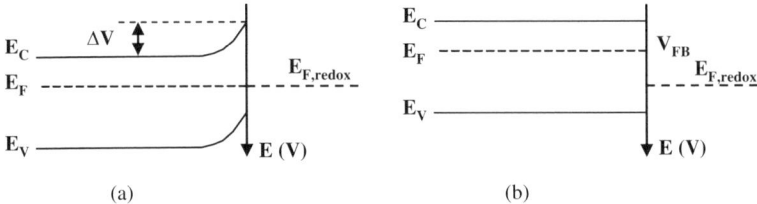

Figure 1.9 Scheme of the energetic levels at the interface semiconductor–electrolyte for an "n" type semiconductor: (a) at equilibrium and (b) flat band potential.

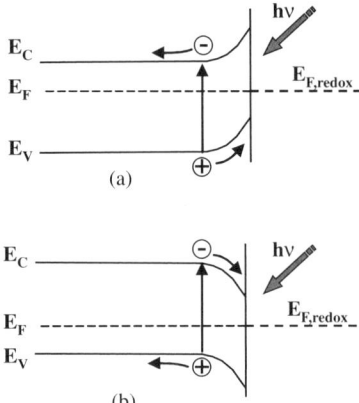

Figure 1.10 Generation of electron–hole pair due to irradiation of (a) an "n" type semiconductor and (b) a "p" type semiconductor.

When a semiconductor is irradiated by radiation of suitable energy – equal to or higher than that of the band gap, E_G – electrons can be promoted from the valence band to the conduction band. Figure 1.10 shows the scheme of electron–hole pair formation due to the absorption of a photon by the semiconductor.

The existence of an electric field in the spatial charge region allows the separation of the photogenerated pairs. In the case of "n" type semiconductors, electrons migrate toward the bulk whereas holes move to the surface (Figure 1.10a). In the case of "p" semiconductors, holes move toward the bulk whereas electrons move to the surface (Figure 1.10b).

Photoproduced holes and electrons, during their migration in opposite directions, can (i) recombine and dissipate their energy as either electromagnetic radiation (photon emission) or more simply as heat, or (ii) react with electron-acceptor or electron-donor species present at the semiconductor–electrolyte interface, thereby reducing or oxidizing them, respectively.

The energy of the conduction band edge, E_C, corresponds to the potential of the photogenerated electrons, whereas the energy of the valence band edge, E_V,

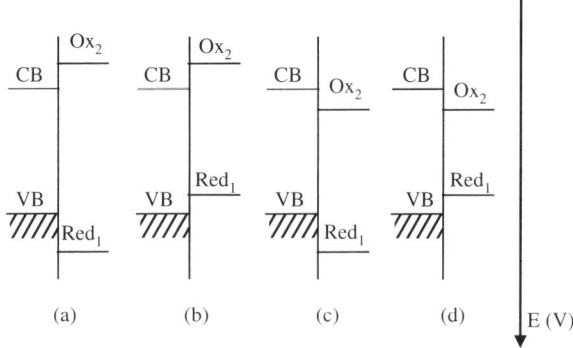

Figure 1.11 Relative positions of the valence and conduction bands and the energies of two redox couples. Only for case (d) are both reduction and oxidation reactions thermodynamically allowed. Ox_2 and Red_1 represent the oxidized and reduced species, respectively, of two different redox couples.

corresponds to the potential of the holes. If E_C is more negative than the potential of a species present in solution, electrons reaching the interface can reduce the oxidized form of the redox couple. Conversely, if the potential of E_V is more positive than that of the redox couple, photoproduced holes can oxidize its reduced form (Figure 1.11). Knowledge of the relative edge positions of the bands and of the energetic levels of the redox couples is essential to establish if thermodynamics allow the occurrence of oxidation and/or reduction of the species in solution.

Figure 1.12 reports the band gaps and the positions of the valence band and conduction band edges for various semiconductors.

1.1.3 Photocatalytic Processes

The photocatalytic properties of a semiconductor depend on the position of the energetic levels, on the mobility and mean lifetime of the photogenerated electrons and holes, on the light absorption coefficient and on the nature of the interface. Moreover, the photoactivity depends on the methods of preparation of the powders, which allows us to vary many physicochemical properties of the semiconductor by controlling the crystalline structure, the surface area and the particle size distribution.

In a photocatalytic system the behavior of each single particle of semiconductor is similar to that of a photoelectrochemical cell constituted by a semiconductor electrode in contact with an electrode of an inert metal.[6] In a photoelectrochemical cell an oxidation or reduction reaction may occur on the semiconductor electrode, whereas in a semiconductor particle immersed in an electrolyte solution both reactions occur simultaneously by hole transfer from the valence band and by electron transfer from the conduction band (Figure 1.13).

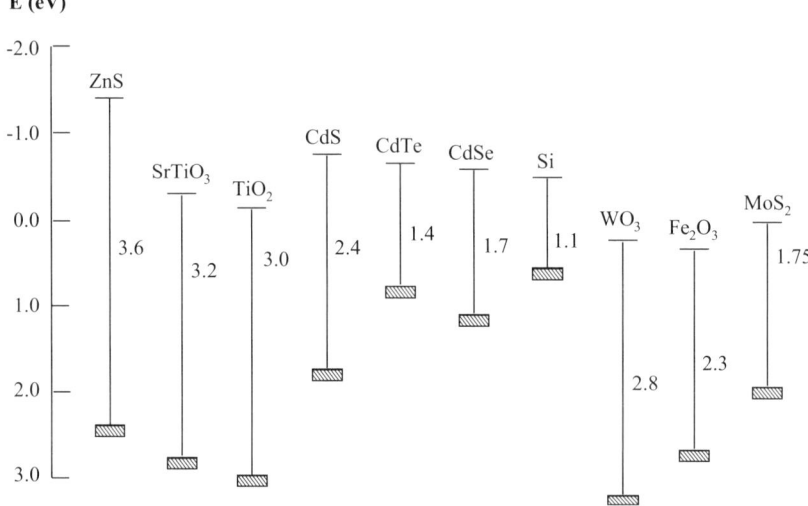

Figure 1.12 Positions of the band edges for some semiconductors in contact with aqueous electrolyte at pH 0.

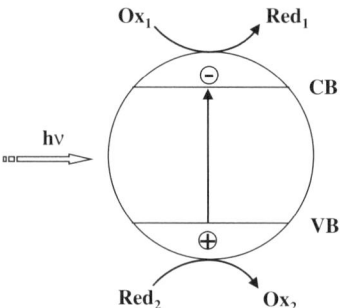

Figure 1.13 Scheme of the photocatalytic process occurring on an illuminated particle in contact with a redox system. The oxidizing agent Ox_1 is oxygen and the reducing agent is an organic substrate.

An advantage in using semiconductor powder suspensions lies in the fact that each particle acts as a small photocell, and in 100 mg of powder, consisting for example of 0.1 mm diameter particles, more than 10^{11} independent particles are present.

Generally, the powders have large surface areas (1–100 m^2 g^{-1}), which favors the occurrence of charge transfer reactions at the interface between the semiconductor surface and the redox electrolyte.

Separation of the electron–hole pairs increases with increasing thickness of the spatial charge region, which depends on the doping of the semiconductor. When the volume of the particles decreases, the effect on charge separation

Fundamentals

becomes a minimum as, sometimes, the particle size is smaller than the thickness of the spatial charge region. The absorption of radiation of suitable wavelength by the semiconductor allows the transformation of light into chemical energy; this phenomenon represents a fundamental step in heterogeneous photocatalysis. In particular, when aqueous suspensions of semiconductor powders are irradiated, at the solid–liquid interface a great variety of photo-induced chemical reactions, able to degrade many organic and inorganic molecules, can occur by means of formation of very reactive radical species that are generated in the presence of O_2 and H_2O. The following scheme shows the events that can occur at the semiconductor–water interface when TiO_2 is used as the photocatalyst:

$$TiO_2 + h\nu \rightarrow TiO_2 \left(e^-_{(CB)} + h^+_{(VB)} \right) \tag{1.18}$$

$$OH^- + h^+_{(VB)} \rightarrow {}^\bullet OH \tag{1.19}$$

$$O_2 + e^-_{(CB)} \rightarrow {}^\bullet O_2^- \tag{1.20}$$

$${}^\bullet O_2^- + H^+ \rightarrow {}^\bullet HO_2 \tag{1.21}$$

$$2 {}^\bullet HO_2 \rightarrow O_2 + H_2O_2 \tag{1.22}$$

$$H_2O_2 + {}^\bullet O_2^- \rightarrow OH^- + {}^\bullet OH + O_2 \tag{1.23}$$

The photocatalytic method allows, moreover, the elimination of many inorganic ionic pollutants present in water by either reducing them to their elemental form on the surface of the catalyst particle:

$$M^{n+}_{(aq)} + ne^-_{(CB)} \rightarrow M_{(surf)} \tag{1.24}$$

or transforming them into less noxious species. For example, the reduction of Cr(vi) to Cr(iii) allows the elimination of chromium by means of its subsequent precipitation as hydroxide.

The deposition of small amounts of noble metals on the particle surface of a semiconductor can be used to increase its photoactivity. Indeed, the photo-oxidation rate of organic compounds is generally limited by the rate of transfer of electrons to oxygen adsorbed on the semiconductor surface. The more electrons there are available to reduce oxygen, the higher the reaction rate.

It is possible to explain the effect of the presence of a noble metal by comparing a photoelectrochemical cell and a process occurring by irradiating an "n" type semiconductor particle partially covered with "islands" of Pt (Figure 1.14).

The photocatalytic activity of a semiconductor powder is based on oxidation and reduction processes that occur continuously on different zones of the same particle. In the presence of metals such as Au, Ag, Pd and Pt physically separated reaction sites for the photogenerated pairs exist, similarly to a cell consisting of a "n" type semiconductor and a platinum counter-electrode, in

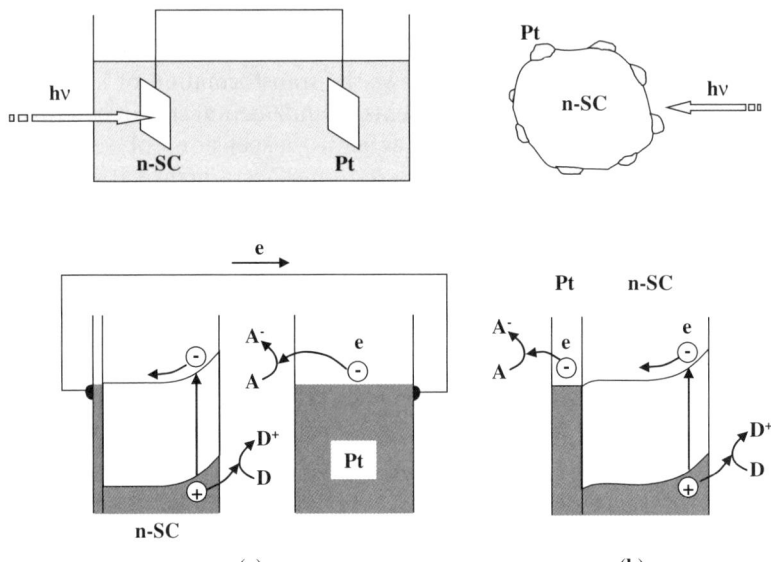

Figure 1.14 Photocatalytic processes in a photoelectrochemical cell (a) and on an irradiated "n" type semiconductor particle covered with "islands" of platinum (b).

which the reduction of oxidized species occurs on the metallic electrode whereas oxidation of the reduced species occurs on the semiconductor.

In the modified particles the photogenerated electrons are transferred onto the metallic islands whereas the holes remain on the semiconductor; this results in an acceleration of the process kinetics, due to the decrease in electron–hole recombination rate.

The same beneficial effect can be obtained by partially covering the particle surface with oxides such as RuO_2 or NiO.[7]

Another way to realize efficient electron–hole pair separation is to use mixtures of two different semiconductor powders, such as CdS and TiO_2[8] or WO_3 and WS_2.[9]

The coupling of two semiconductors with different energies for the valence and conduction bands allows the vectorial transfer of electrons and holes from one semiconductor to the other, thereby decreasing the probability of their recombination and increasing the efficiency of the charge-transfer process at the semiconductor–electrolyte interface.

Figure 1.15 sketches the transfer of electrons from the more cathodic conduction band of a semiconductor to that of another one, and the contemporaneous transfer of holes from the more anodic valence band of the second semiconductor to the valence band of the first one:

$$SC_1(e^- + h^+) + SC_2(\underline{e}^- + \underline{h}^+) \rightarrow SC_1(h^+ + \underline{h}^+) + SC_2(e^- + \underline{e}^-) \quad (1.25)$$

Fundamentals

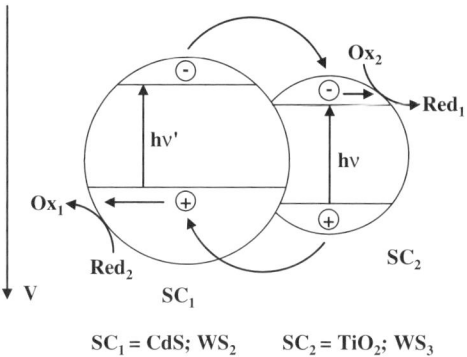

Figure 1.15 Scheme of the charge-transfer processes that occur on two coupled semiconductors.

In the case of CdS and TiO_2 holes in excess in CdS can oxidize the organic substrate, whereas electrons in excess in TiO_2 can react with the adsorbed molecular oxygen.[8]

Depending on the type of photocatalyst used, different reactions can take place with the substrate to be degraded. Clearly, the employed photocatalyst must satisfy some requirements that allow a practical convenient use. First, its absorption spectrum must include radiations in the near-UV region and possibly also in a part of the visible region. Second, it must be a common, (photo)stable and not costly compound that can be reused. Finally, the energy associated with the valence band, E_V, must allow the formation of species able to oxidize most organic molecules and the recombination rate of electrons and holes should be relatively low so that the photogenerated charges can migrate to its surface and give rise to redox processes with appreciable rate.

The semiconductor most frequently used is, undoubtedly, titanium dioxide, which is produced in large amounts as a low cost pigment. The photocatalytic activity of anatase, rutile and brookite polymorphic modifications of TiO_2 is affected by several factors such as the crystalline structure, the surface area, the particle size distribution and the density of surface hydroxyl groups. Although the positions of valence and conduction bands of both anatase and rutile are positive enough to allow the oxidation of many organic molecules, the anatase phase shows a higher activity as the level of its conduction band is more favorable for the electron transfer needed for the complementary reduction reaction. The poor efficiency of rutile is due mainly to the high recombination rate of electron–hole pairs and to its low ability to photoadsorb oxygen.[10] It can be taken into account, however, that the most common TiO_2 photocatalysts, e.g., Degussa P25 (ca. 80% anatase, 20% rutile), contain a mixture of both crystalline modifications.

The high band gap for anatase ($E_G = 3.2$ eV) and rutile ($E_G = 3.0$ eV) phases allows us to utilize only radiations with a wavelength lower than ca. 400 nm, which represents ca. 5% of solar light. A method to increase light absorption in the visible region is to dope TiO_2 by the addition of transition metal ions.

The presence of these dopants introduces energetic levels in the band gap of TiO_2, thus allowing the absorption of visible radiation. This effect is particularly important for powders containing chromium species, which shift light absorption towards the red end of the spectrum.[11–14]

The photocatalytic activity of doped powders is generally lower than that of bare TiO_2, with the exception of samples containing W or Mo in concentrations lower than 1%. The negative effect is due to an increase in the recombination rate of electron–hole pairs and/or a decrease of the free diffusion mean path length of the charge carriers. The higher photoactivity found in the presence of W and Mo can be attributed to a lower recombination rate due to redox processes leading to trapping of photoproduced electrons through the formation of W(v) or Mo(v) species, from which, successively, the electrons are transferred to oxygen molecules.[15,16]

1.2 Kinetics of Photocatalytic Processes

Heterogeneous photocatalytic reactions should be studied by following the same methodology used for heterogeneous catalytic reactions and similar problems must be overcome.

In catalysis the reaction is accelerated by the presence of small amounts of a particular substance called a catalyst. One of the fundamental principles of catalysis is that of catalytic cycle based on the definition of catalysis by Boudart: "A catalyst is a substance that transforms reactants into products, through an uninterrupted and repeated cycle of elementary steps in which the catalyst is changed through a sequence of reactive intermediates, until the last step in the cycle regenerates the catalyst in its original form." Heterogeneous catalysis involves systems in which reactants and catalyst are physically separated phases.

Photocatalysis is a "change in the rate of a chemical reaction or its generation under the action of ultraviolet, visible or infrared radiation in the presence of a substance – the photocatalyst – that absorbs light and is involved in the chemical transformations of the reaction partners." Then a photocatalyst is a "substance able to produce, by absorption of ultraviolet, visible, or infrared radiation, chemical transformations of the reaction partners, repeatedly coming with them into intermediate chemical interactions and regenerating its chemical composition after each cycle of such interactions."

The definition of heterogeneous catalysis implies that at least one reactant remains attached for a significant time on the surface of a solid catalyst.

When a fresh clean surface is exposed to a fluid, the molecules of the fluid or those dissolved in it will concentrate preferentially on the surface: this phenomenon is called "adsorption."[17] The extent of the interactions of these molecules (adsorbate) with the solid (adsorbent) may vary widely for the various systems and it is possible to distinguish two main types of adsorption: physical adsorption and chemisorption. Physical adsorption is due to forces of molecular (van der Waals) interactions such as for instance attractions between

Fundamentals

permanent or induced dipole and quadrupole. Therefore, physical adsorption can be considered as the condensation of a vapor to form a liquid or the liquefaction of a gas on the solid surface.

Chemisorption is due to a rearrangement of electrons of both the fluid and the surface with the consequent formation of chemical bonds. It is possible to distinguish between the two types of adsorption by measuring the "heat of adsorption." The heat exchanged per mole of gas adsorbed generally ranges between 2 and 6 kcal for physical adsorption (but values as large as 20 kcal have been reported) whereas chemisorption can involve a few hundred kcal per mole (rarely are values below 20 kcal found). The amount of fluid adsorbed and the adsorption–desorption equilibrium depends on the nature of the adsorbent and of the adsorbate and also on the pressure and the temperature. The specific surface area of the solid adsorbent, of course, is a parameter of paramount importance. The most convenient method for defining an adsorption equilibrium is by analysis of the "adsorption isotherms." An adsorption isotherm shows the influence of the equilibrium pressure upon the amount of adsorbed gas at constant temperature. Isotherms derived both theoretically and experimentally can be represented by simple equations that correlate directly the concentration of the adsorbed species to the pressure. A selection of adsorption isotherms generally used in various kinetic interpretations of surface reactions is reported here:

$$\frac{V}{V_m} = \theta = \frac{bP}{1+bP} \quad \text{Langmuir} \tag{1.26}$$

$$V = k \cdot P^{1/n}, \; n > 1 \quad \text{Freundlich} \tag{1.27}$$

$$\frac{V}{V_m} = \theta = \frac{1}{a} \ln c_0 P \quad \text{Temkin} \tag{1.28}$$

where V is the volume adsorbed, V_m the volume required to form a monolayer on the adsorbent surface, θ is the fraction of monolayer covered at the equilibrium pressure P, and a, b, k and c_0 are constants.

Photocatalysis on wide band gap solids ($E_G \geq 3\,\text{eV}$) is related to photophysical processes that transform under irradiation the surface of the solid, on which "photoadsorption and/or photodesorption" phenomena can occur. Photocatalysis taking place at the boundaries between two phases can be expressed symbolically as:

$$A \xrightarrow[\text{Cat}]{h\nu} B \tag{1.29}$$

were A and B are the reactant and product, respectively, and Cat is the solid photocatalyst.

Photoadsorption is defined as "adsorption, typically chemisorption, initiated by ultraviolet, visible, or infrared radiation either by the adsorbate or by the adsorbent." It can be represented as:

$$A \xrightarrow[\text{Cat}]{h\nu} A_{ads} \tag{1.30}$$

were A is the adsorbate molecule and A_{ads} represents the photoadsorbed species. Photoadsorption can be seen as the first step of "photoexcitation" of a heterogeneous system, *i.e.*, the act by which an excited state is produced in a molecule or solid catalyst by absorption of UV, Vis or infrared radiation.

For gas–solid and liquid–solid systems the interaction of a species with the solid surface depends on the chemical nature of the species and on the chemical and physical nature of the solid. For non-illuminated surfaces of semiconductor oxides a thermodynamic equilibrium between a species and the solid is established only when the electrochemical potential of the electrons in the entire system is uniform. When the adsorption–desorption equilibrium is established, an aliquot of the species is located in an adsorbed layer, held at the surface by either weak or strong bonding forces.

When a photocatalyst goes from dark conditions to irradiated ones, the radiation absorption causes its surface to undergo the series of changes needed to eventually allow the occurrence of photoprocesses.[18,19] Essentially, the interaction of a photon with a solid semiconductor gives rise to an increase of the vibrational state of the lattice or the number of quasi-free charge carriers, *i.e.*, it generates an excited state of the solid. Illumination with band gap or greater energy creates a perturbation to this adsorption–desorption equilibrium established in the dark; under irradiation, the previous equilibrium is displaced, determining a net photoadsorption or photodesorption of species. A new equilibrium is achieved when the species and the solid acquire the same electrochemical potential.[18,19]

Direct observation of photosorption phenomena at the TiO_2 surface has been reported in several earlier studies, focusing on the role played by surface OH groups in the photoadsorption of oxygen and therefore in regulating the TiO_2 photoactivity. For example, UV illumination stimulates desorption of oxygen and water molecules adsorbed at the TiO_2 surface[20–24] while studies by IR and XPS reveal that the amount of OH groups on the TiO_2 surface increases upon UV illumination.[25–29] These changes are generally fast and reversible, *i.e.*, once irradiation is stopped the surface recovers its previous features, under equal initial conditions. A well-recognized indication that irradiation modifies the surface of wide band gap solids is the post-sorption or memory effect, *i.e.*, the adsorption in the dark caused by pre-irradiation of the solid surface.[30,31]

More recent and clear evidence of this behavior is the phenomenon of induced superhydrophilicity, *i.e.*, the generation of a highly hydrophilic TiO_2 surface by UV illumination.[22,32–37] The superhydrophilic property of the TiO_2 surface allows water to spread completely across the surface rather than remaining as droplets. For a film that consists of only TiO_2, the contact angle of water almost becomes zero during UV irradiation. However, the contact angle increases and is restored comparatively quickly in the dark. One of the most interesting aspects of TiO_2 is that photocatalysis and induced hydrophilicity can take place simultaneously on the same surface even though the mechanisms are different. In the case of photocatalysis UV light excites the catalyst and pairs of electrons and holes are generated; the photogenerated electrons then react with molecular oxygen to produce superoxide radical anions and the photogenerated holes react with water to produce hydroxyl radicals. These two types of radicals work to carry out redox reactions with species adsorbed on the TiO_2 surface. On the other hand, surface hydroxyl groups can trap photogenerated holes and improve the separation of electrons and holes, which results in enhancement of photocatalysis.

Chapter 3 gives a more detailed explanation of superhydrophilicity.

For liquid–solid systems, adsorption phenomena involving ions occur with a transfer of electric charge, causing a significant variation of the electronic band structure of the surface. The charge transfer is responsible for the formation of the so-called space-charge layer where the potential difference between surface and bulk creates an electric field that can play a beneficial role in further adsorption steps. Most photons are absorbed in the superficial layer of the solid where the space-charge exists and the photoelectrons straightforwardly interact with the electric field present there. The consequence is a more efficient separation of the electrons and holes within the space-charge layer, *i.e.*, an increase of their mean lifetime.

Photoadsorbed species can act (i) as surface-hole trapping and photoelectrons can be trapped in the bulk of the solid or (ii) as surface-electron trapping; in the latter case holes can react with OH surface groups and/or H_2O. Both alternatives depend on the chemical nature of the molecule to be adsorbed and on the type of the solid adsorbent. Notably, in the gas–solid regime only gaseous species or lattice ions can be involved, whereas in the liquid phase the interaction with the solvent should also be considered.

All the above considerations allow us to conclude that the electronic surface modifications created under irradiation eventually induce photoadsorption both in gas–solid and in liquid–solid regimes, which are the photoadsorption phenomena strictly related to the photoactivity of the solid photocatalysts. When the light source is switched off, a reversal of the process could occur, although reversibility is rarely complete and the achievement of a new equilibrium depends mainly on kinetics rather than thermodynamic factors.

Photoexcitation of a semiconductor solid is a quite complicated process. Indeed, the absorption of photons is strictly correlated with the nature of the crystal sites, which can be regular or with imperfections such as intrinsic and extrinsic lattice defects. It is possible to distinguish four simplified types of electronic excitations induced by light absorption. In a perfect lattice absorption can produce only intrinsic photoexcitations with (i) the promotion of electrons from the valence band to the conduction band with formation of free electron–hole pairs (separated electrons and holes); (ii) the formation of free bulk excitons (the combination of an electron and a positive hole that is free to move through a non-metallic crystal as a unit). In an imperfect lattice the presence of defects causes extrinsic absorption of light, in particular (iii) photon

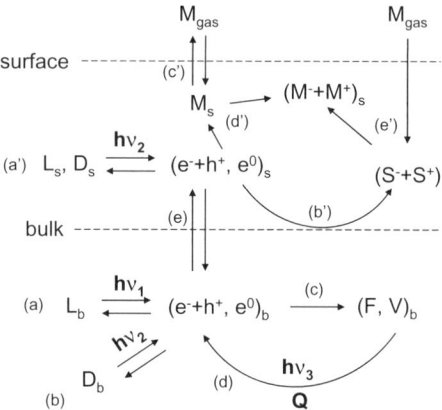

Figure 1.16 Photo-induced processes in a heterogeneous system containing a wide band gap semiconductor.

absorption by defects generating electronically excited defects and bound and/or self-trapped excitons; (iv) photon absorption generating ionization of defects (transitions between localized and delocalized electronic states).

The scheme of Figure 1.16 shows in detail the processes described above. It is possible to distinguish bulk and surface processes.[38,39]

Intrinsic photoexcitation of the lattice, L_b, [(a) in Figure 1.16] occurs by the fundamental absorption of photons of light with energy higher or equal to that of the band gap ($h\nu_1 \geq E_G$) and the consequent formation of free electrons (e^-) and holes (h^+), or excitons (e^0) that can recombine (band to band recombination of free charge carriers).

Pre-existing bulk defects (D_b) can lead to an extrinsic photoexcitation (b) by light with energy $h\nu_2 < h\nu_1$, with the formation of free electrons and holes. The reverse process is the recombination and exciton decay through defects.

Another bulk process is the trapping of free charge carriers and excitons by bulk defects (c) to form color centers or deep traps (F-type: vacancies occupied by electrons, and V-type: vacancies occupied by holes). Photobleaching ($h\nu_3$) or thermo-annealing of color centers (d) affords free electrons and holes.

As far as the surface processes are concerned, light absorption ($h\nu_2$) by regular, L_s, and irregular, D_s, surface states (a′) yields free charge carriers and excitons. Their surface recombination is likely to be accompanied by luminescence, whereas their trapping by surface defects, S, (b′) yields surface active centers (S^- and S^+) for adsorption and catalysis.

Adsorption of molecules in the gas or liquid phase (c′) generates adsorption–desorption equilibria on the solid surface. The interaction of surface charge carriers and excitons with the adsorbed molecules (d′) can promote surface chemical processes by the Langmuir–Hinshelwood mechanism, whereas the interaction of molecules with surface active centers (e′) can initiate surface chemical processes by the Eley–Rideal mechanism. Finally, the exchange of free charge carriers and excitons between the bulk and the surface can occur by diffusion or drift (e).

Fundamentals

In the Eley–Rideal mechanism, proposed in 1938 by D. D. Eley and E. K. Rideal, only one of the reactant molecules adsorbs and the other reacts with it directly from the gas phase, without adsorbing:

$$A_g + S_s \rightleftharpoons AS_{ads} \tag{1.31}$$

$$AS_{ads} + B_g \rightarrow \text{Product} \tag{1.32}$$

The Langmuir–Hinshelwood mechanism proposes that both molecules adsorb and then undergo a bimolecular reaction:

$$A_g + S_s \rightleftharpoons AS_{ads} \tag{1.33}$$

$$B_g + S_s \rightleftharpoons BS_{ads} \tag{1.34}$$

$$AS_{ads.} + BS_{ads.} \rightleftharpoons \text{Product} \tag{1.35}$$

Figure 1.17 shows the two mechanisms.

The Langmuir–Hinshelwood model (LH) is widely applied to liquid- and gas-phase systems (with minor variations) for the degradation of organic substrates on TiO$_2$ surfaces in the presence of oxygen.[40–46] This model successfully explains the kinetics of reactions that occur between two adsorbed species – a free radical and an adsorbed substrate, or a surface-bound radical and a free substrate. The

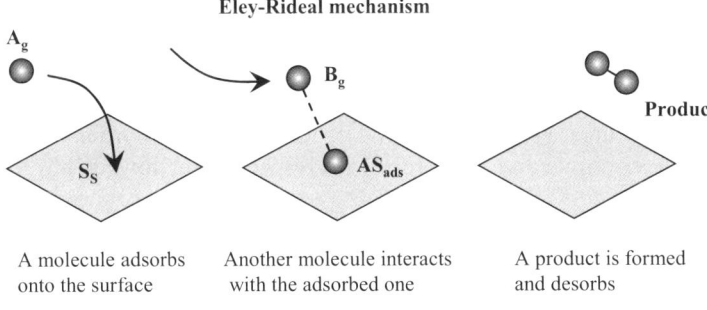

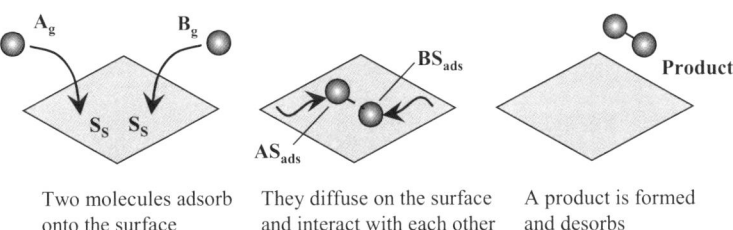

Figure 1.17 Graphic models of the Eley–Rideal and Langmuir–Hinshelwood mechanisms.

main assumptions of the LH model are (i) the adsorption equilibrium is established at all times and the reaction rate is less than the rate of adsorption or desorption; (ii) the reaction is assumed to occur between adsorbed species whose coverage on the catalyst surface is always in equilibrium with the concentration of the species in the fluid phase, so that the rate-determining step of the photocatalytic process is the surface reaction. The concentrations of adsorbed species are determined by adsorption equilibrium by means of a suitable isotherm. Notably, the simple rate form of the LH approach may have origins that take into account different photoreaction mechanisms.[47–49]

The decrease of the amount of a species in a batch photocatalytic system working in a liquid–solid regime is the combined result of photoadsorption and photoconversion processes. To describe this behavior, a molar balance applied to the species at any time[43] can be written as:

$$n_T = n_L + n_S \quad (1.36)$$

where n_T is the total number of moles present in the photoreactor, n_L the number of moles in the fluid phase and n_S the number of moles photoadsorbed on the solid. By dividing Equation (1.36) by the volume of the liquid phase, V, the following relation is obtained:

$$C_T = C_L + \frac{n_S}{V} \quad (1.37)$$

where C_T is the total concentration of the species and C_L the concentration in the liquid phase.

Both substrate and oxygen must be present in the system for the occurrence of the photoreaction; then it is assumed that the total disappearance rate of substrate per unit surface area, r_T, follows second-order kinetics (first order with respect to the substrate coverage and first order with respect to the oxygen coverage):

$$r_T \equiv -\frac{1}{S}\frac{dn_T}{dt} = k''\theta_{Sub}\theta_{Ox} \quad (1.38)$$

where S is the catalyst surface area, t the time, k'' the second order rate constant and θ_{Sub} and θ_{Ox} are the substrate and oxygen fractional coverage of the surface, respectively; θ_{Sub} and θ_{Ox} are defined as follows:

$$\theta_{Sub} \equiv \frac{n_S}{WN_S^*} \quad (1.39)$$

$$\theta_{Ox} \equiv \frac{n_{S,Ox}}{WN_{S,Ox}^*} \quad (1.40)$$

where $n_{S,Ox}$ is the number of moles of oxygen photoadsorbed on the solid, N_S^* and $N_{S,Ox}^*$ are the maximum capacity of photoadsorbed moles of substrate and oxygen, respectively, on the unit mass of irradiated solid and W is the mass of catalyst.

Fundamentals

By considering oxygen that is continuously bubbled into the dispersion, during the experimental runs, its concentration in the liquid phase does not change and it is always in excess. Then the θ_{Ox} term of Equation (1.38) does not depend on time, *i.e.*, it is constant. By defining $k = k''\theta_{Ox}$, Equation (1.38) turns a pseudo-first order rate equation:

$$r_T \equiv -\frac{1}{S}\frac{dn_T}{dt} = k\theta_{Sub} \qquad (1.41)$$

where k is the pseudo-first order rate constant. Introducing the liquid volume, V, and the surface area per unit mass of catalyst, S_S, in Equation (1.41), the following relation is obtained:

$$r_T \equiv -\frac{1}{S}\frac{dn_T}{dt} = -\frac{1}{S_S W} V \frac{dC_T}{dt} = k\theta_{Sub} \qquad (1.42)$$

The kinetic information on a photoprocess consists of knowledge of substrate concentration values in the liquid phase, C_L, as a function of irradiation time. Then the C_T and the θ_{Sub} variables of Equation (1.42) must be transformed as a function of C_L. For this reason it is necessary to choose a suitable adsorption isotherm for θ_{Sub}, *i.e.*, a relationship between θ_{Sub} [and therefore n_S, see Equation (1.39)] and C_L, while Equation (1.37) coupled with the chosen isotherm may be used for C_T. Therefore, Equation (1.42) may be formally written in the following way:

$$-\frac{1}{S_S W} V \frac{d}{dt} C_T(C_L) = k\theta_{Sub}(C_L) \qquad (1.43)$$

For a batch photocatalytic runs, by hypothesizing that the main assumptions of the LH model are valid, the substrate concentration values measured in the liquid phase at a certain time represent the substrate concentration in equilibrium with an (unknown) substrate amount photoadsorbed on the catalyst surface. This feature belongs to all the measured values of substrate concentration except for the initial one. The substrate concentration measured at the start of a photoreactivity run is characteristic of a system without irradiation. Then, during the fitting of a kinetic model to the experimental data, regression analysis must be performed by using the values of concentration measured throughout the run, excluding the initial substrate concentration under dark.[50]

1.2.1 Langmuir Isotherm

The Langmuir isotherm theory[51] assumes that the substrate is adsorbed in a monolayer over a homogenous adsorbent surface. A plateau characterizes the Langmuir isotherm; therefore, a saturation point is reached where no further adsorption occurs. Adsorption is assumed to take place at specific homogeneous sites within the adsorbent. Once a molecule occupies a site, no further adsorption can take place at that site. The Langmuir model assumes that the catalyst surface is completely uniform, *i.e.*, (i) the adsorption onto the surface of each molecule has equal activation energy and (ii) no transmigration of adsorbate in the plane of the surface occurs. These assumptions are hypothesized here to describe photoadsorption. The Langmuir adsorption isotherm is described by the following relationship:

$$\theta_{\text{Sub}} \equiv \frac{n_S}{WN_S^*} = \frac{K_L^* C_L}{1 + K_L^* C_L} \tag{1.44}$$

where K_L^* is the photoadsorption equilibrium constant, related to the free energy of photoadsorption, and N_S^* the monolayer adsorption capacity. K_L^* may be considered a measure of the intrinsic photoreactivity of the catalyst surface. By solving Equation (1.44) with respect to n_S and substituting in Equation (1.37) we obtain:

$$C_T = C_L + \frac{WN_S^*}{V} \frac{K_L^* C_L}{1 + K_L^* C_L} \tag{1.45}$$

Substituting in Equation (1.42) the Langmuir relationship [Equation (1.44)] gives the following relation:

$$-\frac{V}{WS_S} \frac{dC_T}{dt} = k \frac{K_L^* C_L}{1 + K_L^* C_L} \tag{1.46}$$

Taking the derivative of Equation (1.45) with respect to time yields:

$$\frac{dC_T}{dt} = \left[1 + \frac{WN_S^* K_L^*}{V} \frac{1}{(1 + K_L^* C_L)^2}\right] \frac{dC_L}{dt} \tag{1.47}$$

Substituting Equation (1.47) into the left-hand term of Equation (1.46), one obtains:

$$-\frac{V}{WS_S}\left[1 + \frac{WN_S^* K_L^*}{V} \frac{1}{(1 + K_L^* C_L)^2}\right] \frac{dC_L}{dt} = k \frac{K_L^* C_L}{1 + K_L^* C_L} \tag{1.48}$$

and, upon rearranging and separating the variables, the following differential equation is obtained:

$$-\frac{V}{WS_S} \frac{1}{kK_L^*} \frac{dC_L}{C_L} - \frac{V}{WS_S} \frac{1}{k} dC_L - \frac{1}{S_S} \frac{N_S^*}{k} \frac{dC_L}{C_L(1 + K_L^* C_L)} = dt \tag{1.49}$$

Equation (1.49) gives the evolution of concentration in the liquid phase with the irradiation time of a species in photoadsorption equilibrium on the catalyst surface, over which the species undergoes a slow transformation process. On this basis, the integration of Equation (1.49) must be performed with the limiting condition that at $t=0$ the substrate concentration in the liquid phase is that in equilibrium with the initial photoadsorbed amount, $C_{L,0}$; this initial concentration is unknown but it may be determined by regression analysis carried out with the experimental data obtained after the start of irradiation. The integration yields:

$$\frac{V}{WS_S}\frac{1}{kK_L^*}\ln\frac{C_{L,0}}{C_L} + \frac{V}{WS_S}\frac{1}{k}(C_{L,0} - C_L) + \frac{1}{S_S}\frac{N_S^*}{k}\ln\left(\frac{1+K_L^*C_L}{1+K_L^*C_{L,0}}\frac{C_{L,0}}{C_L}\right) = t \quad (1.50)$$

Equation (1.50) contains four unknown parameters, K_L^*, N_S^*, k and $C_{L,0}$, which may be determined by a best fitting procedure.

1.2.2 Freundlich Isotherm

The Langmuir theory of adsorption is an approximation of real surfaces. If the Langmuir theory was universally valid, we should expect the heat of adsorption to remain constant with increasing values of the coverage θ. In some cases the heat of adsorption decreases with coverage. One explanation is that the surfaces of the catalysts and of the solids are energetically heterogeneous. An assumption generally used in describing the physical chemistry of real adsorbed layers is that the surface sites are different, the so-called biographical non-uniformity.[44] By applying the Langmuir adsorption isotherm to a distribution of energies among the sites such that the heat of adsorption decreases logarithmically with coverage, the Freundlich isotherm is derived;[51] it is described by the following relationship:

$$\theta_{Sub} \equiv \frac{n_S}{WN_S^*} = K_F^* C_L^{1/n} \quad (1.51)$$

where K_F^* is the Freundlich isotherm constant, which indicates the relative adsorption capacity of the adsorbent, and n is a dimensionless parameter indicative of the intensity of the adsorption. The parameter n represents the mutual interaction of adsorbed species; a value of n greater than unity indicates the occurrence of repulsion among adsorbed molecules. The Freundlich relation is an exponential equation that assumes that the concentration of adsorbate on the adsorbent surface increases by increasing the adsorbate concentration in the liquid phase. By using the empirical form of the Freundlich isotherm, there is no limit to adsorption; this is not in agreement with the chemisorption theory. Statistical derivation of this isotherm, however, sets a value of the Freundlich maximum adsorption capacity. Solving Equation (1.51) with respect to n_S and substituting in Equation (1.37), gives:

$$C_T = C_L + \frac{WN_S^*}{V}K_F^* C_L^{1/n} \quad (1.52)$$

Substituting the Freundlich relationship [Equation (1.51)] in Equation (1.42) affords the following relation:

$$-\frac{V}{WS_S}\frac{dC_T}{dt} = kK_F^* C_L^{1/n} \qquad (1.53)$$

The derivative of Equation (1.52) with respect to time yields:

$$\frac{dC_T}{dt} = \left[1 + \frac{WN_S^* K_F^*}{V}\frac{1}{n}C_L^{\frac{1-n}{n}}\right]\frac{dC_L}{dt} \qquad (1.54)$$

Substituting Equation (1.54) into the left-hand term of Equation (1.53), one obtains:

$$-\frac{V}{WS_S}\left[1 + \frac{WN_S^* K_F^*}{V}\frac{1}{n}C_L^{\frac{1-n}{n}}\right]\frac{dC_L}{dt} = kK_F^* C_L^{1/n} \qquad (1.55)$$

and by rearranging and separating the variables the following differential equation is obtained:

$$-\frac{V}{WS_S}\frac{1}{kK_F^*}\frac{dC_L}{C_L^{1/n}} - \frac{1}{S_S}\frac{N_S^*}{kn}\frac{dC_L}{C_L} = dt \qquad (1.56)$$

As with Langmuir adsorption isotherm, Equation (1.56) is integrated with the condition that at $t=0$ the substrate concentration in the liquid phase is that in equilibrium with the initial photoadsorbed amount, $C_{L,0}$; this initial concentration is unknown but it may be determined by regression analysis carried out with the experimental data obtained after the start of irradiation. The integration yields:

$$\frac{V}{WS_S}\frac{1}{kK_F^*}\frac{n}{n-1}\left(C_{L,0}^{(n-1)/n} - C_L^{(n-1)/n}\right) + \frac{1}{S_S}\frac{N_S^*}{kn}\ln\frac{C_{L,0}}{C_L} = t \qquad (1.57)$$

Equation (1.57) contains five unknown parameters, K_F^*, N_S^*, n, k, and $C_{L,0}$, which may be determined by a best fitting procedure.

1.2.3 Redlich–Peterson Isotherm

The Redlich–Peterson (R-P) isotherm[52] model combines elements from both the Langmuir and Freundlich equation, and the mechanism of adsorption is a hybrid one and it does not follow ideal monolayer adsorption. The R-P isotherm is an empirical equation, designated the "three parameter equation," that may be used to represent adsorption equilibria over a wide concentration range.[53–56] The R-P adsorption isotherm is described by the following relationship:

$$\frac{n_S}{W} = \frac{K_{R-P}^* C_L}{1 + \alpha_{R-P}^* C_L^\beta} \qquad (1.58)$$

Fundamentals

where K_{R-P}^* and α_{R-P}^* are R-P isotherm constants that refer to the adsorption capacity and the surface energy, respectively. The exponent β is the heterogeneity factor, which lies between 1 and 0. The equation can be reduced to the Langmuir equation as β approaches 1 while it becomes the Henry's law equation when $\beta = 0$. By solving Equation (1.58) with respect to n_S and substituting in Equation (1.37), the result is:

$$C_T = C_L + \frac{W}{V} \frac{K_{R-P}^* C_L}{1 + \alpha_{R-P}^* C_L^\beta} \qquad (1.59)$$

In the cases of Langmuir and Freundlich isotherms it has been assumed that the total disappearance rate of substrate per unit surface area, r_T, follows a pseudo-first order kinetics with respect to the substrate concentration, which is expressed by its fractional coverage. The same assumption is made for R-P isotherm; however, as the R-P isotherm relates an adsorbed amount (and not a fractional coverage) with the equilibrium concentration in the liquid phase, the r_T term is written as:

$$r_T \equiv -\frac{V}{WS_S} \frac{dC_T}{dt} = kC_{\text{Surface}} = k \frac{n_S}{WS_S} \qquad (1.60)$$

where C_{Surface} is the surface concentration of adsorbed species (adsorbed moles/catalyst surface area).[48] Substituting in Equation (1.60) the R-P relationship [Equation (1.58)] produces:

$$-\frac{V}{W} \frac{dC_T}{dt} = k \frac{K_{R-P}^* C_L}{1 + \alpha_{R-P}^* C_L^\beta} \qquad (1.61)$$

Taking the derivative of Equation (1.59) with respect to time yields:

$$\frac{dC_T}{dt} = \left[1 + \frac{WK_{R-P}^*}{V} \frac{1 + \alpha_{R-P}^* C_L^\beta (1 - \beta K_{R-P}^*)}{\left(1 + \alpha_{R-P}^* C_L^\beta\right)^2}\right] \frac{dC_L}{dt} \qquad (1.62)$$

Substituting Equation (1.62) into the left-hand term of Equation (1.61), one obtains:

$$-\frac{V}{W}\left[1 + \frac{WK_{R-P}^*}{V} \frac{1 + \alpha_{R-P}^* C_L^\beta (1 - \beta K_{R-P}^*)}{\left(1 + \alpha_{R-P}^* C_L^\beta\right)^2}\right] \frac{dC_L}{dt} = k \frac{K_{R-P}^* C_L}{1 + \alpha_{R-P}^* C_L^\beta} \qquad (1.63)$$

and, upon rearranging and separating the variables, the following differential equation is obtained:

$$-\frac{V}{WkK^*_{\text{R-P}}}\frac{dC_L}{C_L} - \frac{\alpha^*_{\text{R-P}}V}{WkK^*_{\text{R-P}}}C_L^{\beta-1}dC_L - \frac{1}{k}\frac{dC_L}{C_L\left(1+\alpha^*_{\text{R-P}}C_L^\beta\right)} +$$
$$-\frac{V\alpha^*_{\text{R-P}}(1-\beta K^*_{\text{R-P}})}{WkK^*_{\text{R-P}}}\frac{C_L^{\beta-1}dC_L}{1+\alpha^*_{\text{R-P}}C_L^\beta} = dt \quad (1.64)$$

As with previous adsorption isotherms, Equation (1.64) is integrated with the condition that at $t=0$ the substrate concentration in the liquid phase is that in equilibrium with the initial photoadsorbed amount, $C_{L,0}$; again, this initial concentration is unknown but it may be determined by regression analysis carried out with the experimental data obtained after the start of irradiation. The integration yields:

$$\frac{V}{WkK^*_{\text{R-P}}}\ln\left(\frac{C_{L0}}{C_L}\right) + \frac{\alpha^*_{\text{R-P}}V}{Wk\beta K^*_{\text{R-P}}}\left(C_{L0}^\beta - C_L^\beta\right) + \frac{1}{k\beta}\ln\left(\frac{1+\alpha^*_{\text{R-P}}C_L^\beta}{1+\alpha^*_{\text{R-P}}C_{L0}^\beta}\frac{C_{L0}^\beta}{C_L^\beta}\right)$$
$$+ \frac{V(1-\beta K^*_{\text{R-P}})}{Wk\beta K^*_{\text{R-P}}}\ln\left(\frac{1+\alpha^*_{\text{R-P}}C_{L0}^\beta}{1+\alpha^*_{\text{R-P}}C_L^\beta}\right) = t \quad (1.65)$$

Equation (1.65) contains five unknown parameters, $K^*_{\text{R-P}}$, $\alpha^*_{\text{R-P}}$, β, k and C_{L0}, which may be determined by a best fitting procedure.

1.2.4 Light Intensity Dependence

In most kinetics studies of a photocatalytic reaction, the observed variation of the initial degradation rate of an organic substrate ($-r$) with its initial concentration is generally well described by a Langmuir–Hinshelwood type kinetic equation:

$$(-r) = k\frac{K^*_L C_L}{1+K^*_L C_L} \quad (1.66)$$

Initially, only k was considered to be dependent on the incident light,[26,57–60] which is directly related to the absorbed photons; however, K^*_L is usually greater than the Langmuir adsorption constant under dark conditions and now it is accepted that this parameter is also light intensity dependent.[61–66]

In the following, the different mechanisms proposed for this light-intensity dependence are shown.

Emeline et al. have proposed Eley–Rideal and Langmuir–Hinshelwood based mechanisms:[62–65,67]

$$k_{\text{ER}} = a \cdot I \quad (1.67)$$

$$K_{\text{ER}} = \frac{b}{d} \cdot I \quad (1.68)$$

Fundamentals

and:

$$k_{LH} = \frac{a \cdot I}{e + d \cdot I} \quad (1.69)$$

$$K_{LH} = \frac{(e + d \cdot I)K_{ads}}{d \cdot I} \quad (1.70)$$

where I is the incident light intensity, K_{ads} is the dark Langmuir adsorption constant, a, b, d and e are constants; a plot of $1/K_{LH}$ versus k_{LH} yields a straight line passing from the origin of the axis. This insight was confirmed for the photodegradation of phenol but studies with 4-chlorophenol showed a straight line with a positive intercept.

Ollis[61] has explained this behavior by considering that the adsorption/desorption kinetics of the key substrate, A, are influenced by the oxidation process under pseudo-steady state conditions:

$$A_{liquid} \underset{k_{-1}}{\overset{k_1}{\rightleftarrows}} A_{ads} \quad (1.71)$$

$$A_{ads} \xrightarrow{k_{LH}} products \quad (1.72)$$

Step (1.72) involves oxidation of the organic species by a photogenerated surface hydroxyl radical, which depends upon the incident light intensity. Then:

$$k_{LH} = \alpha \cdot I^\beta \quad (1.73)$$

where α is a proportionality constant and β is a constant whose value is usually equal to 1 and 0.5 under low and high intensity irradiation, respectively. A square-root dependence can be found for high light intensity due to the predominance of electron–hole recombination.

The pseudo-steady state reaction proposed by Ollis[61] involves a kinetic expression in which k_{LH} is given by Equation (1.73) and:

$$K_{LH} = \frac{k_1}{k_{-1} + \alpha \cdot I^\beta} \quad (1.74)$$

This simple model predicts not only the light intensity dependence for k_{LH} and K_{LH} but also the straight line (of gradient $1/k_1$) in the plot of $1/K_{LH}$ versus k_{LH} with a positive intercept in the case of 4-chlorophenol degradation and a zero intercept for phenol photodegradation, as previously described.

It is well known that the reaction rate depends in a complex way on the light intensity; moreover, the reactor geometry influences strongly the ratio between backward reflected and incident photon flow. The apparent napierian extinction coefficient for slurries, which is a parameter related to the absorption photon flow, decreases as the particle size increases.[68]

Evaluation of the local volumetric rate of photon absorption as well as the determination of scattering and absorption coefficient have been studied.[69] In a solid–liquid system, photon absorption by the photocatalyst is strictly related to scattering, which makes the kinetics analysis more difficult.

The rate of backward reflected photon flow can be determined by an extrapolation method[68,70–73] that consists in measuring the rate of transmitted photon flow as a function of the suspension volume, V:

$$\Phi_T = \Phi' \cdot \exp(-E \cdot C_{Cat} \cdot V) \quad (1.75)$$

where Φ' is the rate of photons able to penetrate the suspension, E is the apparent Napierian extinction coefficient of the suspension and C_{Cat} the catalyst concentration.

Application of a macroscopic photon balance on the photoreactor at the limit condition of $V=0$ allows determination of the backward reflected photon flow Φ_r:

$$\Phi_r = \Phi_i - \Phi' \quad (1.76)$$

The absorbed photon flow Φ_a can be obtained by the following photon balance on the system:

$$\Phi_a = \Phi_i - \Phi_T - \Phi_r \quad (1.77)$$

Nevertheless, when the mean particle size is of the same order of magnitude as the wavelength used, inelastic scattering phenomena can occur and this procedure cannot be used.

In photocatalytic reactions, the photons absorbed by the photocatalyst may be considered as a reactant species. Therefore, the spatial distribution of their absorption rate is an important issue in order to interpret experimental results, to develop kinetic models and to optimize or to scale-up a photocatalytic reactor. The Monte Carlo approach can be employed to obtain values of the local volumetric rate of photon absorption (LVRPA or $e^{a,v}$).

In general terms, the Monte Carlo method applied to the radiation field resolution consists of tracking the trajectory of a great number of photons. In doing so, the basic laws of geometric optics are applied (such as Snell refraction law and Fresnel reflectivity equation), as well as the ray tracing technique. The LVRPA can be constructed out of the series of events that is likely to occur for each of the photons that take part in the simulated experiment. The Monte Carlo method has been employed successfully in the analysis and modeling of photocatalytic reactors.[74–80]

1.3 Radiation Sources

The performance of a photoreactor depends strongly on the irradiation source. Six main types of radiation sources may be used: (1) arc lamps, (2) fluorescent lamps, (3) incandescent lamps, (4) lasers, (5) light-emitting diodes (LEDs) and (6) solar radiation.

1.3.1 Arc Lamps

In arc lamps the emission is obtained by activation of a gas by collision with accelerated electrons generated by an electric discharge between two electrodes, typically made of tungsten. The type of lamp is often named according to the gas contained in the bulb, including neon, argon, xenon, krypton, sodium, metal halide and mercury. In particular for mercury lamps, the following classification, based on the Hg pressure, has been made:

1. Low pressure: the lamp contains Hg vapor at a pressure of *ca.* 0.1 Pa at 298 K and it emits mainly at 253.7 and 184.9 nm.
2. Medium pressure: this lamp contains Hg vapor at a pressure from 100 to several hundred kPa. It emits mostly from 310 to 1000 nm with most intense lines at 313, 366, 436, 576 and 578 nm.
3. High pressure: this lamp contains Hg vapor at a pressure equal or higher than 10 MPa and it emits a continuous background between *ca.* 200 and 1000 nm with broad lines.
4. Hg-Xe. This lamp is used to simulate solar radiation as it is an intense ultraviolet, visible and near-IR radiation source. The lamp contains a mixture of Hg and Xe vapors at different concentrations under high pressure. The presence of Hg vapor increases emission in the UV region.

The above lamps are usually cylindrical, with the arc length increasing as the pressure decreases and power increases. The power ranges from a few watts to *ca.* 60 000 W. Generally, medium- and high-pressure mercury lamps need to be cooled by circulating liquids.

1.3.2 Fluorescent Lamps

Fluorescent lamps are filled with gas containing a mixture of low-pressure mercury vapor and argon (or xenon), more rarely argon–neon, sometimes even krypton. The inner surface of the lamp is coated with a fluorescent (often slightly phosphorescent) coating made of varying blends of metallic and rare-earth phosphor salts. The cathode is generally made of coiled tungsten coated with a mixture of strontium, barium and calcium oxides that have a relatively low thermo-ionic emission temperature. When the light is turned on, the cathode is heated sufficiently to emit electrons. These electrons collide with gas atoms, which are ionized to form a plasma by a process of impact ionization. As a result the conductivity of the ionized gas rises, allowing higher currents to flow through the lamp. The mercury, which exists at a stable vapor pressure equilibrium point of about one part per thousand inside the tube (with the noble gas pressure typically being about 0.3% of standard atmospheric pressure), is then likewise ionized, causing it to emit light in the ultraviolet (UV) region of the spectrum, predominantly at wavelengths of 253.7 and 185 nm. Fluorescent lighting is efficient because low-pressure mercury discharges emit about 65% of their total light at the 254 nm line (also about 10–20% of the light emitted in UV is at the 185 nm line). The UV light is absorbed by the bulb's fluorescent coating, which re-radiates the energy at lower frequencies (longer wavelengths: two intense lines of 440 nm and 546 nm wavelengths appear on commercial fluorescent tubes) to emit visible light. The blend of phosphors controls the color of the light and, along with the bulb's glass, prevents escape of the harmful UV light. Actinic fluorescent tubes have an emission in the near-UV region (*ca.* 360 nm) due to their particular fluorescent coating.

1.3.3 Incandescent Lamps

The emission of incandescent lamps is obtained by heating at very high temperature suitable filaments of various substances by current circulation.

One of the most used incandescent lamps is the halogen lamp, wherein a tungsten filament is sealed into a small envelope filled with a halogen gas such as iodine or bromine. A tungsten-halogen lamp creates an equilibrium reaction in which the tungsten that evaporates when giving off light is preferentially re-deposited at the hot-spots, thereby preventing the early failure of the lamp. This also allows halogen lamps to be run at higher temperatures that would cause unacceptably short lamp lifetimes in ordinary incandescent lamps, allowing for higher luminous efficacy, apparent brightness and whiter color temperature. Because the lamp must be very hot to create this reaction, the envelope of the halogen lamp must be made of hard glass or fused quartz, instead of ordinary soft glass which would soften and flow too much at such temperatures.

In applications that require ultraviolet radiation, the lamp envelope is made out of undoped quartz. Thus, the lamp becomes a source of UV-B radiation. A typical halogen lamp is designed to run for about 2000 h, which is twice as long as a typical incandescent lamp.

1.3.4 Lasers

A laser (light amplification by stimulated emission of radiation) is a device that emits light through a specific mechanism. This is a combined quantum-mechanical and thermodynamic process. A typical laser emits light in a narrow and well-defined beam and with a well-defined wavelength (or color). This is in contrast to a light source such as the incandescent light bulb, which emits in almost all directions and over a wide spectrum of wavelengths. All these properties are summarized in the term coherence. A laser consists of a gain medium inside an optical cavity, to supply energy to the gain medium. The gain medium is a material (gas, liquid or solid) with appropriate optical

Fundamentals

properties. In its simplest form, a cavity consists of two mirrors arranged such that light bounces back and forth, each time passing through the gain medium. Typically, one of the two mirrors, the output coupler, is partially transparent. All light that is emitted by the laser passes through this output coupler.

Light of a specific wavelength that passes through the gain medium is amplified (increases in intensity); the surrounding mirrors ensure that most of the light makes many passes through the gain medium. Part of the light that is between the mirrors (*i.e.*, in the cavity) passes through the partially transparent mirror and appears as a beam of light. The process of supplying the energy required for the amplification is called pumping and the energy is typically supplied as an electrical current or as light at a different wavelength. In the latter case, the light source can be a flash lamp or another laser. Most practical lasers contain additional elements that affect properties such as the wavelength of the emitted light and the shape of the beam.

1.3.5 Light Emitting Diodes (LEDs)

LEDs are special p-n junction diodes formed by a thin layer of doped semiconductor material. They are based on semiconductor optical properties to produce photons by the recombination of electron–hole pairs. Electrons and holes are injected in a recombination zone through two parts of the diode doped in different ways, "n" type impurities for electrons and "p" type for holes. Figure 1.18 shows a working scheme. Ultraviolet light is achievable by using GaN as the active layer, with emission in the near-UV region with wavelengths around 350–370 nm. In addition, nitrides containing aluminium, such as AlGaN and AlGaInN, can be used to achieve shorter wavelengths, whereas the use of aluminium nitride enables wavelengths down to 210 nm to be obtained. Figure 1.19 shows the main parts of a LED.

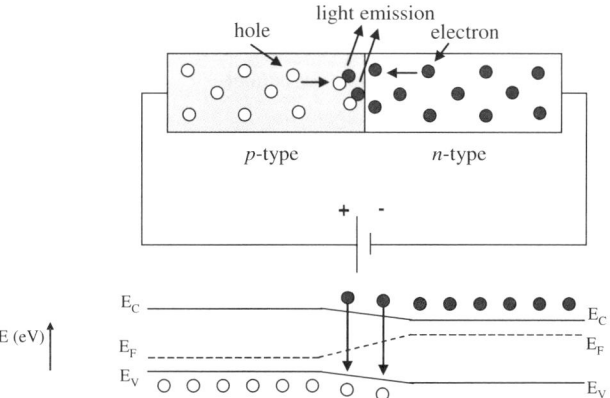

Figure 1.18 Working scheme and energetic levels in a LED.

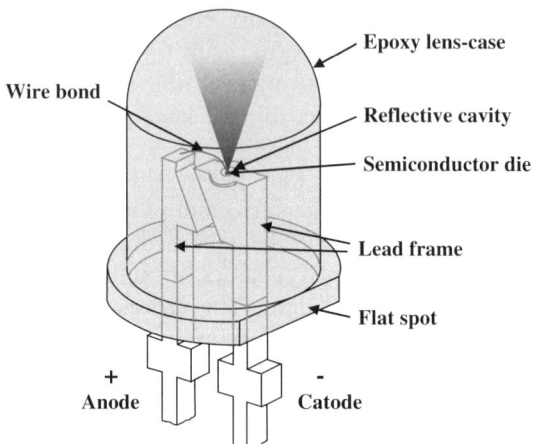

Figure 1.19 Main parts of a LED.

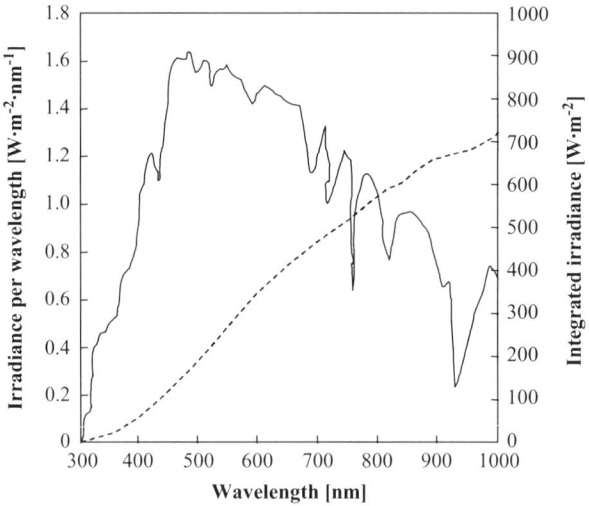

Figure 1.20 ASTM global irradiance standard solar spectrum (AM 1.5) up to a wavelength of 1000 nm, incident on a plane tilted 38° facing south, normalized to 1000 W m^{-2} for the whole spectrum.

1.3.6 Solar Resources

The earth receives about 1.7×10^{14} kW of solar radiation (1.5×10^{18} kWh per year). Extraterrestrial radiation has an intensity of 1367 W m^{-2} in the wavelength range 200–5000 nm, which is reduced to 280–4000 nm when it reaches the ground, due to absorption phenomena by atmospheric compounds such as ozone, oxygen, carbon dioxide and water vapor as well as aerosols, clouds, *etc*.

The solar radiation that reaches the planet without being absorbed or scattered is called direct beam radiation, whereas radiation that reaches the ground but has been dispersed beforehand is called diffuse radiation; the sum of both is called global radiation. Figure 1.20 shows the standard solar spectrum at sea level, where the spectral irradiance data are for the sun at a solar zenith angle of 48.19°. This zenith angle corresponds to an air mass (AM) of 1.5, which is the ratio of the direct-beam solar irradiance path length through the atmosphere at a solar zenith angle of 48.19° to the path length when the sun is in the zenith.

The radiation effectively reaching ground level changes strongly due to several factors, such as date, time, geographic latitude, atmospheric conditions (aerosols, humidity, *etc.*) or cloudiness. To judge the feasibility and profitability of solar applications at a specific site, studies have to be performed to estimate or to measure the amount of radiation actually available at the specific site throughout the year.

References

1. C. Kittel, *Introduction to Solid State Physics*, John Wiley & Sons, New York, 1963.
2. A. J. Dekker, *Solid State Physics*, Prentice-Hall, Englewood Cliffs, NJ, 1957.
3. H. Gerischer, in: *Physical Chemistry, An Advanced Treatise Vol. IX*, H. Eyring, D. Henderson and W. Host, (ed.), Academic Press, New York, 1970.
4. F. Lohmann, Fermi-Niveau und flachbandpotential von molekülkristallen aromatischer kohlenwasserstoffe, *Z. Naturforsch., Teil A*, 1967, **22**, 843–844.
5. Yu. V. Pleskov and Yu. Ye. Gurevich, *Semiconductor Photoelectrochemistry*, Consultants Bureau, New York, 1986, p. 101.
6. A. J. Bard, Photoelectrochemistry and heterogeneous photo-catalysis at semiconductors, *J. Photochem.*, 1979, **10**, 59–75.
7. K. Domen, S. Naito, K. Domen, T. Onishi, K. Tamara and M. Soma, Study of the photocatalytic decomposition of water vapor over a NiO-SrTiO$_3$ catalyst, *J. Phys. Chem.*, 1982, **86**, 3657–3661.
8. N. Serpone, P. Maruthamuthu, P. Pichat, E. Pelizzetti and H. Hidaka, Exploiting the interparticle electron transfer process in the photocatalysed oxidation of phenol, 2-chlorophenol and pentachlorophenol: chemical evidence for electron and hole transfer between coupled semiconductors, *J. Photochem. Photobiol. A: Chem.*, 1995, **85**, 247–255.
9. A. Di Paola, L. Palmisano, A. M. Venezia and V. Augugliaro, Coupled semiconductor systems for photocatalysis. Preparation and characterization of polycrystalline mixed WO$_3$/WS$_2$ powders, *J. Phys. Chem. B*, 1999, **103**, 8236–8244.
10. A. Sclafani, L. Palmisano and M. Schiavello, Influence of the preparation methods of TiO$_2$ on the photocatalytic degradation of phenol in aqueous dispersion, *J. Phys. Chem.*, 1990, **94**, 829–832.

11. L. Palmisano, V. Augugliaro, A. Sclafani and M. Schiavello, Activity of chromium ions doped titania for the dinitrogen photoreduction to ammonia and for the phenol photodegradation, *J. Phys. Chem.*, 1988, **92**, 6710–6713.
12. J. M. Herrmann, J. Disdier and P. Pichet, Effect of chromium doping on the electrical and catalytic properties of powder titania under UV and visible illumination, *Chem. Phys. Lett.*, 1984, **108**, 618–622.
13. M. I. Litter and J. A. Navio, Photocatalytic properties of iron-doped titania semiconductors, *J. Photochem. Photobiol. A: Chem.*, 1996, **98**, 171–181.
14. K. Wilke and H. D. Breuer, The influence of transition metal doping on the physical and photocatalytic properties of titania, *J. Photochem. Photobiol. A: Chem.*, 1999, **121**, 49–53.
15. W. Lee, W. Gao, K. Dwight and A. Wold, Preparation and characterization of titanium(IV) oxide photocatalysts, *Mater. Res. Bull.*, 1992, **27**, 685–692.
16. J. R. Do, W. Lee, K. Dwight and A. Wold, The effect of WO_3 on the photocatalytic activity of TiO_2, *J. Solid State Chem.*, 1994, **108**, 198–201.
17. J. M. Thomas and W. J. Thomas, *Introduction to the Principles of Heterogeneous Catalysis*, Academic Press, London, New York, 1967.
18. R. I. Bickley, in: *Photoelectrochemistry, Photocatalysis and Photoreactors, Fundamentals and Developments*, M. Schiavello, (ed.), Reidel, Dordrecht, 1985, pp. 379–388.
19. R. I. Bickley, in: *Photoelectrochemistry, Photocatalysis and Photoreactors, Fundamentals and Developments*, M. Schiavello, (ed.), Reidel, Dordrecht, 1985, pp. 491–502.
20. P. Meriaudeau and J. C. Vedrine, Electron paramagnetic resonance investigation of oxygen photoadsorption and its reactivity with carbon monoxide on titanium dioxide: The O3-3 species, *J. Chem. Soc., Faraday Trans.*, 1976, **72**, 472–480.
21. W. R. Murphy, T. H. Veerkamp and T.W. Leland, Effect of ultraviolet radiation on zinc oxide catalysts, *J. Catal.*, 1976, **43**, 304–321.
22. N. Sakai, R. Wang, A. Fujishima, T. Watanabe and K. Hashimoto, Effect of ultrasonic treatment on highly hydrophilic TiO_2 surfaces, *Langmuir*, 1998, **14**, 5918–5920.
23. R. Wang, N. Sakai, A. Fujishima, T. Watanabe and K. Hashimoto, Studies of surface wettability conversion on TiO_2 single-crystal surfaces, *J. Phys. Chem. B*, 1999, **103**, 2188–2194.
24. W. C. Wu, L. F. Liao, J. S. Shiu and J. L. Lin, FTIR study of interactions of ethyl iodide with powdered TiO_2, *Phys. Chem. Chem. Phys.*, 2000, **2**, 4441–4446.
25. A. L. Linsebigler, G. G. Lu and J. T. Yates, Photocatalysis on TiO_2 surfaces: principles, mechanisms, and selected results, *Chem. Rev.*, 1995, **95**, 735–758.

26. M. R. Hoffmann, S. T. Martin, W. Y. Choi and D. W. Bahnemann, Environmental applications of semiconductor photocatalysis, *Chem. Rev.*, 1995, **95**, 69–96.
27. N. Asakuma, T. Fukui, M. Toki, K. Awazu and H. Imai, Photoinduced hydroxylation at ZnO surface, *Thin Solid Films*, 2003, **445**, 284–287.
28. A. Ignatchenko, D. G. Nealon, R. Dushane and K. J. Humphries, Interaction of water with titania and zirconia surfaces, *J. Mol. Cat. A: Chem.*, 2006, **256**, 57–74.
29. K. Mori, S. Imaoka, S. Nishio, Y. Nishiyama, N. Nishiyama and H. Yamashita, Investigation of local structures and photo-induced surface properties on transparent Me(Ti, Cr)-containing mesoporous silica thin films, *Microporous Mesoporous Mater.*, 2007, **101**, 288–295.
30. A. N. Terenin and Yu. P. Solonitzyn, Action of light on the gas adsorption by solids, *Disc. Faraday Soc.*, 1959, **28**, 28–35.
31. Yu. P. Solonitzyn, I. M. Prudnikov and V. M. Yurkin, Study of memory effect in photostimulated adsorption at metal oxides adsorbents, *Russ. J. Phys. Chem.*, 1982, **57**, 2028–2030.
32. R. Wang, K. Hashimoto, A. Fujishima, M. Chikuni, E. Kojima, A. Kitamura, M. Shimohigoshi and T. Watanabe, Light-induced amphiphilic surfaces, *Nature*, 1997, **388**, 431–432.
33. T. Watanabe, A. Nakajima, R. Wang, M. Minabe, S. Koizumi, A. Fujishima and K. Hashimoto, Photocatalytic activity and photoinduced hydrophilicity of titanium dioxide coated glass, *Thin Solid Films*, 1999, **351**, 260–263.
34. A. Nakajima, S. Koizumi, T. Watanabe and K. Hashimoto, Effect of repeated photo-illumination on the wettability conversion of titanium dioxide, *J. Photochem. Photobiol. A*, 2001, **146**, 129–132.
35. M. Miyauchi, N. Kieda, S. Hishita, T. Mitsuhashi, A. Nakajima, T. Watanabe and K. Hashimoto, Reversible wettability control of TiO_2 surface by light irradiation, *Surf. Sci.*, 2002, **511**, 401–407.
36. Y. C. Lee, Y. P. Hong, H. Y. Lee, H. Kim, Y. J. Jung, K. H. Ko, H. S. Jung and K. S. Hong, Photocatalysis and hydrophilicity of doped TiO_2 thin films, *J. Colloid Interface Sci.*, 2003, **267**, 127–131.
37. M. Takeuchi, K. Sakamoto, G. Martra, S. Coluccia and M. Anpo, Mechanism of photoinduced superhydrophilicity on the TiO_2 photocatalyst surface, *J. Phys. Chem. B*, 2005, **109**, 15422–15428.
38. A. Emeline, A. Salinaro, V. K. Ryabchuk and N. Serpone, Photoinduced processes in heterogeneous nanosystems. From photoexcitation to interfacial chemical transformations, *Int. J. Photoenergy*, 2001, **3**, 1–16.
39. V. Ryabchuk, Photophysical processes related to photoadsorption and photocatalysis on wide band gap solids: a review, *Int. J. Photoenergy*, 2004, **6**, 95–113.
40. C. S. Turchi and D. F. Ollis, Mixed reactant photocatalysis: intermediates and mutual rate inhibition, *J. Catal.*, 1989, **119**, 483–496.

41. A. V. Vorontsov, E. N. Kurkin and E. N. Savinov, Study of TiO_2 deactivation during gaseous acetone photocatalytic oxidation, *J. Catal.*, 1999, **186**, 318–324.
42. H. Ibrahim and H. De Lasa, Kinetic modeling of the photocatalytic degradation of air-borne pollutants, *AIChE J.*, 2004, **50**, 1017–1027.
43. H. de Lasa, B. Serrano and M. Salaices, *Photocatalytic Reaction Engineering*, Springer, New York, 2005.
44. D. Y. Murzin and T. Salmi, *Catalytic Kinetics*, Elsevier, Amsterdam, 2005.
45. A. Gora, B. Toepfer, V. Puddu and G. Li Puma, Photocatalytic oxidation of herbicides in single-component and multicomponent systems: reaction kinetics analysis, *Appl. Catal. B*, 2006, **65**, 1–10.
46. G. Palmisano, V. Loddo, S. Yurdakal, V. Augugliaro and L. Palmisano, Reaction pathways and kinetics of photocatalytic oxidation of nitrobenzene and phenylamine in aqueous TiO_2 suspensions, *AIChE J.*, 2007, **53**, 961–968.
47. K. Demeestere, A. D. Visscher, J. Dewulf, M. Van Leeuwen and H. Van Langenhove, A new kinetic model for titanium dioxide mediated heterogeneous photocatalytic degradation of trichloroethylene in gas-phase, *Appl. Catal. B*, 2004, **54**, 261–274.
48. C. Minero and D. Vione, A quantitative evaluation of the photocatalytic performance of TiO_2 slurries, *Appl. Catal. B*, 2006, **67**, 257–269.
49. J. Krýsa, G. Waldner, H. Měšťánková, J. Jirkovský and G. Grabner, Photocatalytic degradation of model organic pollutants on an immobilized particulate TiO_2 layer. Roles of adsorption processes and mechanistic complexity, *Appl. Catal. B*, 2006, **64**, 290–301.
50. S. Yurdakal, V. Loddo, G. Palmisano, V. Augugliaro and L. Palmisano, A quantitative method of photoadsorption determination for irradiated catalyst in liquid-solid system, *Catal. Today*, 2009, **143**, 189–194.
51. C. N. Satterfield, *Heterogeneous Catalysis in Practice*, McGraw-Hill, New York, 1980.
52. O. Redlich and D. L. Peterson, A useful adsorption isotherm, *J. Phys. Chem.*, 1959, **63**, 1024.
53. A. P. Matthews and W. J. Weber Jr., Physical, chemical wastewater treatment, *AIChE Symp. Ser.*, 1976, **73**, 91–98.
54. S. J. Allen, G. Mckay and J. F. Porter, Adsorption isotherm models for basic dye adsorption by peat in single and binary component systems, *J. Colloid Interface Sci.*, 2004, **280**, 322–333.
55. M. Ahmaruzzaman and D. K. Sharma, Adsorption of phenols from wastewater, *J. Colloid Interface Sci.*, 2005, **287**, 14–24.
56. K. V. Kumar and S. Sivanesan, Prediction of optimum sorption isotherm: comparison of linear and non-linear method, *J. Hazard. Mat.*, 2005, **126**, 198–201.
57. A. Mills and S. Le Hunte, An overview of semiconductor photocatalysis, *J. Photochem. Photobiol. A*, 1997, **108**, 1–35.

58. D. F. Ollis, E. Pelizetti and N. Serpone, in: *Photocatalysis: Fundamentals and Applications*, N. Serpone and E. Pelizetti, (ed.), Wiley-InterScience, New York, 1989.
59. M. A. Fox and M. T. Dulay, Heterogeneous photocatalysis, *Chem. Rev.*, 1993, **93**, 341–357.
60. C. Turchi and D. F. Ollis, Photocatalytic degradation of organic water contaminants: mechanisms involving hydroxyl radical attack, *J. Catal.*, 1990, **122**, 178–192.
61. D. F. Ollis, Kinetics of liquid phase photocatalyzed reactions: an illuminating approach, *J. Phys. Chem. B*, 2005, **109**, 2439–2444.
62. A. D. Emeline, V. Ryabchuk and N. Serpone, Factors affecting the efficiency of a photocatalyzed process in aqueous metal-oxide dispersions. Prospect of distinguishing between two kinetic models, *J. Photochem. Photobiol. A*, 2000, **133**, 89–97.
63. Y. Xu and C. H. Langford, Variation of Langmuir adsorption constant determined for TiO_2-photocatalyzed degradation of acetophenone under different light intensity, *J. Photochem. Photobiol. A*, 2000, **133**, 67–71.
64. I. Martyanov and E. Savinov, Photocatalytic steady-state methyl viologen oxidation in air-saturated TiO_2 aqueous suspension: Initial photonic efficiency and initial oxidation rate as a function of methyl viologen concentration and light intensity, *J. Photochem. Photobiol. A*, 2000, **134**, 219–226.
65. A. D. Emeline, V. K. Ryabchuk and N. Serpone, Dogmas and misconceptions in heterogeneous photocatalysis. Some enlightened reflections, *J. Phys. Chem. B*, 2005, **109**, 18515–18521.
66. A. Mills and J. Z. Wang, The kinetics of semiconductor photocatalysis: light intensity effects, *Phys. Chem.*, 1999, **213**, 49–58.
67. A. Mills, J. Wang and D. F. Ollis, Dependence of the kinetics of liquid-phase photocatalyzed reactions on oxygen concentration and light intensity, *J. Catal.*, 2006, **243**, 1–6.
68. V. Augugliaro, V. Loddo, L. Palmisano and M. Schiavello, Performance of heterogeneous photocatalytic systems: influence of operational variables on photoactivity of aqueous suspension of TiO_2, *J. Catal.*, 1995, **153**, 32–40.
69. R. J. Brandi, O. M. Alfano and A. E. Cassano, Evaluation of radiation absorption in slurry photocatalytic reactors. 1. Assessment of methods in use and new proposal, *Environ. Sci. Technol.*, 2000, **34**, 2623–2630.
70. V. Augugliaro, L. Palmisano and M. Schiavello, Photon absorption by aqueous TiO_2 dispersion contained in a stirred photoreactor, *AIChE J.*, 1991, **37**, 1096–1100.
71. M. Schiavello, V. Augugliaro and L. Palmisano, An experimental method for the determination of the photon flow reflected and absorbed by aqueous dispersions containing polycrystalline solids in heterogeneous photocatalysis, *J. Catal.*, 1991, **127**, 332–341.

72. V. Augugliaro, V. Loddo, L. Palmisano and M. Schiavello, Heterogeneous photocatalytic systems: influence of some operational variables on actual photons absorbed by aqueous dispersions of TiO_2, *Solar Energy Mater. Solar Cells*, 1995, **38**, 411–419.
73. V. Augugliaro, V. Loddo, L. Palmisano and M. Schiavello, in: *Photochemie Konzepte, Methoden, Experimente*, D. Wöhrle, M. W Tausch and W. D. Stohrer, (ed.), Wiley-VCH Verlag, Weinheim, 1998, pp. 459–465.
74. G. Spadoni, E. Bandini and F. Santarelli, Scattering effects in photosensitized reactions, *Chem. Eng. Sci.*, 1978, **33**, 517–524.
75. M. Pasquali, F. Santarelli, J. F. Porter and P. L. Yue, Radiative transfer in photocatalytic systems, *AIChE J.*, 1996, **42**, 532–536.
76. R. Changrani and G. B. Raupp, Monte Carlo simulation of the radiation field in a reticulated foam photocatalytic reactor, *AIChE J.*, 1999, **45**, 1085–1094.
77. R. Changrani and G. B. Raupp, Two-dimensional heterogeneous model for a reticulated-foam photocatalytic reactor, *AIChE J.*, 2000, **46**, 829–842.
78. Q. Y. Yang, P. Ling Ang, M. B. Ray and S. O. Pehkonen, Light distribution field in catalyst suspensions within an annular photoreactor, *Chem. Eng. Sci.*, 2005, **60**, 5255–5268.
79. A. Alexiadis, 2-D radiation field in photocatalytic channels of square, rectangular, equilateral triangular and isosceles triangular sections, *Chem. Eng. Sci.*, 2006, **61**, 516–525.
80. G. E. Imoberdorf, O. M. Alfano, A. E. Cassano and H. A. Irazoqui, Monte Carlo model of UV-radiation interaction with TiO_2-coated spheres, *AIChE J.*, 2007, **53**, 2688–2703.

CHAPTER 2
Powders versus Thin Film Preparation

2.1 Introduction

Numerous papers have been published in last decade on the preparation, characterization and use in photocatalytic studies of the most common TiO_2 polymorphs, *i.e.*, anatase and rutile phases, both as powdered forms or films onto various types of supports, mainly glasses. Few studies have been concerned with the brookite phase, due to the difficulty in obtaining it straightforwardly in the absence of the other phases. Little information is available on the other rarer phases, reported below in Table 2.1, and they generally are not used to carry out heterogeneous photocatalytic reactions.

Important manufacturing processes applied to prepare TiO_2 powders, which are used mainly as base pigments in the anatase or rutile phase, are the sulfate, chloride and flame pyrolysis processes.

The sol-gel, hydrothermal, solvothermal, sol, laser pyrolysis and microwave methods reported in the literature are generally more suitable for scientific purposes.[1]

The main techniques used to prepare, instead, photocatalytic films are dip-coating, spin-coating, flow coating, (plasma) spray drying and spray-pyrolysis methods, physical vapor deposition (PVD), chemical vapor deposition (CVD), chemical bath deposition, thermal or anodic oxidation, and electrophoretic deposition.

Any explanation of the photoactivity of TiO_2 powders must carefully consider not only the type of phase(s) present and its (their) intrinsic electronic properties but also its (their) morphological, structural and surface physico-chemical properties. Surface hydroxylation, crystallinity, size of particles and porosity are of paramount importance among the parameters to be evaluated. Note, for instance, the light absorption blue-shift (quantum size effect) observed for TiO_2 nanoparticles and the influence of particle size in the anatase→rutile transition temperature.

Clean by Light Irradiation: Practical Applications of Supported TiO_2
By Vincenzo Augugliaro, Vittorio Loddo, Mario Pagliaro, Giovanni Palmisano and Leonardo Palmisano
© V. Augugliaro, V. Loddo, M. Pagliaro, G. Palmisano and L. Palmisano 2010
Published by the Royal Society of Chemistry, www.rsc.org

Table 2.1 Natural and synthetic TiO_2 forms.

Polymorphic form	Crystal system	Synthesis
Rutile	Tetragonal	
Anatase	Tetragonal	
Brookite	Orthorhombic	
$TiO_2(II)$-(α-PbO_2-like form)	Orthorhombic	
$TiO_2(OI)$	Orthorhombic	
Baddeleyite-like form (seven-coordinate Ti)	Monoclinic	
Cubic form	Cubic	
$TiO_2(OII)$, cotunnite; ($PbCl_2$)-like	Orthorhombic	
$TiO_2(B)$	Monoclinic	Hydrolysis of $K_2Ti_4O_9$ followed by heating
$TiO_2(H)$, hollandite-like form	Tetragonal	Oxidation of the related potassium titanate bronze, $K_{0.25}TiO_2$
$TiO_2(R)$, ramsdellite-like form	Orthorhombic	Oxidation of the related lithium titanate bronze $Li_{0.5}TiO_2$

The same considerations apply to films, and in addition it is essential to consider their thickness, transparency and roughness. Transparency, for instance, is an essential property for superhydrophilic mirrors or glasses that give rise to an antifouling effect without formation of steam on their surface. The superhydrophilicity of a TiO_2 film, which usually increases under irradiation, can be verified by measuring the so-called contact angle of a drop of water onto its surface. The higher the angle, the higher is the hydrophilicity.

The existing scientific literature does not satisfactorily highlight the differences between powders and films as far as their scientific and practical utilizations are concerned.

Notably, the surface area of the photoactive films is smaller than that of the powders and consequently the photodegradation rate of organic pollutants is generally lower than that observed in the presence of powders. Despite this, to-date practical applications of films are more numerous, as it can be noticed by reading other chapters of this book. The main reason for this is that it is not necessary to separate the photocatalyst from the reaction ambient in liquid–solid systems and films of TiO_2 on various supports can be used to successfully reduce the indoor pollution, while also killing some kinds of bacteria, if present.

This chapter is confined to the main preparation methods of pure polycrystalline or nanostructured anatase, rutile and brookite TiO_2 polymorphs mainly used in photocatalytic studies and to the most popular techniques employed to prepare TiO_2 films. Only a few examples, with essential information from the literature, are given; interested readers are invited to scrutinize the cited papers for more details.

2.2 Structures and some Properties of the various TiO$_2$ Polymorphs

TiO$_2$ occurs in nature as rutile, anatase and brookite minerals. Rutile is the most abundant TiO$_2$ polymorph, is an important minor constituent in natural rocks and is isostructural with stishovite. For general nanostructured samples, the transformation sequence between anatase and brookite depends on the initial particle sizes since particle size determines the thermodynamic phase stability.[2,3] All the allotropic phases contain six-coordinated titanium atoms.

Rutile is the most stable phase for particle size above 35 nm, anatase for particle size below *ca.* 11 nm, whereas brookite has been found to be the most stable phase for nanoparticles in the 11–35 nm range. Figure 2.1 shows schematic structures and photographs of the main TiO$_2$ phases.

In addition to rutile, anatase and brookite there are five high-pressure forms: α-PbO$_2$-like form, TiO$_2$(OI), baddeleyite-like form, cubic form, and TiO$_2$(OII) cotunnite-like form.[4,5] Moreover, three forms are produced synthetically, *i.e.*, monoclinic TiO$_2$(B),[6] tetragonal TiO$_2$(H) hollandite-like form[7] and TiO$_2$(R) ramsdellite-like orthorhombic form.[8] As far as the high-pressure forms are concerned, the α-PbO$_2$-like is a dense post-rutile TiO$_2$ phase with Raman bands identical to those of the α-PbO$_2$-structured polymorph of TiO$_2$ synthesized in a dynamic laboratory experiment (2% denser than rutile). The baddeleyite-like is an ultradense post-rutile polymorph of TiO$_2$ (11% denser than rutile) with monoclinic a lattice, isostructural with the baddeleyite ZrO$_2$ polymorph, and the titanium cation is coordinated with seven oxygen anions. The cotunnite-like polymorph is a cotunnite-structured titanium oxide that represents the hardest oxide known. This is a new polymorph of TiO$_2$, where titanium is nine-coordinated to oxygen in the cotunnite (PbCl$_2$) structure.[9] Table 2.1 summarizes the above-cited polymorphic forms.

TiO$_2$ is used in a wide range of applications in ceramics and in electric devices and moreover as a gas sensor, a white pigment (*e.g.*, in paints and cosmetic products), a corrosion-protective coating and an optical coating. It plays an important role in earth sciences, in biocompatibility of bone implants, is spacer materials, and finds applications in nanostructured form in Li-based batteries and electrochromic devices. TiO$_2$ is also used in heterogeneous photocatalysis both in powdered form and as film on various types of supports and in solar cells for the production of hydrogen and electric energy.[10]

2.3 Preparation Methods of Powdered TiO$_2$

Many methods have been used to prepare powdered TiO$_2$, mainly in the anatase or in the rutile phase. In the following, after a short presentation of some industrial processes, the preparations more frequently found in the scientific literature will be presented. They will be examined in terms of the photocatalytic applications of TiO$_2$ by describing also briefly some papers reporting

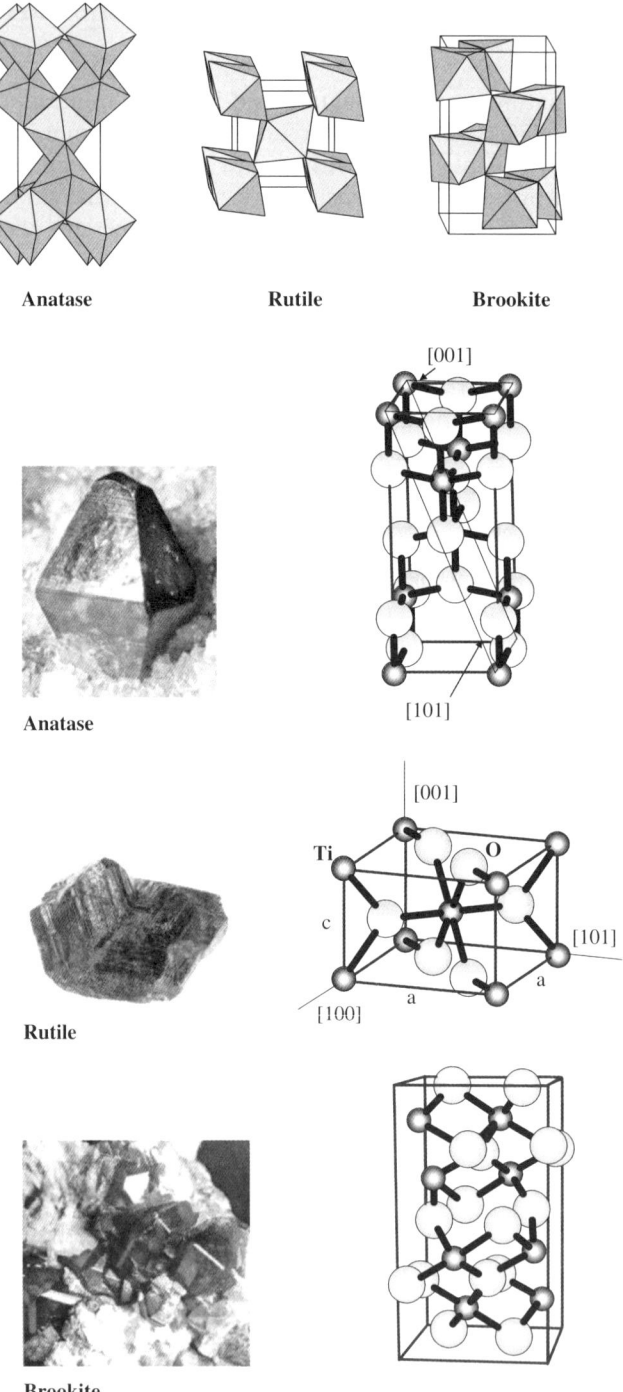

Figure 2.1 Structures and photographs of crystals of the main polymorphs of TiO$_2$. (Photograph courtesy of http://www.probertencyclopaedia.com.)

the formation under particular experimental conditions of nanorods, nanotubes and nanowires.

2.3.1 Sulfate Process

The sulfate process was the first commercial process for manufacturing powdered TiO_2. The first used starting raw material was ilmenite ($FeTiO_3$), but subsequently other ores with a much higher TiO_2 content have been chosen.

The ore is subjected to drying and grinding and then it is reacted with concentrated sulfuric acid (exothermic digestion reaction). The experimental conditions are controlled (minimum amount of acid should be used) to maximize the conversion of TiO_2 into water-soluble titanyl sulfate. The resultant dry, green-brown cake of metal sulfates is dissolved in a weak acidic aqueous solution, the temperature is lowered to avoid premature hydrolysis, and the solution clarified by settling and chemical flocculation. The resultant clear solution is then further cooled to crystallize the by-product $FeSO_4 \cdot 7H_2O$, which is separated.

The insoluble mud so-obtained is washed to recover the titanyl sulfate liquor, which is filtered to remove the insoluble impurities. The solution is evaporated to a precise composition and hydrolyzed to obtain a pulp consisting mainly of clusters of colloidal hydrous titanium oxide. Precipitation is induced by a seeding or nucleation technique. The pulp is then separated from the mother liquor and washed to remove residual traces of metallic impurities, using sometimes chelating agents. The washed pulp is treated with chemicals that adjust the physical texture and act as catalysts in the calcination step. This process can produce either anatase or rutile crystal forms depending on the additives used before the calcination process.

2.3.2 Chloride Process

The starting material for the chloride process is generally a mineral rutile containing over 90% TiO_2. An ore blend mixed with a source of carbon is reacted in a fluidized bed with chlorine at *ca.* 1173 K. The exothermic reaction, which needs accurate temperature control, yields $TiCl_4$ together with the chlorides of all the impurities present. The mixed chlorides are cooled and the low-volatile chloride impurities (*e.g.*, manganese, chromium and iron) are separated by condensation and removed from the gas stream along with any unreacted solid starting materials. After condensation of the $TiCl_4$ vapor, an extremely pure, colorless, mobile liquid $TiCl_4$ intermediate product, freezing at 249 K and boiling at 409 K, is produced by fractional distillation. The success of the chloride process is due to this stable intermediate, which can be purified, tested, stored, reprocessed and handled as a liquid or vapor. Moreover, potentially discoloring trace contaminants can be eliminated with benefit to pigment color because the distillation process is a vapor-phase process. The second stage in the chloride process is the oxidation of $TiCl_4$ to TiO_2. $TiCl_4$ is reacted with O_2 (exothermic reaction) to form TiO_2 with liberation of chlorine, which is recycled to the chlorination stage. Owing to the high temperature, only rutile phase is produced. After cooling, the gas stream passes through a separator to collect the TiO_2 particles and it is treated to remove adsorbed chlorine.

Notably, in both sulfate and chloride processes the commercial TiO_2, usually sold as a pigment, is subjected to a surface treatment with a combination of Al_2O_3, SiO_2 and/or ZrO_2. Sometimes, organic species are also used, such as for instance polyols, which are alcohols containing multiple hydroxyl groups. The aim of the treatment is to improve the wetting, stability of the dispersion in different media, compatibility with the binder, color stability, and durability.

It is imperative to know the nature and the extent of additives and impurities when commercial powdered TiO_2 samples are used as photocatalysts, as their presence could significantly influence the photoactivity.

2.3.3 Flame Pyrolysis Process

This process has been used to prepare the most popular commercial sample used in photocatalytic studies, *i.e.*, Degussa P25, but some research papers[11–13] also report its use to prepare TiO_2 soft agglomerates (or just agglomerates) and hard agglomerates (aggregates).

In ref. 13, for instance, titanium tetraisopropoxide, used as the precursor, diluted (0.8 M) with xylene, was injected (5 ml min^{-1}) by a syringe pump into the reactor through the center capillary of the flame synthesis pyrolysis nozzle. Oxygen (5 ml min^{-1}) was fed through the surrounding annulus, dispersing the solution into droplets. The pressure drop at the nozzle was held constant at 1.5 bar by adjusting the annulus gap width. A supporting premixed CH_4/O_2 (ratio: 1.5/3.2) flame surrounding the dispersion oxygen annulus ignited and stabilized the spray flame. The extent of agglomeration or aggregation was a function of the applied dispersion pressure.

2.3.4 Preparation of Powdered and Nanocrystalline TiO_2 by the Sol-gel Method

The sol-gel technique enables one to synthesize highly pure and homogeneous materials by starting from a colloidal suspension (sol) obtained from liquid solutions of organometallic or inorganic precursors. The stability of a sol suspension is due to van der Waals and coulombic repulsive strengths among very small particles negligibly subjected to gravitational forces. The transformation of a sol into another phase (the gel) can be induced under various experimental conditions. The gel is a macromolecular and highly viscous solid that encloses molecules of solvent. By drying the gel it is possible to obtain porous solids that usually are used to produce advanced materials (*e.g.*, ceramics).

In a typical sol-gel process carried out to prepare TiO_2, the sol is obtained by hydrolysis and polymerization reactions of the TiO_2 precursors that are generally organic titanium alkoxides (also metal inorganic salts can be used) as tetra-*n*-butoxide or tetraisopropoxide dissolved in alcohol (2-propanol, 1-pentanol, ethanol, *etc*.). The latter are added to water [Equation (2.1)] to induce hydrolysis to titanium hydroxide and the size of the TiO_2 particles depends on the experimental conditions used, such as for instance the ratio between precursor and water, initial pH, reaction time, presence of foreign ions and temperature:

$$Ti(iso-OC_3H_7)_4 + 4H_2O \rightarrow Ti(OH)_4 + 4C_3H_7OH \qquad (2.1)$$

Polymerization and loss of solvent gives rise to the transition from liquid sol into a solid gel phase. The formation of the Ti-O-Ti chains is facilitated by a low content of water, low hydrolysis rate and excess of precursor in the reaction ambient. The sol-gel method is a versatile approach to preparing TiO_2 with tailored morphological properties and the removal of the organic counterpart from the TiO_2 gels plays an important role in the preparation of TiO_2 samples with physicochemical features strictly related to their photocatalytic properties.[14-22] A gelling process can be carried out by treating the reactant mixtures under reduced pressure[23] and/or by using various additives, such as HCOOH, HO_2CCO_2H, CH_3COOH, HCl, $SnCl_2$, *etc*.[24-28] Some additives can play the role of chelating agents, others the role of hydrolysis catalysts (HCl or $SnCl_2$). In particular, the presence of appropriate chelating agents that modifies the titanium alkoxide precursor allows one to control the hydrolysis and condensation reaction rate.

An example of the synthesis of TiO_2 samples is reported in the following. Titanium isopropoxide in 2-propanol was allowed to react with a stoichiometric amount of a chelating ligand (HCOOH or CH_3COOH). Subsequently it was hydrolyzed under strong acidic catalysis (HCl) or with water using $SnCl_2$ as the catalyst. The gelling process was carried out by either leaving the reactant mixture in air or treating the pre-hydrolyzed solution under reduced pressure. The addition of water gave rise to fine emulsions that turned into a clear solution after *ca*. 1 day. The pre-hydrolyzed solution was divided in two parts: the first was treated under reduced pressure until a granular gel was obtained, while the second was left in the air at 298 K until a monolithic gel was obtained. The latter was dried in the air for 7 days and then under vacuum for 2 days. The samples obtained from ligand-modified Ti-alkoxide precursors were white powders, whereas those prepared in the presence of $SnCl_2$ were pale yellow materials. The samples gelled under reduced pressure presented low values of BET specific surface areas compared with those spontaneously gelled in the air, due to the forced solvent evaporation from incompletely hydrolyzed titanium oligomers. Both types of crude gels (vacuum or air gelled) were not satisfactorily crystallized and they were also subjected to a water reflux treatment or to heating in air to obtain anatase (688–808 K), rutile ($T > 923$ K) or mixed anatase–rutile phases. In some cases brookite phase appeared together with anatase.

Notably, the photocatalytic activity of such TiO_2 nanocrystalline samples depended not only on the structural, intrinsic electronic and surface physicochemical properties of the samples but also on the possible presence of adsorbed organic impurities, which had detrimental effects by poisoning the catalytic sites. The most photoactive samples for photodegradation reaction were those prepared in the presence of formic acid gelled spontaneously in air.

The reflux conditions reported in ref. 29, which describes the synthesis of anatase nanoparticles, were a 16 h reflux process performed after precipitation of a white precipitate consisting of amorphous particles. Subsequent centrifugation, evaporation under reduced pressure at 313 K, and redispersion in toluene gave rise to a sol of *ca.* 20 wt% of nanoparticles of titania with a size of a few nanometers. The separated nanoparticles were mixed with 3,3,4,4,5,5,6,6,7,7,8,8,8,-tridecafluorooctyl-1,1,1,-triethoxysilane (FTS) in dry methyl ethyl ketone to obtain a transparent binder, which is stable in a refrigerator for at least 2 months, that can be used for the preparation of photocatalytic coatings.

Other papers report the preparation of polycrystalline or nanostructured TiO_2 by using inorganic salts [$TiCl_3$, $TiCl_4$, $Ti_2(SO_4)_3$, $Ti(SO_4)_4$] as the precursors and by carrying out the hydrolysis in water and ammonia[30,31] or in pure water.[32,33]

Pioneering studies on bare and iron-loaded TiO_2 photocatalysts by Bickley *et al.*[30,31] have reported the preparation of polycrystalline TiO_2 in the anatase, rutile or anatase and rutile mixed phases using $TiCl_3$ as precursor. The preparation started with an aqueous solution of $TiCl_3$ to which ammonia was added dropwise at room temperature. The reaction products were allowed to stand for 24 h in a vessel and subsequently the resulting solid was filtered off, washed thoroughly with distilled water until it was free of residual Cl^- ions and finally dried at 373 K for 24 h. Various portions of the obtained solid were calcined for 24 h at 773, 923, 1073 and 1273 K. The presence of rutile phase increased with increasing calcination temperature up to *ca.* 1273 K, at which point only rutile phase was detected by X-ray diffraction measurements.

Another fundamental study[32] deals with the preparation of some TiO_2 samples both in the anatase and rutile phases by using $TiCl_3$ or $TiCl_4$ (ex $TiCl_3$ or ex $TiCl_4$) as the precursors. The method used to obtain the samples ex $TiCl_3$ is virtually identical to that employed in 30 and 31, whereas when $TiCl_4$ was used as the precursor it was allowed to react with pure water at *ca.* 278 K and the pH was adjusted by NaOH up to 4.5. Notably, caution is needed in the last case because the reaction is strongly exothermic and gaseous HCl is released. The subsequent procedure was the same as described above for samples ex $TiCl_3$.[30,31]

Both samples ex $TiCl_3$ and ex $TiCl_4$ were tested to carry out the photodegradation of phenol in aqueous medium as a probe reaction. The physicochemical features as determined by the preparation methods affect the photocatalytic behavior in addition, of course, to the semiconducting properties. Rutile phase was active or inactive according to the preparation conditions; in particular, it was completely inactive only when prepared by calcining at relatively high temperatures ($T>973$ K) and for long times ($t>3$ h). The most photoactive samples were those ex $TiCl_4$ that were prepared by calcining 336 h at 823 K, or 24 h at 873 K.

The improved crystallinity beneficially influenced the photoreactivity, but the presence of surface hydroxyls and the particle size also appeared to be essential factors. A compromise between these properties seems to be essential to guarantee an appreciable photoactivity. Irreversible dehydroxylation, the extent of which depends on the temperature and time of calcination, indeed, does not favor the adsorption of O_2 onto the TiO_2 surface. The photoreduction of $O_{2(ads)}$ by electrons produced under illumination in the conduction band is in principle possible on both TiO_2 modifications. $O_2^-{}_{(ads)}$ species may evolve in various ways, producing further oxygenated species involved in the photo-oxidation of the organic molecules. On the other hand, the oxidative processes, which involve the photogenerated holes and adsorbed phenolic species as well as OH^- groups, have similar driving forces on anatase and rutile. Therefore, the photooxidative degradation processes, from a thermodynamic point of view, should occur for both anatase and rutile.

Consequently, the differences in photoactivity reported in the literature often do not depend very strongly on the intrinsic electronic characteristics of the two phases, although the electron–hole recombination rate of rutile phase has been reported to be higher than that of anatase,[32,34] but mainly on the surface physicochemical properties of the samples.

The influence of one or many of these surface properties can be more or less significant, depending on the type of molecule to be photodegraded, on the pH of the medium, the irradiation intensity, *etc.* Consequently, it is not easy to straightforwardly correlate all these factors with the photoreactivity, since interplay between them and the redox reactions takes place, which sets the final level of photoreactivity.

2.3.5 Preparation of Nanoparticles, Nanorods, Nanotubes and Nanowires of TiO_2 by the Hydrothermal Method

Hydrothermal syntheses of TiO_2 nanoparticles[35–44] or nanorods[45] are carried out in water or in mixtures water/ethanol and, generally, steel pressure autoclaves are used, with the possibility of controlling temperature and pressure.

In ref. 44 the TiO_2 nanoparticles were prepared by adding dropwise titanium tetraisopropoxide to a water/ethanol solution at pH 0.7 (adjusted with HNO_3) and allowing the reaction to occur at 153 K for 4 h. The size of the anatase particles obtained in such a way ranged between 7 and 25 nm, depending on the concentration of the precursor and the composition of the mixed water/ethanol solvent. Nanorods were obtained, instead, by treating a dilute aqueous solution of $TiCl_4$ at 606–696 K for 12 h in the presence of acids or inorganic salts.[45] Some authors tune the morphology of the resulting nanorods by using additives or changing the experimental conditions.[46–48]

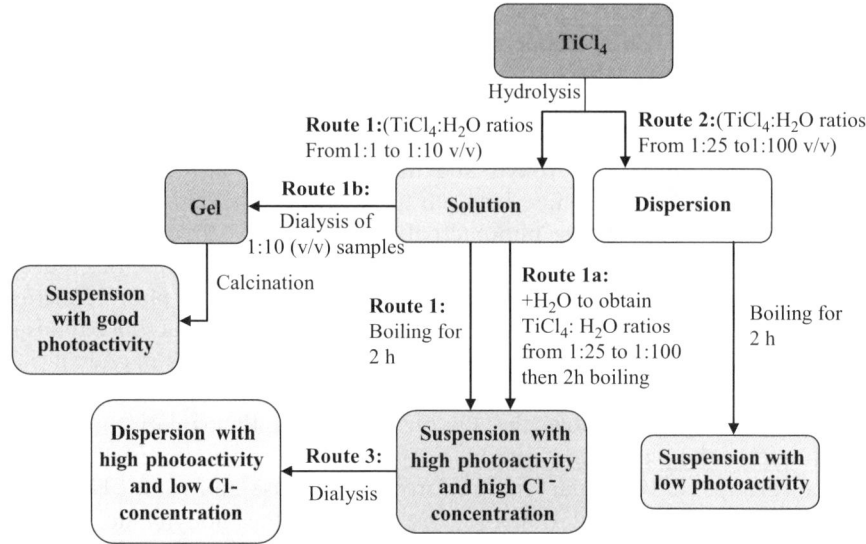

Scheme 2.1 Preparation procedure of nanostructured TiO$_2$ samples obtained by hydrolysis of TiCl$_4$, as described in ref. 50.

Other papers[49–51] report the use of TiCl$_4$ as precursor to obtain nanoparticles of TiO$_2$ under experimental milder conditions than those reported above that employed hydrothermal processes in a closed vessel at higher pressures than 1 atm. Scheme 2.1 sketches the procedure followed to obtain the samples described in ref. 50, and Figure 2.2 shows XRD diffractograms of some samples.

In these papers the photoactivity mainly depended on a compromise between crystallinity and particle size. In fact, the beneficial role of an increased crystallinity contrasts with the detrimental effect due to the corresponding increase of particle size. The most active samples prepared by using a TiCl$_4$/H$_2$O ratio equal to 1 : 50 just matched the highest crystallinity and the lowest particle size. Notably, the nanostructured TiO$_2$ catalysts were more efficient than Degussa P25 at pH 2. The activity of the catalysts was enhanced by dialyzing the suspensions against water. The technique presented seems an economical and fast way to prepare active photocatalysts. Limitations are expected for their use in solar reactors due to the slight shift (quantum size effect) in the band gap of the samples whose diameter range between *ca.* 10 and 45 nm.

More recently nanocrystalline anatase, rutile and brookite TiO$_2$ powders have been prepared by thermohydrolysis starting from TiCl$_4$ in HCl or NaCl solutions.[52,53] Reference 52 reports that rutile, mixtures of brookite and rutile or mixtures of anatase, brookite and rutile were obtained depending on the acidity of the medium. Mixtures of brookite and rutile were prevalently obtained when the hydrolysis of TiCl$_4$ occurred in diluted HCl. Ternary mixtures of anatase, brookite and rutile were always formed in NaCl solutions.

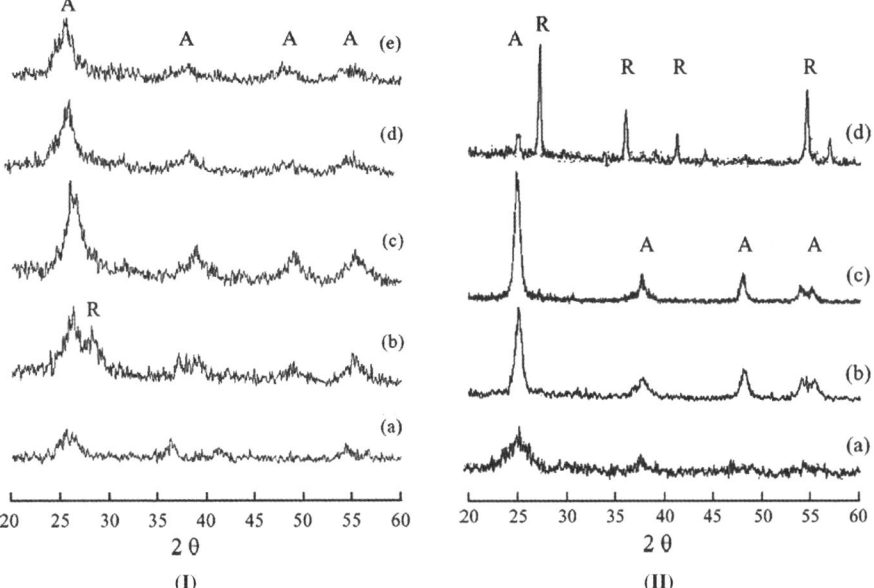

Figure 2.2 (I) XRD patterns of samples prepared with various TiCl$_4$/H$_2$O ratios: (a) 1 : 1, (b) 1 : 10, (c) 1 : 50, dialyzed, (d) 1 : 50 and (e) (1:100). A: anatase, R: rutile. (II) XRD patterns of powders obtained by drying or calcining the gel (Route 1b in Scheme 2.1): (a) 298, (b) 673, (c) 773 and (d) 873 K. A: anatase, R: rutile. (Reproduced with permission from ref. 50).

Pure brookite nanoparticles were separated from rutile by simple peptization with water by slightly modifying the method reported in ref. 54, where the peptization was performed by using HNO$_3$. 4-Nitrophenol photodegradation was used to evaluate the photocatalytic activity of the various samples. The highest activity corresponded to powders consisting of more than one crystalline phase. The effects of phase composition and characterization of the photocatalysts were studied in more detail in ref. 53. The most efficient samples consisted of a ternary mixture of anatase, brookite and rutile and the presence of brookite phase did not seem to negatively influence the photoreactivity. Some powders were more active than Degussa P25.

Some photoelectrochemical measurements[53] performed according to the method reported in ref. 55 have allowed determination of the position of the flat band potential and the valence and conduction bands of the three polymorphic phases (Figure 2.3).

Notably, the experimental flat-band potential of pure brookite (−0.46 V), reported for the first time in ref. 53, is very close to the −0.44 V calculated by Grätzel and Rotzinger by means of extended Hückel molecular orbital (MO) calculations.[56]

TiO$_2$ brookite phase has been used less frequently in photocatalysis than the anatase and rutile phases because it is obtained as a mixture with one or both of

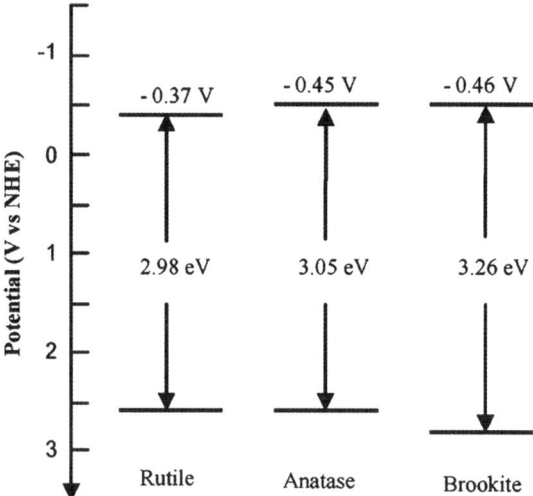

Figure 2.3 Band edges (*versus* NHE) at pH 7 of rutile, anatase and brookite samples. (Reproduced with permission from ref. 53.)

the other phases and its separation implies an additional peptization process as above reported.[53,57]

Moreover, the interpretation of XRD diffractograms is not easy when the brookite phase is present together with the anatase. In fact the main (101) diffraction peak of anatase at $2\theta = 25.28°$ overlaps with the (120) and (111) peaks of brookite at $2\theta = 25.34°$ and $25.69°$, respectively. Therefore, it is advisable to use Raman spectroscopy to demonstrate unequivocally the complete absence of anatase phase, by checking the absence of anatase characteristic peaks at 391 and 510 cm^{-1}.[57]

Figure 2.4 shows some XRD diffractograms of brookite, and brookite–rutile mixtures before their separation, while a representative Raman spectrum of the pure brookite phase is given in Figure 2.5.

Very recently a set of papers has reported the preparation under very mild conditions of badly crystallized nanocrystalline TiO$_2$ anatase, brookite and rutile phases that were more selective than Degussa P25 (*ca.* 30% rutile, 70% anatase) and Merck (anatase) commercial samples for the photooxidation in water of some aromatic alcohols (benzyl, 4-methylbenzyl, 4-methoxybenzyl and 4-nitrobenzyl alcohols) to the corresponding aldehydes.[57–63] The photocatalysts were prepared by hydrolysis of TiCl$_4$ in water at room temperature. Scheme 2.2 reports the synthesis of brookite and rutile TiO$_2$ by hydrolysis of TiCl$_4$ in HCl aqueous solution.

The most selective samples to aldehyde (up to 72% mol with respect to the transformed alcohol) were rutile samples prepared under very mild conditions at room temperature from TiCl$_4$.[61]

A comparative ATR-FTIR study of selected laboratory prepared photocatalysts and Degussa P25 has been reported[63] together with the determination

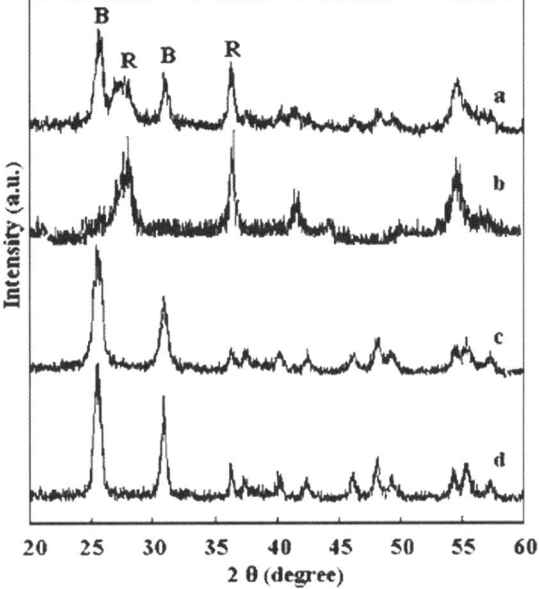

Figure 2.4 XRD diffractograms of (a) brookite–rutile mixture, (b) pure rutile, (c) pure brookite and (d) calcined brookite (723 K, 2 h). (Reproduced with permission from ref. 57.)

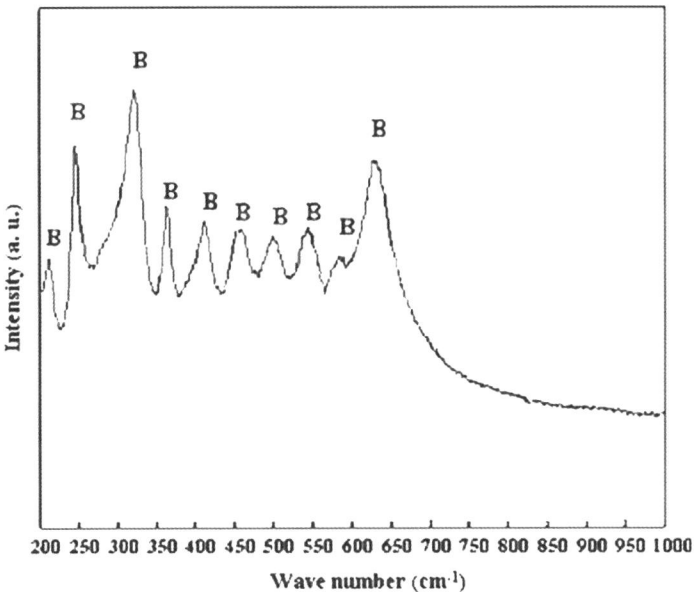

Figure 2.5 Raman spectrum of a representative sample of pure TiO_2 brookite. (Reproduced with permission from ref. 57.)

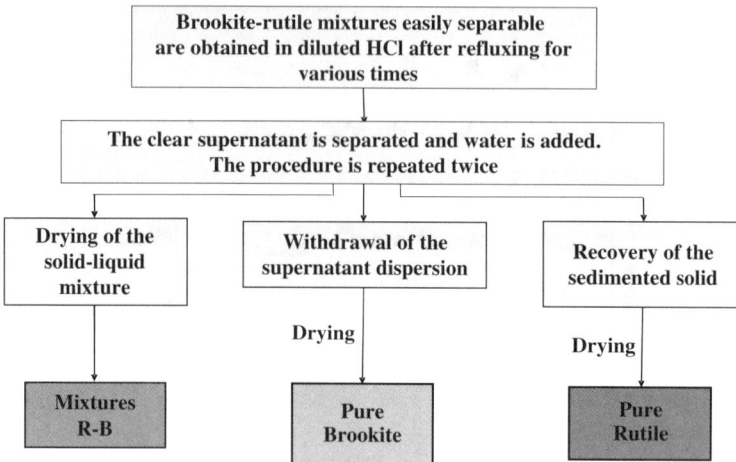

Scheme 2.2 Synthesis of brookite and rutile TiO$_2$ by hydrolysis of TiCl$_4$ in HCl aqueous solution.[57–63]

of their main photoelectrochemical properties (band gap, conduction and valence band edges) by means of the method reported in ref. 55. The band gap, valence band and conduction band edges were almost identical for all of the laboratory prepared samples in which anatase phase was predominant, although the samples showed a different selectivity to aldehydes. Conversely, appreciable differences were noticed for laboratory-prepared samples containing other phases. The ATR-FTIR results showed a very dissimilar hydrophilicity (Degussa was more hydrophobic than the laboratory-prepared samples) and different ability for adsorbing the organic compounds deriving from benzyl alcohol photocatalytic oxidation. These results indicate that for selective photooxidation of alcohol to aldehyde, analogously to the photomineralization reactions, the surface physicochemical properties should be considered because they could play a more important role than the intrinsic electronic properties.

Preparation of mesoporous TiO$_2$ microspheres with sizes ranging between 5 and 30 µm starting from Ti$_2$(SO$_4$)$_3$ industrial grade (16.9 wt%) was also reported.[64] The microspheres could be used as photocatalyst in fluidized bed reactors to treat waste water. They were prepared as follows: blue Ti(OH)$_3$ was obtained by injecting contemporaneously Ti$_2$(SO$_4$)$_3$ and NH$_4$HCO$_3$ solutions *via* two syringe pumps into a reactor containing distilled water under vigorous stirring. The pH value was maintained at 6.5–7.0 by adjusting the flux of ammonium hydrogen carbonate to guarantee the complete precipitation of Ti^{3+} ions. The precipitate was filtered off, washed with distilled water and dried at 373 K for 12 h to be transformed into TiO$_2$ microspheres. Different fractions of them were finally subjected to calcinations at 773, 873, 973, 1073 or 1173 K for 2 h in air. XRD data indicated that the microspheres were all amorphous when the calcination temperature was below 773 K, while relatively broad diffraction peaks due to anatase phase were present for the materials calcined at

873–1073 K. The crystallite sizes, determined by using the Scherrer equation, increased by increasing the calcination temperature, and they ranged between *ca.* 7 and 35 nm. The samples calcined at 1173 K showed the presence of the rutile phase in addition to the anatase phase.

According to calculations, the TiO_2 sample calcined at 1173 K was a mixture of rutile 29.6% and anatase 70.4%, which is similar to Degussa P25. In addition, the phase transformation from anatase into rutile in TiO_2 materials did not occur during calcination below 1073 K and, consequently, only the anatase phase was present at 1073 K, some 100–200 K higher than the reported phase transformation temperatures.[65,66] This shows that the thermodynamic phase stability can be dependent on particle size, and in particular the anatase structure proved to be thermodynamically more stable than the rutile one for particle sizes of *ca.* 14 nm.[67] The latter value is similar to the crystal sizes of the sample calcined at 1073 K.

The hydrothermal method has also been used to prepare TiO_2 nanorods (Figure 2.6),[45] nanotubes (Figure 2.7)[68–73] and nanowires (Figure 2.8).[74,75]

TiO_2 nanotubes have been reported to be obtained from TiO_2 powders held at 293–383 K for 20 h in an autoclave in an aqueous solution of NaOH (2.5–20 M).[68] It is, however, necessary to wash the products with HCl aqueous solution and water. It has been hypothesized that some of the Ti–O–Ti bonds break in the presence of NaOH, with resultant formation of Ti–O–OH and Ti–ONa bonds and the subsequent formation of nanotubes after the acid treatment.

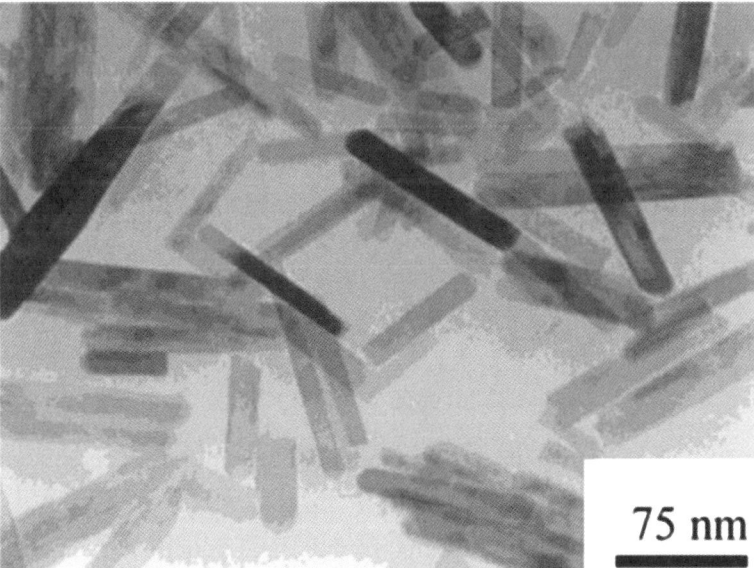

Figure 2.6 TEM photograph of TiO_2 nanorods obtained by the hydrothermal method. (Reproduced with permission from ref. 45.)

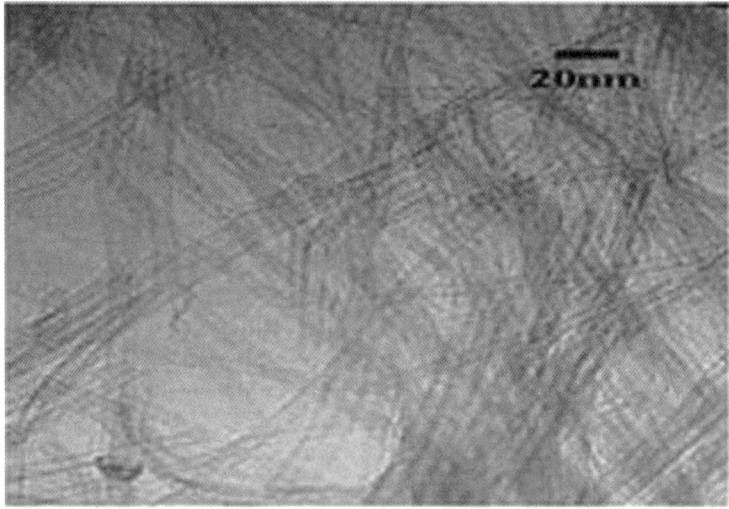

Figure 2.7 TEM photograph of TiO_2 nanotubes. (Reproduced with permission from ref. 71.)

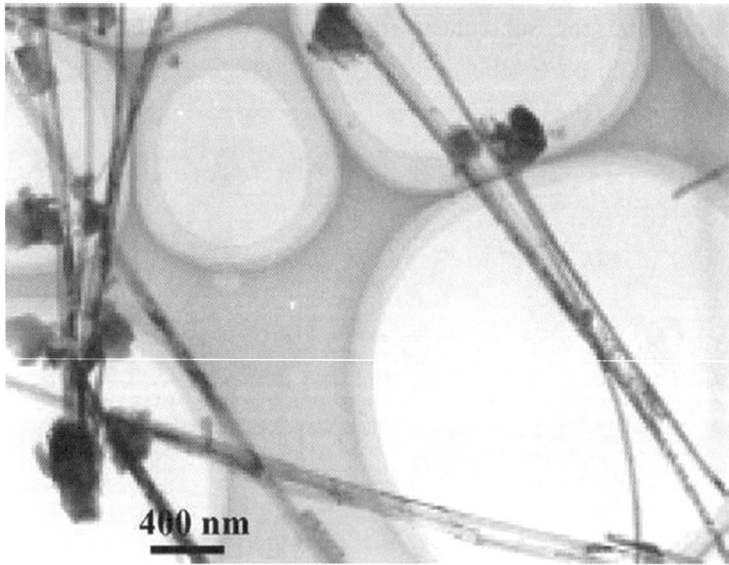

Figure 2.8 TEM photograph of TiO_2 nanowires. (Reproduced with permission from ref. 76.)

TiO_2 nanowires have been prepared from layered titanate particles using the hydrothermal method.[75] The layered structured $Na_2Ti_3O_7$ was treated with HCl (0.05–0.1 M) at 413–443 K for 3–7 days in an autoclave and subsequently the product was washed with water. Before the formation of nanowires, steps involving exfoliation of layered $Na_2Ti_3O_7$ and the formation of nanosheets occurred.

2.3.6 Solvothermal Method

The solvothermal method differs from the hydrothermal method in that a non-aqueous solvent it is used. Higher temperatures can be achieved because many organic solvents with boiling points higher than that of water can be chosen. Their use allows a better control of size and shape distributions of the TiO_2 nanoparticles, nanorods or nanowires.[76–80]

In ref. 80 titanium tetraisopropoxide was mixed with toluene (weight ratio 1–3 : 10) for 3 h at 523 K. The average particle size of TiO_2 powders increased by increasing the composition of titanium tetraisopropoxide, but no TiO_2 was produced at 1 : 20 and 2 : 5 weight ratios.

Surfactants have been used to synthesize TiO_2 nanoparticles or nanorods with narrow size distribution.[76,79] In ref. 79 titanium tetraisopropoxide, dissolved in anhydrous toluene in the presence of oleic acid as the surfactant, was kept for 20 h at 523 K in an autoclave without stirring. Nanorods were formed in the presence of a sufficient amount of titanium tetraisopropoxide or surfactant. The diameter and length of these nanorods were 3–5 nm and 18–25 nm, respectively, depending on the ratio precursor/surfactant and precursor/solvent.

The solvothermal method was also used to prepare nanowires in the same way as with the hydrothermal method.[81,82] In ref. 81 a TiO_2 suspension in 5 M NaOH water–ethanol solution was kept in an autoclave at 443–473 K for 24 h. TiO_2 nanowires were obtained after the product was cooled at room temperature, washed with a dilute HCl aqueous solution and dried at 333 K for 12 h in air. The type of solvent influenced dramatically the crystal morphology. In the absence of ethanol, for instance, flake-like structures of TiO_2 were obtained, whereas in the presence of chloroform TiO_2 nanorods were produced.

2.3.7 Sol Method

This method is a nonhydrolytic sol-gel process and generally involves the reaction of $TiCl_4$ (or other halides) with different oxygen donor species, such as organic ethers or metal alkoxides:[83,84]

$$TiCl_4 + 2ROR \rightarrow TiO_2 + 4RCl \quad (2.2)$$

$$TiCl_4 + Ti(OR)_4 \rightarrow 2TiO_2 + 4RCl \quad (2.3)$$

In ref. 83 metal alkoxides were allowed to react for no longer than 5 min with a hot solution of $TiCl_4$ mixed with trioctylphosphine oxide in heptadecane at 573 K under inert gas atmosphere. The size of the particles was not affected by the type of alkyl substituent, R, whereas the reaction rate increased upon increasing the branching of R.

Various titanium halides were tested in addition to $TiCl_4$ and the particles size ranged from 9.2 nm for TiF_4 to 3.8 nm for TiI_4, indicating that the size of the formed nanocrystals of anatase decreased upon increasing the size of the halide.

In ref. 84 the reaction between $TiCl_4$ and benzyl alcohol was performed at 313–423 K for 1–21 days. The product was washed and subsequently calcined at 723 K for 5 h. The particle size depended strongly on temperature and on the concentration of $TiCl_4$. In particular, particles in the range 4–8 nm were obtained by choosing appropriate thermal conditions and the concentration of $TiCl_4$. Moreover, an almost uniform size and shape of the particles was observed at low temperatures (*e.g.*, 313 K).

In the presence of surfactants, TiO_2 nanorods of different size and shape were synthesized.[85–90] The preparation started with the hydrolysis under nitrogen flow of titanium tetraisopropoxide in oleic acid at 353–373 K for 5 min. To this solution a 0.1–2 M aqueous base solution was added rapidly and then the mixture was mixed for 6–12 h. Some of the bases used were trimethylamine, trimethylamino-*N*-oxide, triethylamine, tributylamine, tetramethylammonium hydroxide and tetrabutylammonium hydroxide.

2.3.8 Laser Pyrolysis Method

IR laser pyrolysis has been employed successfully as a single-step reaction route for the synthesis of anatase/rutile mixed nanopowders. Sensitized mixtures containing $TiCl_4$ (vapor) and air as gas-phase precursors were used and mixed crystallographic phases of TiO_2 (rutile and anatase, 20 and 30 nm mean diameter, respectively) were obtained. TiO_2 nanopowders with *ca.* 90% anatase content were obtained at a high concentration of air.[91,92]

Another paper[93] reports the synthesis of TiO_2 powders with an average diameter as low as 8 nm by laser pyrolysis by using an aerosol of titanium tetraisopropoxide as the main precursor sensitized by C_2H_4. The possibility of controlling the anatase/rutile phase ratio was demonstrated over a large range by tuning the experimental parameters.

Samples prepared by using laser radiation usually appear to be less porous materials.

2.3.9 Microwave Method

It is possible to use microwave radiation to prepare TiO_2 nanomaterials.[94–97] In ref. 94 high quality rutile TiO_2 nanorods were prepared by means of a microwave hydrothermal method, while ref. 95 describes the preparation of nanotubes between 8 and 12 nm diameter and up to 200–1000 nm long. TiO_2 nanotubes formed solely by microwave irradiation of NaOH aqueous solutions contained anatase, rutile or both phases.

Corradi *et al.*[96] have reported the rapid preparation (5 min to 1 h) of nanoparticle suspensions of TiO_2 with microwave radiation, whereas 32 h was needed for a conventional hydrothermal synthesis at 468 K.

More recently, Addamo et al. have presented[97] a very simple procedure to prepare anatase, rutile or mixed phase TiO_2 photocatalysts by microwave irradiation at low temperature. The samples were active for 4-nitrophenol photodegradation in aqueous solution and 2-propanol photooxidation in a gas–solid system. The starting materials were unusual TiO_2 gels formed from $TiCl_4$. The use of microwaves for very short times enhanced the TiO_2 crystallinity, preventing an increase of particle size and minimizing the decrease of specific surface area. The particle size ranged between 16 and 25 nm, depending on the treatment temperature (393 or 423 K) and time (15, 30 or 60 min). The most active sample was prepared by irradiating for 30 min at 423 K the gel obtained from a 1 : 10 $TiCl_4$: H_2O (v/v) solution. For this last sample the observed initial reaction rate of 4-nitrophenol photodegradation was 93×10^{-9} versus 50×10^{-9} mol l^{-1} s^{-1} determined when TiO_2 Degussa P25 was used. A relatively low efficiency for the photooxidation of gaseous 2-propanol was obtained, though, due to the large amount of water present in the gel, which reduces the number of active TiO_2 surface sites.

2.4 Preparation of Films

Research on the preparation, characterization, and testing of films (thickness ranging from a few nanometers to ca. 100 μm) constituted by many kinds of insulator and semiconducting materials has attracted great attention among the scientific community because these devices are widely used in micro- and optoelectronic industries.

Some types of films can be used in devices (sensors) that provide information on the occurrence of some phenomena by exploiting thermal, biological, optical, magnetic, etc. effects. Many, for instance, are very important in the field of environmental chemistry, allowing the detection of noxious gases. This chapter does not aim to report general information on the numerous typologies of existing films;[98–102] instead, only some methods suitable for preparing photocatalytic TiO_2 films are presented.

2.4.1 Wet Coating Technologies to Prepare TiO_2 Films

These technologies have been developed for various types of supports, but mainly for glasses. Both the wet coating materials and the equipment are not standardized and therefore they have been tested and developed by the users. Transparent and non-transparent coated materials can be obtained, depending on the experimental conditions employed. Examples are glasses with antifogging, conductive and photocatalytic properties, although to choose the wet coating technique for large area or wide volume applications is not opportune because other techniques that will be illustrated below (see for instance magnetron sputtering) are more convenient. Wet coating techniques, however, often appear cost-effective and can be divided basically into two categories: the first implies a heating treatment after the coating step at temperatures generally ranging between 423 and 823 K, depending on the support used and on the TiO_2 phase that it is expected to form, while the second involves an UV or IR treatment. In the second case, unlike the first, the chemical structures obtained in the liquid coating material do not change.

Notably, the coating step has to be performed onto a cleaned surface and under clean room conditions.

In the following, only dip-coating, spin-coating, flow coating, (plasma) spray drying and spray-pyrolysis methods will be presented as they have been selected very often for the preparation of TiO_2 photocatalytic films. In most cases the obtainment of TiO_2 photoactive films requires a heating post-treatment because the precursor species – which are generally dissolved (usually as titanium alkoxides or titanium chlorides) in alcoholic solution or are present as a sol or a not highly viscous gel in aqueous or alcoholic suspension (due to the occurrence of a pre-hydrolysis process) – need to be transformed into nanocrystalline TiO_2 after their contact with the surface of the support.

2.4.1.1 Dip Coating

Dip-coating is carried out by first immersing the support in a liquid, in which the precursor is present, and then withdrawing it at controlled speed and temperature under atmospheric conditions (Figure 2.9).

This process has been developed both for flat and curved surfaces, but a drawback is the difficulty encountered in coating large surfaces. An old coating process that has been used to produce colored glasses in the crystal industry implies the dipping of an uncolored glass into a melt colored glass with similar chemical composition before it is blown to the desired shape.

Dip-coating technology has been used for plate glass, solar energy systems, anti-reflective coatings on windows, optical filters, dielectric mirrors, photocatalytic films, *etc*.[103–111]

It is possible to obtain coating thicknesses from few nanometers to 1 micron with high precision by selecting a suitable viscosity of the coating liquid and withdrawal rate, in accord with the Landau–Levich equation:[112]

$$t = \frac{0.94(\mu \cdot U)^{2/3}}{\gamma_{LV}^{1/6}(\rho \cdot g)^{1/2}} \quad (2.4)$$

where t is the coating thickness, μ the viscosity, U the withdrawal rate, γ_{LV} the liquid–vapor surface tension and ρ the solution density.

The Landau–Levich equation can be applied only when $U > 1 \text{ mm s}^{-1}$; it cannot be applied to systems in which the particles repel each other.

The atmosphere controls the evaporation of the solvent and the subsequent transformation of the sol into the film. Moreover, the thickness of the layer depends on the dipping angle between the support and the liquid surface and different thicknesses of the layer can be produced at the top and bottom sides of the surface that has been coated.[113]

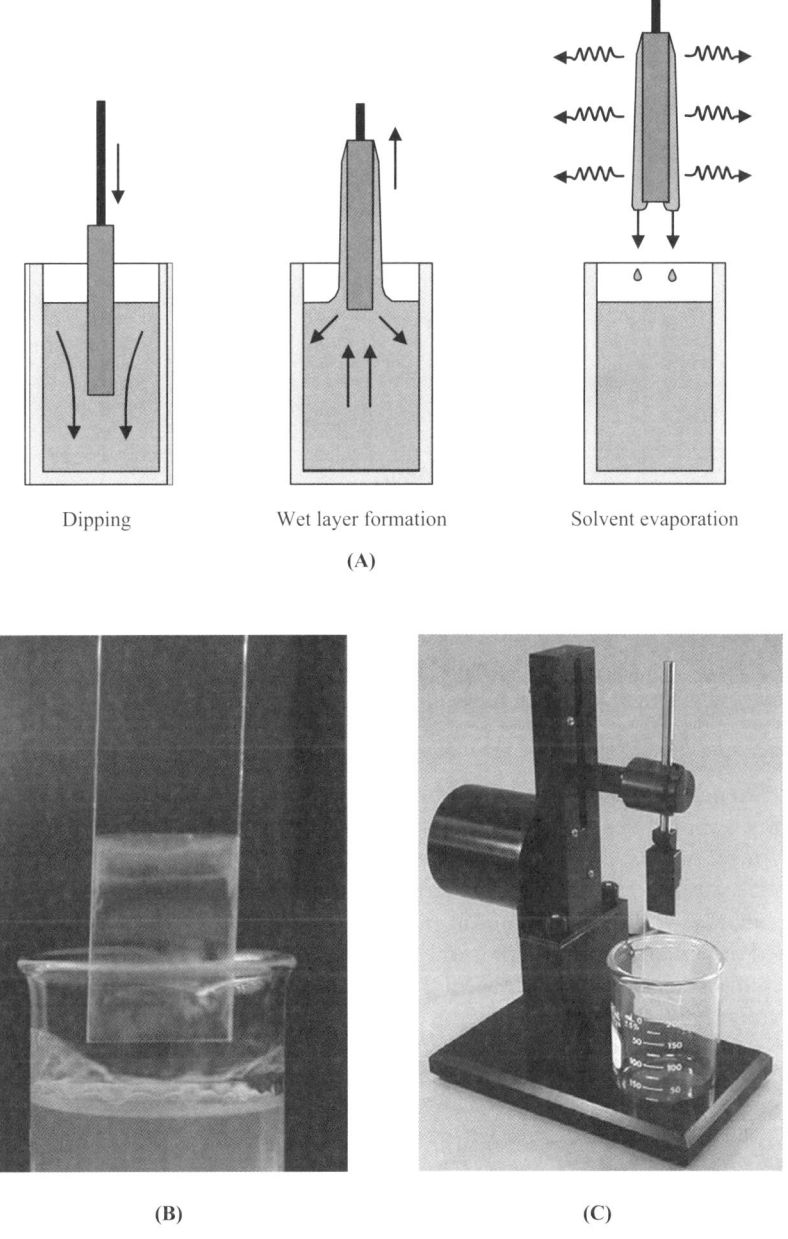

Figure 2.9 (a) Steps of the dip coating process: (i) dipping of the support in the coating liquid, (ii) formation of the wet layer by withdrawing and (iii) evaporation of the solvent; (b) formation of the film on the substrate; (c) apparatus for dip coating. [(b) Photograph courtesy of www.advanced-materials.at/rd/rd_seg_en.html. (c) Photograph courtesy of http://www.engineerlive.com/ChemicalEngineer/Materials/Precision_dip_coating_for_the_fabrication_of_thin_films/21225/.]

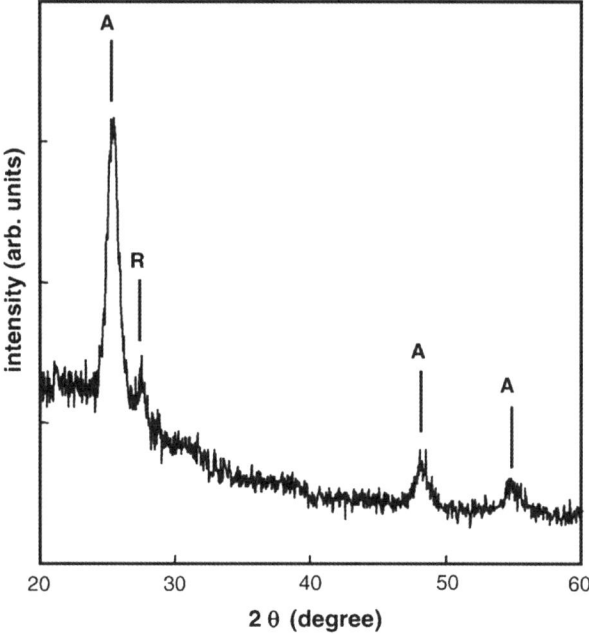

Figure 2.10 Low-angle XRD diffractogram of a 15 cycle deposited TiO_2 film. (Reproduced with permission from ref. 107.)

Reference 107 reports the preparation of photocatalytic TiO_2 films by the dip-coating technique by using sols derived from titanium tetraisopropoxide on glass substrates previously covered by a SiO_2 layer obtained from a tetraethyl orthosilicate sol. The films, after thermal treatment at 673 K, mainly consisted of TiO_2 anatase with a small amount of rutile (Figure 2.10). The calcinations were carried out either step-by step or only at the end of the various dip coating steps. The films subjected only to a final post-deposition calcination were thinner, probably due to a sintering effect involving the whole mass of TiO_2 (154 *versus* 252 nm for the 15 cycle deposited TiO_2 film). The adherence was excellent, unlike that of a similar film (easily detached by wiping) prepared by coating with TiO_2 Degussa P25 suspended in water. The samples were tested under UV illumination by using as probe reaction the photooxidation of 2-propanol in the gas–solid regime in the presence of water vapor; the results indicated a satisfactory efficiency (initial reaction rate, r_0, in the range *ca.* $6–24 \times 10^{-9}\,\text{mol}\,\text{l}^{-1}\,\text{s}^{-1}$), with CO_2 and propanone being the only compounds detected.

The laboratory-made experimental set-up used to carry out the experimental runs is sketched in Figure 2.11, while Figure 2.12 shows photoreactivity results of a selected experiment.

Photoactive films consisting of pure anatase, brookite or rutile have been prepared[108] by dip coating from water dispersions obtained by using $TiCl_4$ as the precursor (Figure 2.13). Also in this case, the photodegradation of 2-

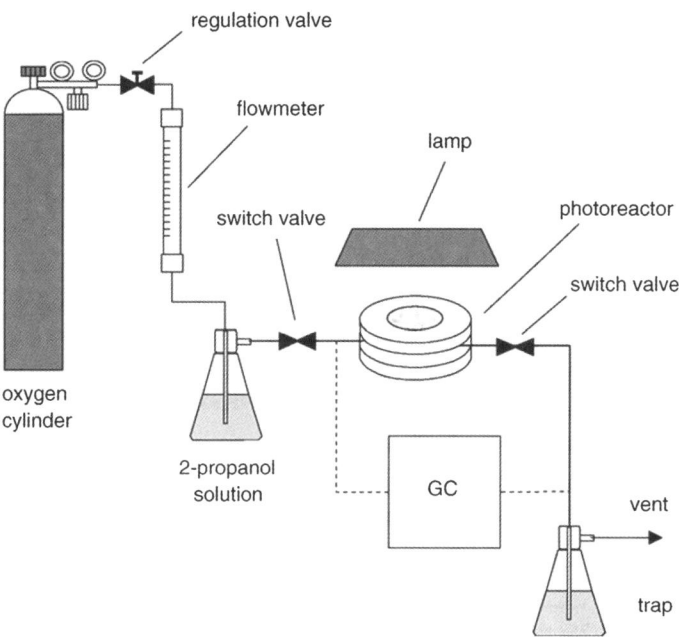

Figure 2.11 Set-up used for the photocatalytic degradation of 2-propanol. (Reproduced with permission from ref. 107.)

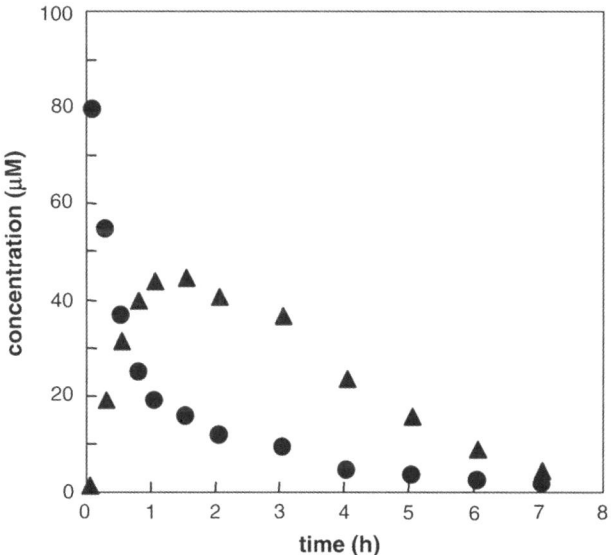

Figure 2.12 2-Propanol (●) and propanone (▲) concentrations *versus* irradiation time. Sample: TiO_2 film deposited by 15 cycles and calcined after each deposition. Reaction temperature: 310 K; water vapor concentration: 1.4 mM; irradiance: 1.3 mW cm^{-2}. (Reproduced with permission from ref. 107.)

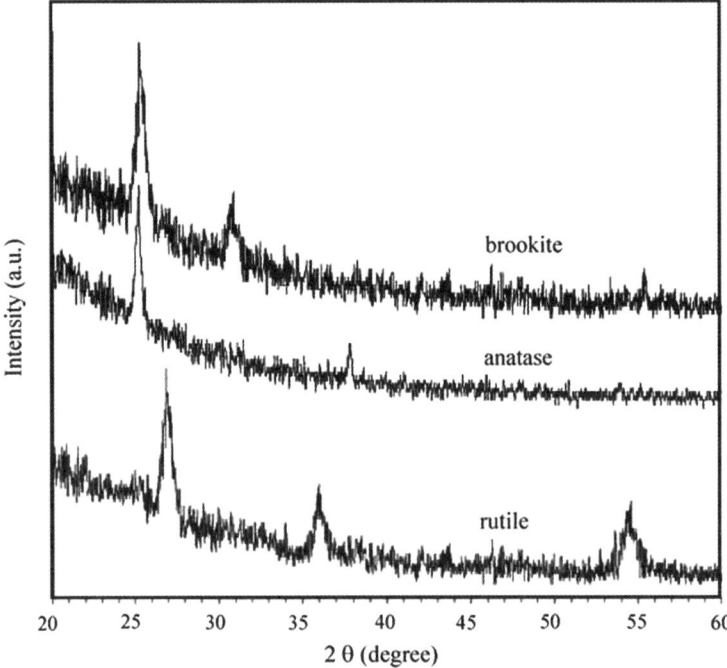

Figure 2.13 XRD patterns of TiO_2 films, deposited in ten steps, constituted of pure brookite, anatase and rutile phases. (Reproduced with permission from ref. 108.)

propanol was used as probe reaction and the brookite film was shown to be more efficient (Figure 2.14) than the anatase one, although a strict comparison between the photoreactivity results cannot be performed because the morphology and the thickness of the films were different.

2.4.1.2 Spin Coating and Flow Coating

The spin coating technique is usually employed to prepare films onto flat surfaces, although a homogeneous coating can also be obtained with non-planar substrates. The solution or the sol is placed onto the substrate to be coated and then rotated at high speed to spread it by centrifugal force. During the spinning step, evaporation of the solvent and gelation can occur. The thickness of the coating varies between some hundreds nanometres and *ca.* 10 µm, depending on the duration and the rate (usually ranging between 500 and 10 000 rpm) of the spinning (Figure 2.15).

The quality of the coating depends on the rheological parameter of the fluid used to coat. In principle, one should operate in the Newtonian regime and an important parameter to be considered is the Reynolds number of the

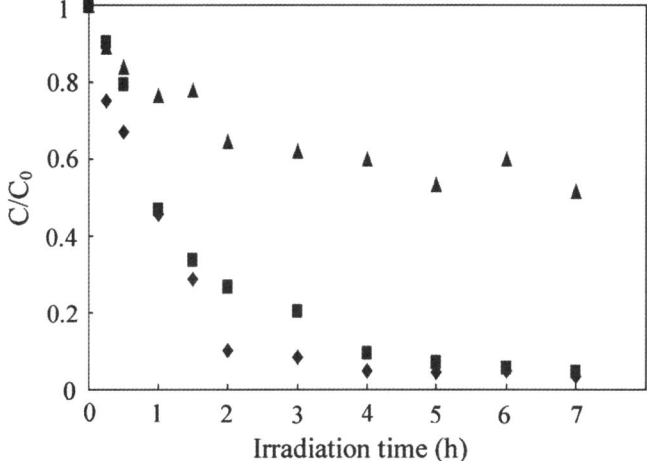

Figure 2.14 2-Propanol degradation *versus* irradiation time in the presence of ten-step deposited films: (■) pure anatase, (♦) pure brookite and (▲) pure rutile. Initial 2-propanol concentration: 60 mM. (Reproduced with permission from ref. 108.)

atmosphere around the sample. A high Reynolds number could be responsible for turbulence and bad quality of the layer.

The film thickness can be correlated with the angular velocity, the solvent evaporation rate and the viscosity of the fluid by using the following Meyer-hofer equation:[114]

$$t = \left(1 - \frac{\rho_A}{\rho_{A0}}\right) \cdot \left(\frac{3\mu m}{2\rho_{A0}\omega^2}\right)^{1/3} \quad (2.5)$$

where t is the final thickness, ρ_A the density of the volatile solvent, ρ_{A0} the initial value of ρ_A, μ the viscosity, m the evaporation rate of the solvent and ω the angular speed.

A simpler equation can be used [Equation (2.6)]:

$$t = A \cdot \omega^B \quad (2.6)$$

where A and B must be empirically determined. Results reported in 115 and 116 concerning the thickness of films obtained by spin coating at different angular speeds fit Equation (2.6) well and B ranges between 0.4 and 0.7.

Examples of preparation of TiO_2 or TiO_2 composite films have been reported in the literature.[117–120] Reference 117, for instance, deals with the preparation and testing of TiO_2–methylcellulose (MC) nanocomposite films. A sol suspension was prepared by adding titanium tetraisopropoxide (TTP) to a mixture of ethanol and HCl (molar ratio TTP : HCl : EtOH : H_2O = 1 : 1.1 : 10 : 10) and

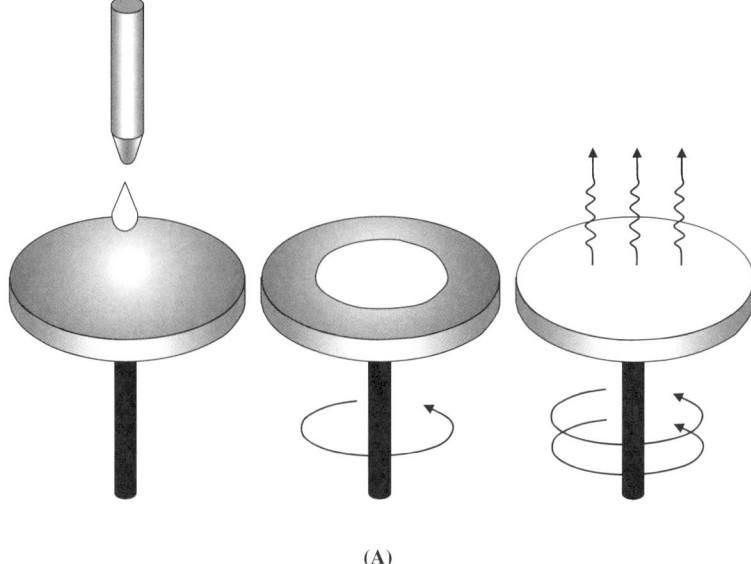

(A)

(B)

Figure 2.15 (a) Steps of the spin coating process: deposition of the fluid, spinning and evaporation/gelation of the solvent; (b) typical apparatus for spin coating. (Photograph courtesy of www.mbraun.com/oled-process-spincoater.htm.)

then adding a 2 wt% solution of methylcellulose. A TiO_2 nanopowder was dispersed in the sol and a microscope glass slide was coated with the mixture by using the spin coating technique. The presence of MC, acting as a dispersant agent, avoided problems due to film inhomogeneity and defects that could cause peeling and cracking during calcination carried out up to 773 K. Photocatalytic activity of the composite film was evaluated by degrading a textile dye, *i.e.* Light Yellow X6G (C.I. Reactive Yellow 2), as a model pollutant, and it was compared with the activity of (i) a similar composite film without MC and (ii) a TiO_2 nanopowder. The good mechanical integrity makes this composite film an interesting candidate for practical applications.

Nanoscale composite TiO_2–SnO_2 thin films with different ratios have been deposited from sol-gel TiO_2–SnO_2 mixtures on quartz substrates by spin coating.[119] XRD and AFM experiments showed that smooth and uniform anatase, anatase–rutile as well as rutile structural thin films were formed at 273 K, depending on the Sn ratio. Nanocomposite films with a Ti : Sn ratio of 3 : 1 showed excellent photocatalytic activity in the UV–Vis region.

The flow coating process is not usually employed for the preparation of TiO_2 films because it is not easy to obtain uniform layers by this method (the coating thickness increases from the top to the bottom of the substrate), unless it is not coupled with a spinning process. The coating thickness depends on the inclination angle of the substrate, on the solvent evaporation rate and on the viscosity of the fluid used (Figure 2.16).

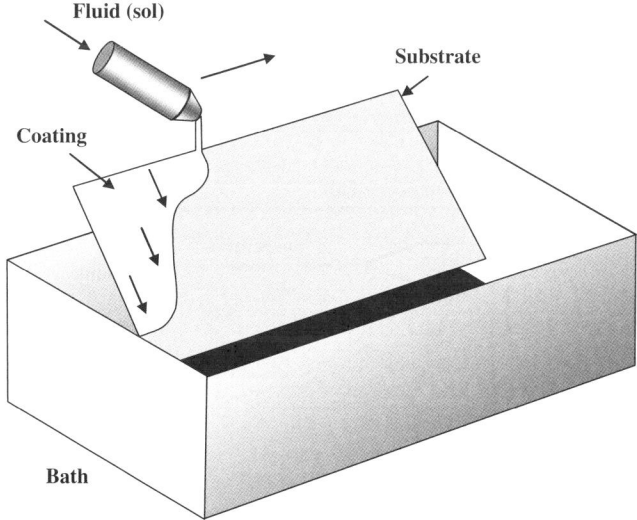

Figure 2.16 Sketch of a typical flow-coating apparatus.

2.4.1.3 (Plasma) Spray Drying

The use of this process allows films to be obtained by spraying and drying a slurry that contains dispersed TiO_2 particles.[121–127] This method, which is suitable to be industrially scaled up, offers advantages since it is *ca.* ten-times faster than the dip coating technique, although the films usually show poor mechanical properties. When plasma spray drying is used, a feedstock material is melted and sprayed onto a substrate where its solidification occurs with the consequent formation of a film.

TiO_2 coatings with a mixture of anatase and rutile phases have been obtained by plasma spraying using a spray-dried powder as feedstock material.[124,125] The characteristic of the formed layer depended not only on the parameters of the process but also on the type of feedstock material. Figure 2.17 shows a plasma-spray experimental set-up.

In ref. 126 nano-TiO_2 photocatalytic coatings were obtained on stainless steel by an atmospheric plasma spraying process. The starting nano-TiO_2 powder, consisting of 100% anatase phase, was agglomerated by the spray drying process. The photodecomposition efficiency of the TiO_2 coatings was evaluated by the kinetics of decomposition of the methylene blue in aqueous solution. The best photodecomposition efficiency was observed when TiO_2 coatings were prepared under a lower heat input, resulting in higher anatase phase fraction and smaller anatase grain size.

In a comprehensive comparative study[127] on the microstructure and photocatalytic performances of TiO_2 coatings obtained by various thermal spraying methods it is reported that the crystalline structure depends strongly on the technique of thermal spraying deposition. Agglomerated spray dried anatase TiO_2 powder was used as feedstock material for spraying. The photocatalytic behavior of the TiO_2-base surfaces was evaluated from the conversion rate of gaseous nitrogen oxides (NO_x). A high amount of anatase was suitable for the photocatalytic degradation of the pollutants. Suspension plasma spraying allowed the retention of the original anatase phase.

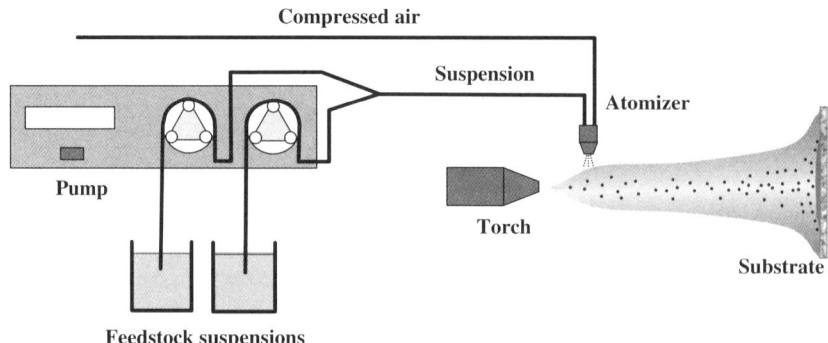

Figure 2.17 Example experimental set-up used for the plasma-spray technique.

2.4.1.4 Spray Pyrolysis

The spray pyrolysis method is based on the pyrolytic decomposition of an inorganic or organic precursor of the film dissolved in solution. A spray nozzle, with the help of carrier gases, gives rise to the formation of fine drops (aerosol) from the solution. The temperature of the substrate is maintained at a constant value by using a furnace or a hot plate. The best conditions for the preparation of the film occur when the droplets approach the substrate just before complete elimination of the solvent. Films grown at a substrate temperature less than 573 K are generally amorphous or badly crystallized. To obtain polycrystalline films, one needs to employ either higher temperatures or a post-annealing treatment.[128–130]

Figure 2.18 shows a sketch of a spray pyrolysis set-up.

TiO_2 films have also been prepared and characterized by spray pyrolysis.[131–136] In some cases the photocatalytic activity has been tested.

The photoelectrochemical behavior of type 304 stainless steel (SUS 304) coated with a TiO_2 thin film with the aim of photoprotecting the material from cathodic corrosion has been studied under different experimental conditions.[131] The photo-potential of TiO_2-coated SUS 304 was, for example, -350 mV vs. Ag/AgCl in an aerated aqueous solution containing 3 wt% NaCl (pH 7) under illumination with 10 mW cm^{-2} ultraviolet (UV) light, which was more negative

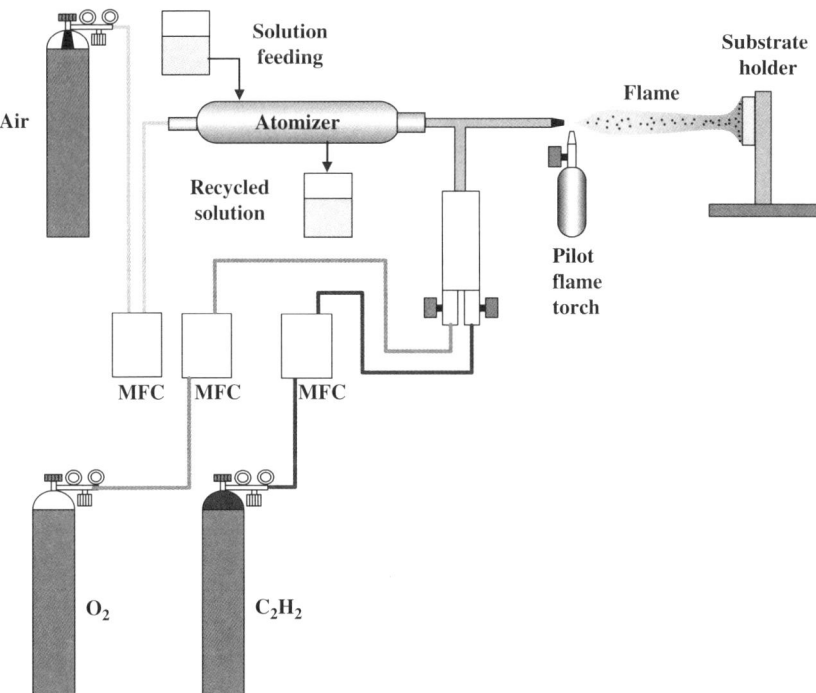

Figure 2.18 Sketch of a spray pyrolysis set-up.

than the corrosion potential of the bare steel (-100 mV). Moreover, bleaching of methylene blue dye on TiO_2 thin film-coated SUS 304 was observed under $1\,\text{mW cm}^{-2}$ UV illumination, indicating the possibility of photocatalytic self-cleaning effects.

Hydrophilic TiO_2 films have been synthesized from $TiCl_4$ by spraying a sol on glass substrates. The hydrophilic properties of the films were investigated by contact angle measurements under UV light illumination.[133]

Transparent TiO_2 thin films prepared on glass substrates were amorphous up to a deposition temperature of 723 K. The crystallization process to anatase phase started at *ca.* 823 K. These quite homogeneous and porous films were photocatalytically active and the photooxidation of methanol to methanal in water was successfully carried out.[134]

TiO_2 thin films have been formed both by spray pyrolysis and DC magnetron sputtering (see below) methods onto glasses that were coated with fluorine-doped tin oxide (FTO).[136] The films were characterized by X-ray diffraction, scanning electron microscopy, atomic force microscopy and UV–visible spectroscopy. The influence of the total pressure of an $Ar-O_2$ mixture on the deposition rate, phase composition, crystallinity, surface morphology and photocatalytic properties was investigated for films deposited using the sputtering technique. Transparent films were prepared by spray pyrolysis at low concentrations of titanium precursor. The photocatalytic properties of TiO_2 thin films were tested by using the degradation of methylene blue under UV light irradiation as probe photoreaction. The highest degradation rates were observed for films prepared by spray pyrolysis with a substrate temperature close to 673 K.

2.4.2 Physical Vapor Deposition (PVD) Techniques

This coating method involves purely physical processes rather than a chemical reaction at the surface to be coated such as in chemical vapor deposition (CVD, see next section). Physical vapor deposition (PVD) includes methods that allow deposition of thin films by condensation of a vaporized form of the material onto various surfaces.

Some examples of PVD techniques are (i) vacuum evaporation–deposition, (ii) electron beam physical vapor deposition, (iii) sputtering deposition, (iv) cathode arc (plasma) deposition and (v) pulsed laser deposition.

2.4.2.1 Vacuum Evaporation–Deposition

The vacuum evaporation–deposition technique is a convenient, simple and widely used deposition method. The method implies the evaporation in low vacuum of a material that condenses onto a substrate. The material to be deposited is electrically heated to obtain the vapor pressure (generally *ca.* 10^{-5} Torr) needed for the evaporation process. The substrate is kept at a temperature that depends on the required properties of the film. The distance between the substrate and the source is *ca.* 10–50 cm. After transition into the gaseous

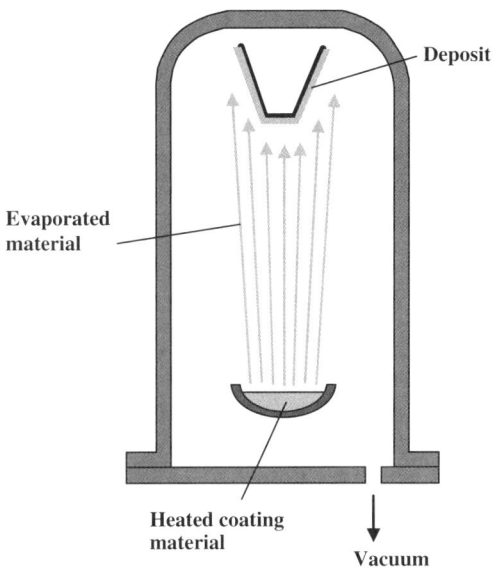

Figure 2.19 Sketch of a vacuum evaporation–deposition set-up.

state of the solid or liquid material to be deposited and the transport of vapor from the source to the substrate, the condensation of the vapor occurs with formation of the film (Figure 2.19).

Amorphous TiO_2 films were vacuum-deposited on rock salt at room temperature by evaporating small quantities (4–6 mg) of powdered anatase TiO_2. The films crystallized in anatase and rutile forms when they were irradiated by an intense electron beam.[137] Electron microscopy and electron diffraction studies were performed on these films: rutile crystallites were small granules, whereas anatase crystals grew to occupy larger areas of several microns in linear dimension. The structural difference of the crystallized films seemed to depend on the intrinsic properties of the amorphous films. The generation of dislocations that constituted small angle grain boundaries was observed in growing anatase crystals. In these crystals, twins were formed on the (112) planes and dislocations were also observed in the individual grains.

In another paper[138] anatase TiO_2 thin films were prepared with deposition rates equal to 16 nm min^{-1} by a vacuum arc-plasma evaporation (VAPE) method using sintered TiO_2 pellets as the source material (Figure 2.20). The films appeared to be highly transparent at substrate temperatures from room temperature to 673 K, regardless of whether oxygen (O_2) gas was introduced during the deposition process.

The photocatalytic activity of the deposited TiO_2 thin films was evaluated by immersing the samples (size 2.5 cm^2) in an aqueous methylene blue solution and irradiating them with a black light lamp. The photo-induced hydrophilicity was studied by contact angle measurements under UV irradiation. The highest

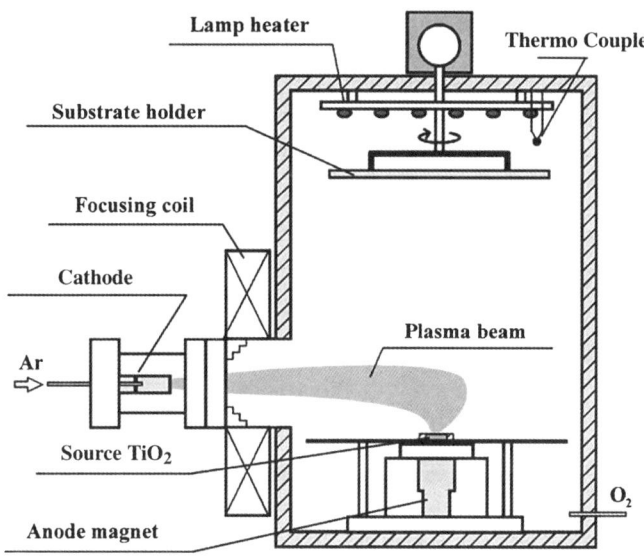

Figure 2.20 Sketch of the VAPE set-up. (Reproduced with permission from ref. 138.)

photocatalytic activity and photo-induced hydrophilicity were obtained in anatase TiO_2 thin films prepared at 573 K. Moreover, a transparent and conductive anatase TiO_2 thin film with a resistivity of $2.6 \times 10^{-1}\,\Omega\,cm$ was prepared at a substrate temperature of 673 K without the introduction of O_2 gas.

2.4.2.2 Electron Beam Physical Vapor Deposition

In this technique the material to be deposited is heated to a high vapor pressure by electron bombardment in high vacuum. Reference 139 reports the preparation of *ca.* 1.2-μm thick TiO_2 films by electron-beam evaporation. Figure 2.21 shows a sketch of the set-up. The chamber was evacuated by a mechanical pump, the distance between the electron-beam evaporation system and the substrate holder was 550 mm, and deposition occurred in an O_2 atmosphere using rutile TiO_2 as the source material. The temperature of the substrate was maintained at 523 K by a quartz lamp. Various films consisting of mixtures of amorphous or anatase TiO_2 were prepared at various O_2 pressures that influenced both the crystallinity and the hydrophilicity (which was evaluated by contact angle measurements).

2.4.2.3 Sputtering Deposition: Reactive Direct Current (DC) and Radio Frequency (RF) Magnetron Sputtering

Sputtering consists in removing atomized material from a solid due to energetic bombardment of its surface layers by ions or neutral particles.

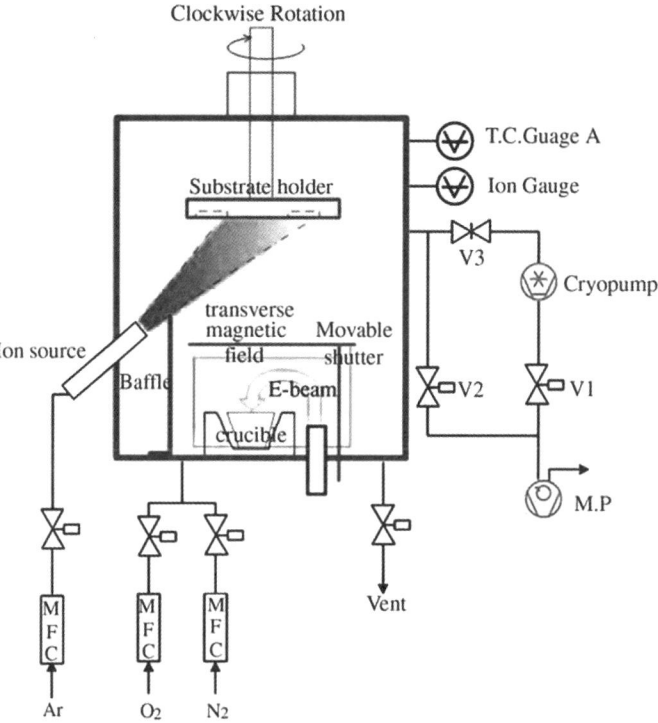

Figure 2.21 Sketch of an electron-beam (E-beam) evaporation set-up. (Reproduced with permission from ref. 139.)

Magnetron sputtering is a powerful and flexible vacuum ($<10^{-6}$ atm) coating process that can be used for a wide range of materials and is especially suitable for large-scale film deposition, due to its low-cost and easy controllability. An inert gas such as, for instance, argon is introduced to reach the minimum value of pressure needed to operate the magnetrons, which are a class of cold cathode discharge devices used generally in a diode mode. The difference between a magnetron cathode and a conventional diode cathode is the presence of a magnetic field. A surface atom becomes sputtered if the energy transferred is more than three times the surface binding energy (approximately equal to the heat of sublimation). Sputtering of a target atom is just one of the possible results of ion bombardment of a surface. The second important process is the emission of secondary electrons from the target surface that enable the discharge to be maintained. The target materials range from pure metals, where a DC power supply can be used, to semiconductors and isolators, which require a RF power supply or pulsed DC. A magnetic field is used to trap secondary electrons close to the target; these electrons follow helical paths around the magnetic field lines, increasing the number of the ionizing collisions with neutral gas molecules. This phenomenon enhances the ionization of the

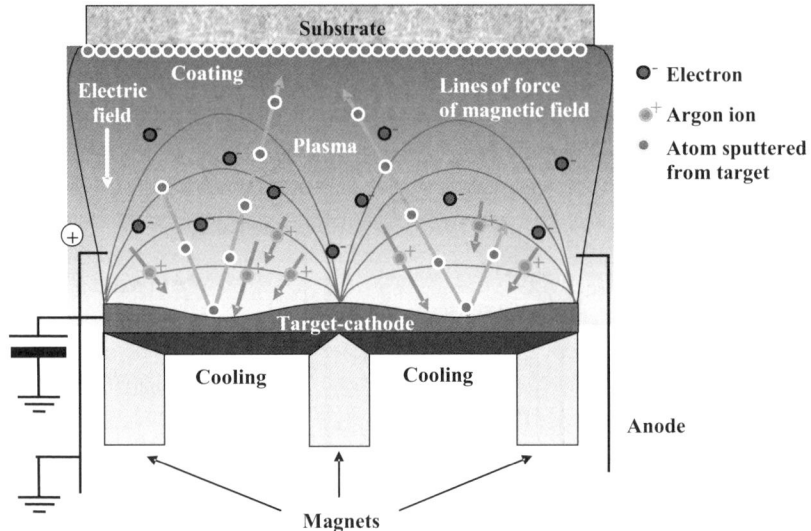

Figure 2.22 Typical RF sputtering working scheme.

plasma near the target, producing a higher sputter rate. The sputtered atoms are not charged and are not affected by the magnetic field.

The set-up used for RF magnetron sputtering is sketched in Figure 2.22.

Among various types of films, TiO_2 films have been prepared by this technique.[140–148]

Reference 140 reports the preparation of TiO_2 films starting from a Ti metal target and their photocatalytic behavior, although films prepared by spray pyrolysis are sometimes reported to be more photoactive.[136] The films were deposited on an unheated non-alkali glass (AN, Asahi Glass) by RF reactive magnetron sputtering. Depositions were carried out under various RF substrate biases and the total gas pressure during deposition was 1.0 or 3.0 Pa. The oxygen flow ratio $[O_2/(O_2 + Ar)]$ and RF sputtering power were kept constant at 60% and 200 W, respectively. The photocatalytic decomposition of acetaldehyde under UV illumination was carried out to test the samples.

Transmission electron microscopy has been used to study the structure, morphology and orientation of thin TiO_2 films prepared by reactive RF magnetron sputtering on glass slides at different substrate temperatures (373–673 K).[143] The microstructure and photocatalytic reactivity of TiO_2 films depended on the deposition temperature. In the temperature range examined, all film samples exhibited a porous nanostructure and the dimension of the particles grew upon increasing the deposition temperature. Films were amorphous at 373 K, and anatase phase was found at $T \geq 473$ K. A preferred orientation was noticed for films deposited between 473 and 573 K, whereas films obtained at 673 K showed a random orientation. The most photoactive films to abate Rhodamine B present in waste water were those deposited at

523 K, due probably to the high degree of preferred orientation and to the satisfactory crystallinity and nanoporosity of the anatase phase.

DC magnetron sputtering has been used to form transparent TiO_2 thin films (200–300 nm thick) on glass substrates.[145] A semiconducting TiO_{2-x} target in pure Ar was used at pressures ranging between 0.1 and 1.0 Pa. Post-deposition heat treatment carried out at different temperatures allowed achievement of the crystallization of anatase TiO_2. Both the as-deposited and heat-treated films were subjected to UV–VIS, SEM and XRD investigation; it was found that the pressure of argon during the deposition process influenced the structure modification that occurred throughout the heat treatment. In addition, decomposition of ethanol in synthetic air was used as a probe reaction to test the photocatalytic activity, with the aim of relating it to the microstructures of the samples.

The existence of a solid–solid interface in TiO_2 films produced by the reactive DC magnetron sputtering[147] explained the high photoactivity of mixed phase TiO_2 catalysts. The sputter deposition allows easy control of the phase and interface formation. Moreover, the influence of oxygen partial pressure, target power, substrate bias, deposition incidence angle and post-annealing treatment on the structural and functional characteristics of the catalysts has been studied. The photocatalytic activity of the films for the photooxidation of acetaldehyde under UV illumination in the gas phase was higher (normalized for surface area) than that showed by mixed phase TiO_2 samples prepared by other methods (flame hydrolysis powders and sol-gel deposited TiO_2 films), although the comparison between films prepared by means of very different techniques or between films and powders is difficult, due to the many variables that could influence the photoactivity to differing extents.

TiO_2 thin films have been deposited very recently on three different unheated substrates by DC magnetron sputtering.[148] The effects of sputtering current and deposition time on the crystallization of TiO_2 thin films were studied. The TiO_2 thin films were deposited at three sputtering currents, 0.50, 0.75 and 1.00 A, with deposition times of 25, 35 and 45 min, respectively. Anatase films were obtained at low sputtering current values and under all deposition conditions. The crystallinity of the anatase phase with grain size in the range 10–30 nm increased as the sputtering current and deposition time increased.

2.4.2.4 Cathode Arc (Plasma) Deposition

In this case a high power arc directed at the target material blasts away some material into a vapor sputter. Detailed information on cathode arcs can be found in ref. 149.

In direct cathode arc deposition, a flow of a highly ionized metal plasma from a cathode arc spot transfers coating material to the substrate surface. A disadvantage of this method is the formation of droplets (also called macroparticles) in the cathode arc flow that could negatively influence the critical properties of the coatings. Figure 2.23 shows a sketch of cathode arc deposition.

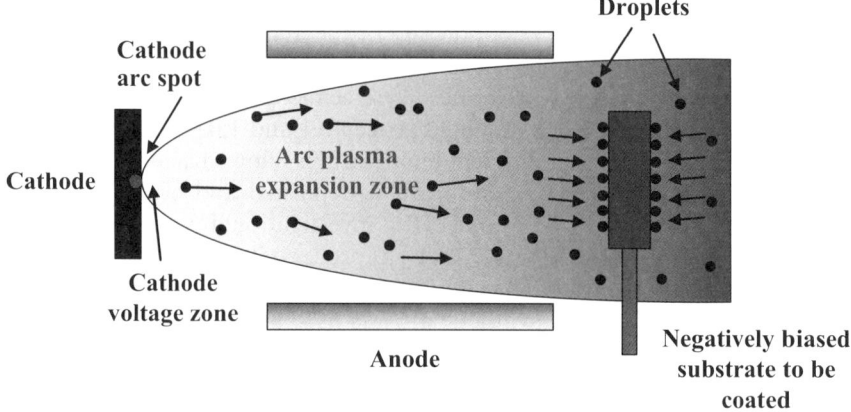

Figure 2.23 Sketch of cathode arc deposition process.

Few applications of this technique can be found for the preparation of TiO_2 thin films, although one report[150] deals with the obtainment of TiO_2 films on glass substrates at various temperatures (from room temperature to 673 K) using an unfiltered cathode arc device. The films deposited at temperatures lower than 573 K were amorphous, whereas those obtained at higher temperatures grew as a crystalline anatase phase. The phase transition amorphous-to-anatase was observed after post-annealing at 673 K. The average transmittance was always higher than 80% and it was slightly higher for films obtained at the highest temperatures. The size of the grains of the as-deposited crystalline films was *ca.* 20 nm and the surface roughness *ca.* 2 nm.

2.4.2.5 *Pulsed Laser Deposition (PLD)*

The use of this technique allows production of multicomponent materials. Ablation of the target occurs by means of a high-power laser and deposition of the vapor phase onto the substrate affords stoichiometric layers that match the target material.

The PLD technique, although industrial uses are known, is generally recognized as a well-established laboratory coating technology, because industrial usage is restricted mainly by the need for large-area deposition. The combination of PLD with other techniques such as magnetron sputtering has been suggested with the aim of potential industrial application.[151]

Figure 2.24 shows a schematic of a typical PLD system.

Some TiO_2 films prepared by PLD have been reported, as for instance in refs 152–156, although photoactivity tests appear only rarely.[156]

A colloidal TiO_2 nanoparticle film has been deposited on a quartz substrate and the morphological and optical properties determined.[153] Atomic force microscopy demonstrated that a good uniformity of deposition was obtained.

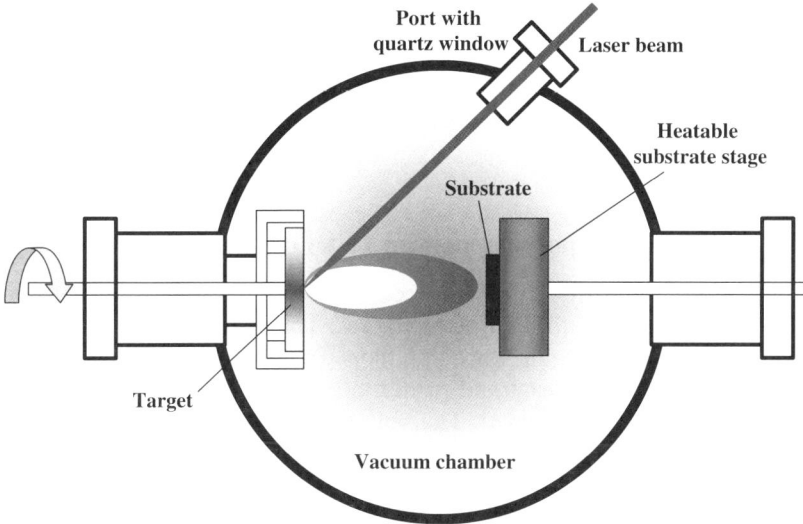

Figure 2.24 Scheme of pulsed laser deposition.

Agglomerates with dimensions of *ca.* 1 µm in size were observed. The optical constants, the energy gap and the film thickness were determined by UV–Vis transmission spectra, recorded in the 200–800 nm range. The optical constants obtained agree with values reported in literature for TiO_2 nanoparticle thin films.

Reference 154 reports the growth of TiO_2 nanocrystalline films on Si substrates by PLD. Rutile sintered targets were used and were irradiated by a KrF excimer laser ($\lambda = 248$ nm, pulse duration ~ 30 ns) in a controlled oxygen environment at a constant substrate temperature of 923 K. The structural and morphological properties of the films were studied by varying the oxygen partial pressure (0.05–5 Pa) and the laser fluence (2–4 J cm^{-2}). Both rutile and anatase phases formed, but the latter disappeared at the highest laser fluences. X-Ray photoelectron spectroscopy (XPS) measurements allowed the determination of the stoichiometry of the grown films. The surface morphology of the deposits, studied by scanning electron (SEM) and atomic force (AFM) microscopies, indicated that they were nanostructured films. The dimensions and density of the nanoparticles observed at the surface depended on the partial pressure of oxygen during growth. The smallest particles (*ca.* 40 nm diameter) were obtained at the highest pressures of inlet gas.

The photocatalytic properties of different crystalline titanium dioxide films have been evaluated and compared.[156] The films grew directly on titanium substrates by surface anodization (see anodic oxidation in the following) and by PLD on titanium and silicon substrates, followed by thermal annealing. Not only were the structure and morphology of the films studied by means of SEM observations and X-ray diffraction analyses, but also photocatalytic tests on stearic acid mineralization were performed. The results indicated higher photocatalytic efficiency for PLD layers. This finding can be explained by considering the different microstructured/microporous morphology of the surfaces. Notably, moreover, PLD TiO_2 films with a relatively high content of rutile phase showed lower photocatalytic efficiency than films containing mainly the anatase phase.

2.4.3 Chemical Vapor Deposition (CVD)

Chemical vapor deposition (CVD) is a chemical process used to produce high-purity solid materials with high performance. The process is often used in the semiconductor industry to produce thin films. In a typical CVD process, one or more volatile precursors react and/or decompose on the surface of a substrate where the desired film is formed. The volatile by-products can be removed by means of a gas flow.

Chemical vapor synthesis (CVS) is a modified CVD method where the parameters of the process are adjusted to produce nanoparticles instead of films. Chemical vapor condensation (CVC), chemical vapor reaction (CVR) and chemical vapor precipitation (CVP) are synonyms used frequently in the literature. Some precursors both in CVD and CVS are hydrides, carbonyls, metallorganics, chlorides and other volatile compounds in gaseous, liquid or solid state. A limitation of the CVS process is the difficulty in finding suitable precursors. The energy necessary to convert the reactants into nanoparticles is supplied by means of flame, hot wall, plasma, and laser reactors. Important parameters to be controlled to obtain good reproducibility are the total pressure, partial pressure of the precursor, carrier gas, temperature or power of the energy source, and reactor geometry. Modifications of the precursor delivery system and the reaction zone allow the synthesis of pure and doped oxides, coated and functionalized nanoparticles, and granular films. Figure 2.25 gives a schematic representation of physical vapor deposition, chemical vapor deposition and chemical vapor synthesis.

TiO_2 films have been prepared by CVD.[157–164] In some cases the photocatalytic activity of the films was tested.

The synthesis of anatase TiO_2 supported on porous solids by coating anatase TiO_2 onto three different particle supports, *i.e.*, activated carbon, γ-Al_2O_3 and silica gel, was studied in ref. 158. The presence of H_2O (vapor) during or before the CVD process was found essential to obtain anatase TiO_2 on the surface of supports. High inflow concentration of the precursor (titanium tetraisopropoxide), high CVD reactor temperature and long coating time gave rise to blockage problems, due to deposition of particles that occurred together with the vapor deposition. Macroporous silica gel with a high density of surface hydroxyl groups and high surface area was found to be the best support and TiO_2/SiO_2 samples showed the highest activity in the photodegradation of phenol in water.

The selective deposition of TiO_2 films starting from titanium isopropoxide as the precursor was performed by using a KrF laser CVD technique on a quartz

Powders versus Thin Film Preparation

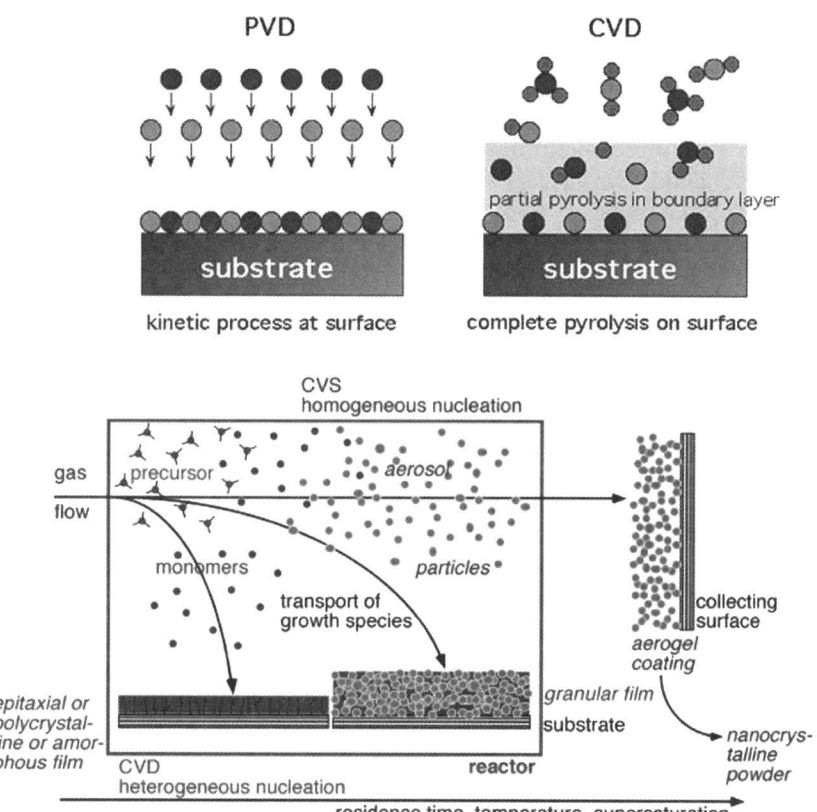

Figure 2.25 Schematic representation of physical vapor deposition (PVD), chemical vapor deposition (CVD) and chemical vapor synthesis (CVS) of nanocrystalline powders (to form a film on a collecting surface). (Courtesy of http://python.rice.edu/~arb/Courses/Images/360_02_handout9.jpg http://www.uni-due.de/ivg/nano/synthesis_nppt.shtml.)

substrate under the fixed laser fluence of 450 J m^{-2}.[159] The substrate temperatures, laser repetition rates and supply rates of the precursor ranged between 328 and 573 K, 10 and 100 Hz, 60 and 300 mg h^{-1}, respectively. Rutile and/or anatase phases were found in the film for all experimental conditions used. Low substrate temperature, low laser repetition rate and high supply rate of the precursor are beneficial for rutile formation.

TiO_2 photocatalytic films prepared by low-pressure metal–organic (LPMO) CVD, by using titanium tetraisopropoxide as the precursor and various reaction temperatures and deposition times, have been characterized.[162] The results indicated that the film thickness was linearly proportional to the deposition time, while their structure depended not only on the deposition time but also on the reaction temperature. Moreover, the photocatalytic decomposition of methylene blue in aqueous solution was chosen as probe reaction to test the photoactivities of the films. The anatase and rutile TiO_2 films showed the highest photocatalytic activity, whereas the amorphous TiO_2 films showed lower activities. The photocatalytic activity depended on film thickness non-linearly and the optimum thickness is located between 3 and 5 µm.

Pure brookite phase TiO_2 crystals with high specific surface area have been deposited at room temperature on silicon(100) substrates by means of plasma-enhanced chemical vapor deposition. Titanium tetraisopropoxide in the vapor phase and argon–oxygen mixtures, under an applied substrate DC bias voltage of –250 V, were used.[163] X-Ray diffraction analysis for the as-deposited TiO_2 confirmed the presence of a crystalline brookite phase.

Thin TiO_2 films have been deposited using a RF plasma-enhanced chemical vapor deposition.[164] The optical properties and thickness, determined by means of UV–Vis absorption spectrophotometry, indicated a good optical quality and their potential application both as interference optical filters and photocatalytic surfaces. The structural analysis, carried out by Raman shift spectroscopy, showed that the coatings posses an amorphous structure. However, Raman spectra of the same films subjected to thermal annealing at 723 K revealed the appearance of the anatase crystalline form. Surface morphology of the films was also investigated by atomic force microscopy, which indicated the presence of granular, broccoli-like topography.

2.4.4 Chemical Bath Deposition

The chemical bath deposition (CBD) technique is the cheapest known technique for the preparation of films of various compounds (oxides, sulfides, *etc.*). It is based on direct deposition onto the chosen support of a compound produced from precursor species by means of chemical reaction(s). It is essential to check the temperature, pH and experimental conditions under which the compound of interest must be produced before its deposition. The films could be subjected to a heating treatment as a second step.

Figure 2.26 shows a CBD set-up.

Some papers report the preparation of TiO_2 films by this method[165,166] but, notably, the adhesion of TiO_2 to the support is not satisfactory in most cases, particularly for films intended to be used in photocatalytic reactions. Consequently, not many relevant papers can be found in the literature.

TiO_2 films on glass substrates have been prepared at room temperature by this method starting from an aqueous peroxotitanate solution.[165] The films were 117 nm thick and consisted of closely packed particles (10–15 nm in diameter) aggregated into large grains (50–100 nm). Anatase TiO_2 thin films were obtained by heating the as-deposited thin films at 773 K for 1 h in air.

Dense, transparent rutile TiO_2 films 2.5 µm thick have been obtained on a F-doped SnO_2-covered glass substrate.[166] The starting solution contained peroxotitanate complex ions, which were relatively stable under the experimental conditions chosen. Kinetic studies suggested that the precipitation started from the formation of amorphous solids, followed by crystallization through a dissolution–recrystallization process. Anatase was detected only for powders

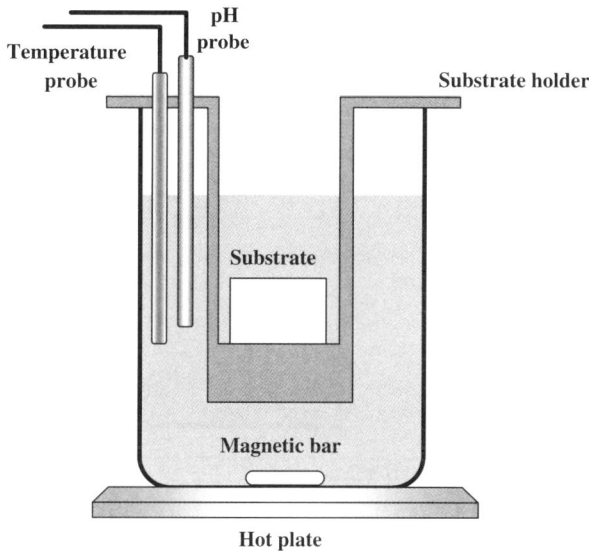

Figure 2.26 Chemical bath deposition set-up.

collected from solutions after film preparation, not for films, and the transformation from amorphous into anatase was hypothesized to occur before further transformation of anatase into rutile.

2.4.5 Thermal Oxidation and Anodic Oxidation

Thermal oxidation is one way to produce a thin layer of oxide on the surface of a wafer. The technique forces an oxidizing agent, which could be water vapor (wet oxidation) or molecular oxygen (dry oxidation), to diffuse into the wafer at high temperature and react with it. Thermal oxidation of silicon, for instance, is usually performed at between 1073 and 1473 K, resulting in the so-called high-temperature oxide (HTO) layer. Wet oxidation, generally, is preferred to dry oxidation for growing thick oxides, because of the higher growth rate.[167]

Thermal oxidation has been sometimes applied to metallic titanium to form a film of TiO_2 on the metallic surface that can then be used to carry out photochemical water splitting or can work as anode in an electrochemical photocell where hydrogen is produced.[168,169]

Reference 168 describes the preparation of chemically modified n-type TiO_2 by controlled combustion of Ti metal in a natural gas flame. This material, in which carbon substituted some lattice oxygen atoms, absorbs light at wavelengths below 535 nm and has a lower band-gap energy than rutile (2.32 instead of 3.00 eV). By applying a potential of 0.3 V, water splitting occurred with a total conversion efficiency of 11% and a maximum photoconversion efficiency of 8.35% (illumination equal to 40 mW cm^{-2}).

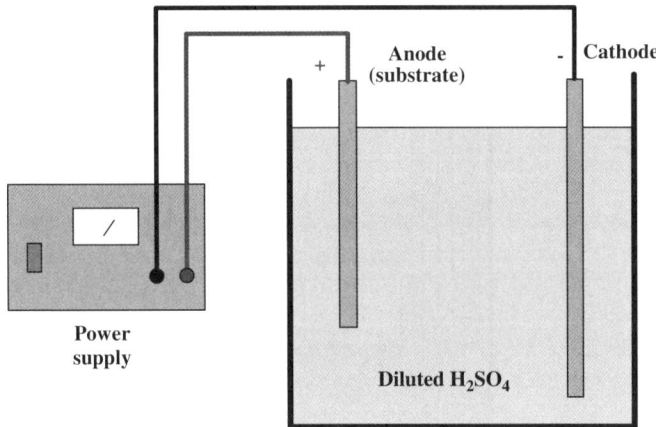

Figure 2.27 Set-up for anodic oxidation.

The anodic oxidation method allows the generation of an oxide film having protective, decorative or functional properties on a surface of the corresponding metal.[170]

This technique has been used to prepare TiO_2 films in a diluted H_2SO_4 solution, and the parameters of their formation have been studied in depth.[171] The set-up of this technique (Figure 2.27), which can be used quite easily, consists of a power generator, an electrolyte, titanium samples that are connected to the positive pole (anode) where oxidation takes place, and another metal (generally titanium) working as cathode connected to the negative pole of the generator, where reduction reactions take place.

The electrochemically enhanced oxidation of a metal under some circumstances yields solid oxide nanotubes. Anodic oxidation of Ti in electrolytes containing a small amount of fluoride has been used to convert Ti into TiO_2 nanotubes.[172] Ordered nanotube arrays have been obtained by anodization of titanium in fluoride-based baths.[173,174] The nanotubes showed various pore sizes (22–110 nm), lengths (200–6000 nm) and wall thicknesses (7–34 nm). Nanotubes with different shape (cylindrical, tapered), length and wall thickness have been produced by varying the anodization parameters such as electrolyte concentration, pH, voltage and bath temperature. They are of great interest for use in water photoelectrolysis, photocatalysis, heterojunction solar cells and gas sensing.

Very recently nitrogen-doped TiO_2 films with amorphous nanotubular morphology and enhanced specific surface area were produced, inducing the growth of the oxide layer directly on titanium substrates.[175] Anodization was performed in fluoride-containing electrolytes and the doping was achieved by means of thermal treatments in a nitrogen atmosphere. The anodically formed N-doped TiO_2 layers showed high efficiency in the photodecomposition of Rhodamine B under visible radiation.

2.4.6 Electrophoretic Deposition

In the electrophoretic deposition process (EDP) colloidal charged particles suspended in a liquid are forced by a direct current (DC) electric field to move toward an oppositely charged electrode, onto which they are deposited.[176] This low-cost method not only allows us to easily control the thickness of the coatings, but also to coat surfaces of conductive substrates with large areas. EDP has been applied very frequently to produce ceramic layers from powders dispersions.[177–179] Figure 2.28 shows a sketch of the EDP set-up, which consists of an electrochemical cell.

TiO_2 coatings have been produced on glasses,[180] stainless steel,[181,182] SiC and carbon fiber.[183]

The TiO_2 coatings were produced on glass foam substrates by EDP and sol-gel method with the aim of imparting to the surface antibacterial and photocatalytic properties.[180] Charged TiO_2 nanometric particles TiO_2 suspended in acetylacetone were used to obtain coatings on the glass foam substrates. Best results were obtained by applying 25 V for 4 min and a subsequent thermal treatment (723 K for 1 h). In addition, multilayer sol-gel TiO_2 coatings on glass foam were also prepared, which were subjected to the same thermal treatment. Both techniques led to partially microcracked coatings that were, however, well adhered to the glass foam substrate.

TiO_2 films produced on various substrates have been tested for photocatalytic degradation of malic acid (hydroxybutanedioic).[181] Two TiO_2 samples supported on stainless steel by EPD showed a lower catalytic activity than those supported on quartz by a dip-coating procedure. The results were explained by considering the negative role played by some cationic impurities (Si^{4+}, Na^+, Cr^{3+} and Fe^{3+}) that were present in the TiO_2 layer as a

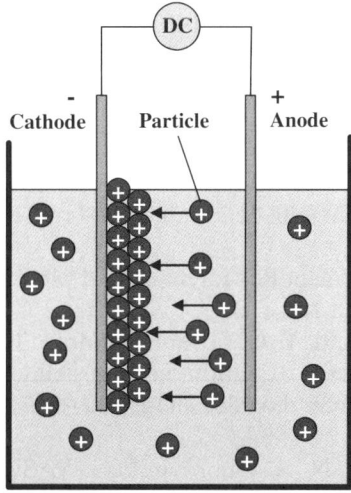

Figure 2.28 Schematic of electrophoretic deposition process set-up.

consequence of the thermal treatment necessary to improve the cohesion and adhesion to the support. TiO_2 Degussa P25 suspended in acetone (1 g in 100 ml) was used to coat the stainless steel foils. The suspension was first homogenized by sonication and stirred magnetically during the deposition. A counter-electrode of Pt acting as the cathode was placed in the suspension just in front of the stainless steel substrate with the same size and shape. The applied potential between the two electrodes was either 125 V (for 4 min) or 200 V (for 2 min), depending on the sample. TiO_2 particles, due to their natural surface charge, moved to the stainless foils, forming a layer. Subsequently, the two samples were heated in a N_2 flow at 973 K for 4 or 8 h.

Crack-free rough anatase TiO_2 coatings have also been prepared on stainless mesh by an EDP technique, and they were found active for the photocatalytic oxidation of 2-propanol to acetone.[182]

Finally, ref. 184 describes the preparation of an electrophoretically deposited film – without the use of a surfactant or any post-thermal treatments – that is potentially useful for application in flexible dye-sensitized solar cells. The resultant film was composed of wide band gap nanocrystalline TiO_2.

References

1. X. Chen and S. S. Mao, Titanium dioxide materials: synthesis, properties, modifications, and applications, *Chem. Rev.*, 2007, **107**, 2891–2959.
2. H. Zhang and J. F. Banfield, Understanding polymorphic phase transformation behavior during growth of nanocrystalline aggregates: insights from TiO_2, *J. Phys. Chem. B*, 2000, **104**, 3481–3487.
3. A. Fujishima, X. Zhang and D. A. Tryk, TiO_2 photocatalysis and related surface phenomena, *Surf. Sci. Rep.*, 2008, **63**, 515–582.
4. A. El Goresy, M. Chen, L. S. Dubrovinsky, P. Gillet and G. Graup, An ultradense polymorph of rutile with seven-coordinated titanium from the Ries crater, *Science*, 2001, **293**, 1467–1470.
5. A. El Goresy, M. Chen, P. Gillet, L. Dubrovinsky, G. Graup and R. Ahuja, A natural shock-induced dense polymorph of rutile with α-PbO_2 structure in the suevite from the Ries crater in Germany, *Earth Plan. Sci. Lett.*, 2001, **192**, 485–495.
6. R. Marchand, L. Brohan and M. Tournoux, A new form of titanium dioxide and the potassium octatitanate $K_2Ti_8O_{17}$, *Mater. Res. Bull.*, 1980, **15**, 1129–1133.
7. M. Latroche, L. Brohan, R. Marchand and M. Tournoux, New hollandite oxides: $TiO_2(H)$ and $K_{0.06}TiO_2$, *J. Solid State Chem.*, 1989, **81**, 78–82.
8. J. Akimoto, Y. Gotoh, Y. Oosawa, N. Nonose, T. Kumagai, K. Aoki and H. Takei, Topotactic oxidation of ramsdellite-type $Li_{0.5}TiO_2$, a new polymorph of titanium dioxide: $TiO_2(R)$, *J. Solid State Chem.*, 1994, **113**, 27–36.
9. L. S. Dubrovinsky, N. A. Dubrovinskaia, V. Swamy, J. Muscat, N. M. Harrison, R. Ahuja, B. Holm and B. Johansson, Materials science: the hardest known oxide, *Nature*, 2001, **410**, 653–654.

10. U. Diebold, Structure and properties of TiO$_2$ surfaces: a brief review, *Appl. Phys. A*, 2003, **76**, 681–687.
11. H. K. Kammler, L. Madler and S. E. Pratsinis, Flame synthesis of nanoparticles, *Chem. Eng. Technol.*, 2004, **20**, 5933–5939.
12. R. N. Grass, S. Tsantilis and S. E. Pratsinis, Design of high-temperature, gas-phase, synthesis of hard or soft TiO$_2$ agglomerates, *AIChE J.*, 2006, **52**, 1318–1325.
13. A. Teleki, R. Wengeler, L. Wengeler, H. Nirschl and S. E. Pratsinis, Distinguishing between aggregates and agglomerates of flame-made TiO$_2$ by high-pressure dispersion, *Powder Technol.*, 2008, **181**, 292–300, and references therein.
14. C. J. Brincker and G. W. Scherer, *Sol-Gel Science: The Physics and Chemistry of Sol-Gel Processing*, Academic Press, San Diego, 1990.
15. D. Robert and J. V. Weber, Titanium dioxide synthesis by sol gel methods and evaluation of their photocatalytic activity, *J. Mater. Sci. Lett.*, 1999, **18**, 97–98.
16. R. Campostrini, G. Carturan, L. Palmisano, M. Schiavello and A. Sclafani, Sol-gel derived anatase TiO$_2$: morphology and photoactivity, *Mater. Chem. Phys.*, 1994, **38**, 277–283.
17. P. R. Mishra and O. N. Srivastava, On the synthesis, characterization and photocatalytic applications of nanostructured TiO$_2$, *Bull. Mater. Sci.*, 2008, **31**, 545–550.
18. W. W. So, S. B. Park and S. J. Moon, The crystalline phase of titania particles prepared at room temperature by a sol-gel method, *J. Mater. Sci. Lett.*, 1998, **17**, 1219–1222.
19. G. Marcì, L. Palmisano, A. Sclafani, A. M. Venezia, R. Campostrini, G. Carturan, C. Martin, V. Rives and G. Solana, Influence of tungsten oxide on structural and surface properties of sol-gel prepared TiO$_2$ employed for 4-nitrophenol photodegradation, *J. Chem. Soc., Faraday Trans.*, 1996, **92**, 819–829.
20. G. Facchin, G. Carturan, R. Campostrini, S. Gialanella, L. Lutterotti, L. Armelao, G. Marcì, L. Palmisano and A. Sclafani, Sol-gel synthesis and characterisation of TiO$_2$-anatase powders containing nanometric platinum particles employed as catalysts for 4-nitrophenol photodegradation, *J. Sol-Gel Sci. Technol.*, 2000, **18**, 29–59.
21. C. Sanchez, P. Toledano and F. Ribot, Molecular structure of metal alkoxide precursors, *Mater. Res. Soc. Symp. Proc.*, 1990, **180**, 47–59.
22. T. J. Boyle, T. M. Alam, C. J. Tafoya and B. L. Scott, Formic acid modified Ti(OCHMe$_2$)$_4$. Syntheses, characterization, and X-ray structures of Ti$_4$(μ_4-O)(μ-O)(OFc)$_2$(μ-OR)$_4$(OR)$_6$ and Ti$_6$(μ_3-O)$_6$(OFc)$_6$(OR)$_6$ (OFc = O$_2$CH; OR = OCHMe$_2$), *Inorg. Chem.*, 1998, **37**, 5588–5594.
23. A. P. Hagan, M. G. Lofthouse, F. S. Stone and M. A. Trevethan, in *Preparation of Catalysts II*, ed. B. Delmon, P. Grange, P. Jacobs and G. Poncelet, Studies in Surface Science and Catalysis, Elsevier, Amsterdam, 1979, p. 417.

24. R. Campostrini, M. Ischia and L. Palmisano, Pyrolysis study of sol-gel derived TiO_2 powders Part I. TiO_2-anatase prepared by reacting titanium(IV) isopropoxide with formic acid, *J. Therm. Anal. Cal.*, 2003, **71**, 997–1009.
25. R. Campostrini, M. Ischia and L. Palmisano, Pyrolysis study of sol-gel derived TiO_2 powders. Part II TiO_2-anatase prepared by reacting titanium(IV) isopropoxide with oxalic acid, *J. Therm. Anal. Cal.*, 2003, **71**, 1011–1021.
26. R. Campostrini, M. Ischia and L. Palmisano, Pyrolysis study of sol-gel derived TiO_2 powders Part III. TiO_2-anatase prepared by reacting titanium(IV) isopropoxide with acetic acid, *J. Therm. Anal. Cal.*, 2004, **75**, 13–24.
27. R. Campostrini, M. Ischia and L. Palmisano, Pyrolysis study of sol-gel derived TiO_2 powders Part IV. TiO_2-anatase prepared by reacting titanium(IV) isopropoxide without chelating agents, *J. Therm. Anal. Cal*, 2004, **75**, 25–34.
28. M. Ischia, R. Campostrini, L. Lutterotti, E. García-López, L. Palmisano, M. Schiavello, F. Pirillo and R. Molinari, Synthesis, characterisation and photocatalytic activity of TiO_2 powders prepared under different gelling and pressure conditions, *J. Sol-Gel Sci. Technol.*, 2005, **33**, 201–213.
29. H. Schmidt, M. Naumann, T. S. Müller and M. Akarsu, Application of spray techniques for new photocatalytic gradient coatings on plastics, *Thin Solid Films*, 2006, **502**, 132–137.
30. R. I. Bickley, J. S. Lees, R. J. D. Tilley, L. Palmisano and M. Schiavello, Characterisation of iron/titanium oxide photocatalysts Part 1. Structural and magnetic studies, *J. Chem. Soc., Faraday Trans.*, 1992, **88**, 377–383.
31. R. I. Bickley, T. Gonzalez-Carreño, A. R. Gonzalez-Elipé, G. Munuera and L. Palmisano, Characterisation of iron/titanium oxide photocatalysts Part 2. Surface studies, *J. Chem. Soc., Faraday Trans.*, 1994, **90**, 2257–2264.
32. A. Sclafani, L. Palmisano and M. Schiavello, Influence of the preparation methods of TiO_2 on the photocatalytic degradation of phenol in aqueous dispersion, *J. Phys. Chem.*, 1990, **94**, 829–832.
33. H. Kawaguchi, Photocatalytic decomposition of phenol in the presence of titanium dioxide, *Environ. Technol. Lett.*, 1984, **5**, 471–474.
34. V. Augugliaro, L. Palmisano, A. Sclafani, C. Minero and E. Pelizzetti, Photocatalytic degradation of phenol in aqueous titanium dioxide dispersion, *Toxicol. Environ. Chem.*, 1988, **16**, 89–109.
35. Z. Zhang, C. C. Wang, R. Zakaria and J. Y. Ying, Role of particle size in nanocrystalline TiO_2-based photocatalysts, *J. Phys. Chem. B*, 1998, **102**, 10871–10878.
36. H. Kominami, J. Kato, S. Y. Murakami, Y. Kera, M. Inoue, T. Inui and B. Ohtani, Synthesis of titanium(IV) oxide of ultra-high photocatalytic activity: high-temperature hydrolysis of titanium alkoxides with water liberated homogeneously from solvent alcohols, *J. Mol. Catal. A: Chem*, 1999, **144**, 165–171.

37. Q. H. Zhang, L. Gao and J. Guo, Effects of calcination on the photocatalytic properties of nanosized TiO_2 powders prepared by $TiCl_4$ hydrolysis, *Appl. Catal. B: Environ.*, 2000, **26**, 207–215.
38. K. Hashimoto, K. Wasada, N. Toukai, H. Kominami and Y. Kera, Photocatalytic oxidation of nitrogen monoxide over titanium(IV) oxide nanocrystals large size areas, *J. Photochem. Photobiol. A: Chem.*, 2000, **136**, 103–109.
39. A. J. Maira, K. L. Yeung, J. Soria, J. M. Coronado, C. Belver, C. Y. Lee and V. Augugliaro, Gas-phase photo-oxidation of toluene using nanometer-size TiO_2 catalysts, *Appl. Catal. B: Environ.*, 2001, **29**, 327–336.
40. H. Kominami, H. Kumamoto, Y. Kera and B. Ohtani, Immobilization of highly active titanium(IV) oxide particles: A novel strategy of preparation of transparent photocatalytic coatings, *Appl. Catal. B: Environ.*, 2001, **30**, 329–335.
41. S. D. Burnside, V. Shklover, C. Barbè, P. Comte, F. Arendse, K. Brooks and M. Grätzel, Self-organization of TiO_2 nanoparticles in thin films, *Chem. Mater.*, 1998, **10**, 2419–2425.
42. H. Yin, Y. Wada, T. Kitamura, S. Kambe, S. Murasawa, H. Mori, T. Sakata and S. Yanagida, Hydrothermal synthesis of nanosized anatase and rutile TiO_2 using amorphous phase TiO_2, *J. Mater. Chem.*, 2001, **11**, 1694–1703.
43. K. Yanagisawa and J. Ovenstone, Crystallization of anatase from amorphous titania using the hydrothermal technique: effects of starting material and temperature, *J. Phys. Chem. B*, 1999, **103**, 7781–7787.
44. S. Y. Chae, M. K. Park, S. K. Lee, T. Y. Kim, S. K. Kim and W. I. Lee, Preparation of size-controlled TiO_2 nanoparticles and derivation of optically transparent photocatalytic films, *Chem. Mater.*, 2003, **15**, 3326–3331.
45. Q. Zhang and L. Gao, Preparation of oxide nanocrystals with tunable morphologies by the moderate hydrothermal method: insights from rutile TiO_2, *Langmuir*, 2003, **19**, 967–971.
46. X. Feng, J. Zhai and L. Jiang, The fabrication and switchable superhydrophobicity of TiO_2 nanorod films, *Angew. Chem. Int. Ed.*, 2005, **44**, 5111–5118.
47. Q. Huang and L. Gao, A simple route for the synthesis of rutile TiO_2 nanorods, *Chem. Lett.*, 2003, **32**, 638.
48. S. Yang and L. Gao, Fabrication and characterization of nanostructurally flowerlike aggregates of TiO_2 via a surfactant-free solution route: effect of various reaction media, *Chem. Lett.*, 2005, **34**, 1044 and references therein.
49. M. Addamo, V. Augugliaro, A. Di Paola, E. García-López, V. Loddo, G. Marcì, R. Molinari, L. Palmisano and M. Schiavello, Preparation, characterization and photoactivity of polycrystalline nanostructured TiO_2 catalysts, *J. Phys. Chem. B*, 2004, **108**, 3303–3310.
50. M. Addamo, V. Augugliaro, A. Di Paola, E. García-López, V. Loddo, G. Marcì and L. Palmisano, Preparation and photoactivity of

nanostructured TiO$_2$ particles obtained by hydrolysis of TiCl$_4$, *Colloids Surf., A: Phys. Eng. Aspects*, 2005, **265**, 23–31.
51. M. Addamo, M. Del Arco, M. Bellardita, D. Carriazo, A. Di Paola, E. García-Lüpez, G. Marcì, C. Martín, L. Palmisano and V. Rives, Photoactivity of nanostructured TiO$_2$ catalysts in aqueous system and their surface acid-base, bulk and textural properties, *Res. Chem. Intermed.*, 2007, **33**, 465–479.
52. A. Di Paola, G. Cufalo, M. Addamo, M. Bellardita, R. Campostrini, M. Ischia, R. Ceccato and L. Palmisano, Photocatalytic activity of nanocrystalline TiO$_2$ (brookite, rutile and brookite-based) powders prepared by thermohydrolysis of TiCl$_4$ in aqueous chloride solutions, *Colloid Surf., A: Phys. Eng. Aspects*, 2008, **317**, 366–376.
53. A. Di Paola, M. Bellardita, R. Ceccato, L. Palmisano and F. Parrino, Highly active photocatalytic TiO$_2$ powders obtained by thermohydrolysis of TiCl$_4$ in water, *J. Phys. Chem. C*, 2009, **113**, 15166–15174.
54. A. Pottier, C. Chanéac, E. Tronc, L. Mazerolles and J. P. Jolivet, Synthesis of brookite TiO$_2$ nanoparticles by thermolysis of TiCl$_4$ in strongly acidic aqueous media, *J. Mater. Chem.*, 2001, **11**, 1116–1121.
55. A. M. Roy, G. C. De, N. Sasmal and S. S. Bhattacharyya, Determination of the flatband potential of semiconductor particles in suspension by photovoltage measurement, *Int. J. Hydrogen Energy*, 1995, **20**, 627–630.
56. M. Grätzel and F. P. Rotzinger, The influence of the crystal lattice structure on the conduction band energy of oxides of titanium (IV), *Chem. Phys. Lett.*, 1985, **118**, 474–477.
57. M. Addamo, V. Augugliaro, M. Bellardita, A. Di Paola, V. Loddo, G. Palmisano, L. Palmisano and S. Yurdakal, Environmentally friendly photocatalytic oxidation of aromatic alcohol to aldehyde in aqueous suspension of brookite TiO$_2$, *Catal. Lett.*, 2008, **126**, 58–62.
58. G. Palmisano, S. Yurdakal, V. Augugliaro, V. Loddo and L. Palmisano, Photocatalytic selective oxidation of 4-methoxybenzyl alcohol to aldehyde in aqueous suspension of home-prepared titanium dioxide catalyst, *Adv. Synth. Catal.*, 2007, **349**, 964–970.
59. S. Yurdakal, G. Palmisano, V. Loddo, V. Augugliaro and L. Palmisano, Nanostructured rutile TiO$_2$ for selective photocatalytic oxidation of aromatic alcohols to aldehydes in water, *J. Am. Chem. Soc.*, 2008, **130**, 1568–1569.
60. V. Augugliaro, T. Caronna, V. Loddo, G. Marcì, G. Palmisano, L. Palmisano and S. Yurdakal, Oxidation of aromatic alcohols in irradiated aqueous suspensions of commercial and home-prepared Rutile TiO$_2$: a selectivity study, *Chem. Eur. J.*, 2008, **14**, 4640–4646.
61. S. Yurdakal, G. Palmisano, V. Loddo, O. Alagöz, V. Augugliaro and L. Palmisano, Selective photocatalytic oxidation of 4-substituted aromatic alcohols in water with rutile TiO$_2$ prepared at room temperature, *Green Chem.*, 2009, **11**, 510–516.
62. V. Augugliaro, H. Kisch, V. Loddo, M. J. Lüpez-Muñoz, C. Márquez-Álvarez, G. Palmisano, L. Palmisano, F. Parrino and S. Yurdakal,

Photocatalytic oxidation of aromatic alcohols to aldehydes in aqueous suspension of home-prepared titanium dioxide 1. Selectivity enhancement by aliphatic alcohols, *Appl. Catal. A: General*, 2008, **349**, 182–188.
63. V. Augugliaro, H. Kisch, V. Loddo, M. J. Lüpez-Muñoz, C. Márquez-Álvarez, G. Palmisano, L. Palmisano, F. Parrino and S. Yurdakal, Photocatalytic oxidation of aromatic alcohols to aldehydes in aqueous suspension of home-prepared titanium dioxide 2. Intrinsic and surface features of catalysts, *Appl. Catal. A: General*, 2008, **349**, 189–197.
64. J. Mao-Xiang, J. Xue-Qin, L. Wang-Xing, L. Dong-Hong and W. Zhou, Preparation and photocatalytic activity of mesoporous TiO_2 microspheres, *Micro Nanosys.*, 2009, **1**, 12–16 and references therein.
65. J. H. Xua, W. L. Daia, J. X. Lia and Y. Caoa, Novel core-shell structured mesoporous titania microspheres: preparation, characterization and excellent photocatalytic activity in phenol abatement, *J. Photochem. Photobiol. A: Chem.*, 2008, **195**, 284–294.
66. D. S. Kim, S. J. Han and S. Y. Kwak, Synthesis and photocatalytic activity of mesoporous TiO_2 with the surface area, crystallite size, and pore size, *J. Colloid Interface Sci.*, 2007, **316**, 85–91.
67. G. A. Waychunas, C. S. Kim, B. Gilbert, H. Zhang and J. F. Banfield, Nanoparticle structure, surface structure and crystal chemistry, *Geochim. Cosmochim. Acta*, 2006, **70**, A692–693.
68. T. Kasuga, M. Hiramatsu, A. Hoson, T. Sekino and K. Niihara, Formation of titanium oxide nanotube, *Langmuir*, 1998, **14**, 3160–3163.
69. D. V. Bavykin, E. V. Milsom, F. Marken, D. H. Kim, D. H. Marsh, D. J. Riley, F. C. Walsh, K. H. El-Abiary and A. A. Lapkin, A novel cation-binding TiO_2 nanotube substrate for electro- and bioelectro-catalysis, *Electrochem. Commun.*, 2005, **7**, 1050–1058.
70. D. V. Bavykin, A. A. Lapkin, P. K. Plucinski, J. M. Friedrich and F. C. Walsh, Reversible storage of molecular hydrogen by sorption into multilayered TiO_2 nanotubes, *J. Phys. Chem. B*, 2005, **109**, 19422–19427.
71. S. H. Lim, J. Luo, Z. Zhong, W. Ji and J. Lin, Room-temperature hydrogen uptake by TiO_2 nanotubes, *Inorg. Chem.*, 2005, **44**, 4124–4126.
72. Y. Q. Wang, G. Q. Hu, X. F. Duan, H. L. Sun and Q. K. Xue, Microstructure and formation mechanism of titanium dioxide nanotubes, *Chem. Phys. Lett.*, 2002, **365**, 427–431.
73. B. D. Yao, Y. F. Chan, X. Y. Zhang, W. F. Zhang, Z. Y. Yang and N. Wang, Formation mechanism of TiO_2 nanotubes, *Appl. Phys. Lett.*, 2004, **82**, 281–282.
74. A. R. Armstrong, G. Armstrong, J. Canales, R. Garcia and P. G. Bruce, TiO_2-B nanowires, *Angew. Chem. Int. Ed.*, 2004, **43**, 2286–2288.
75. M. Wei, Y. Konishi, H. Zhou, H. Sugihara and H. Arakawa, A simple method to synthesize nanowires titanium dioxide from layered titanate particles, *Chem. Phys. Lett.*, 2004, **400**, 231–234.
76. X. L. Li, Q. Peng, J. X. Yi, X. Wang and Y. D. Li, Near monodisperse TiO_2 nanoparticles and nanorods, *Chem. Eur. J.*, 2006, **12**, 2383–2391.

77. X. Wang, J. Zhuang, Q. Peng and Y. D. Li, A general strategy for nanocrystal synthesis, *Nature*, 2005, **437**, 121–124.
78. B. Wen, C. Liu and Y. Liu, Bamboo-shaped Ag-doped TiO_2 nanowires with heterojunctions, *Inorg. Chem.*, 2005, **44**, 6503–6505.
79. C. S. Kim, B. K. Moon, J. H. Park, B. C. Choi and H. J. Seo, Solvothermal synthesis of nanocrystalline TiO_2 in toluene with surfactant, *J. Cryst. Growth*, 2003, **257**, 309–315.
80. C. S. Kim, B. K. Moon, J. H. Park, S. T. Chung and S. M. Son, Synthesis of nanocrystalline TiO_2 in toluene by a solvothermal route, *J. Cryst. Growth*, 2003, **254**, 405–410.
81. B. Wen, C. Liu and Y. Liu, Solvothermal synthesis of ultralong single-crystalline TiO_2 nanowires, *New J. Chem.*, 2005, **29**, 969–971.
82. B. Wen, C. Liu and Y. Liu, Depositional characteristics of metal coating on single-crystal TiO_2 nanowires, *J. Phys. Chem. B*, 2005, **109**, 12372–12375.
83. T. J. Trentler, T. E. Denler, J. F. Bertone, A. Agrawal and V. L. Colvin, Synthesis of TiO_2 nanocrystals by nonhydrolytic solution-based reactions, *J. Am. Chem. Soc.*, 1999, **121**, 1613–1614.
84. M. Niederberger, M. H. Bartl and G. D. Stucky, Benzyl alcohol and titanium tetrachloride – a versatile reaction system for the nonaqueous and low-temperature preparation of crystalline and luminescent titania nanoparticles, *Chem. Mater.*, 2002, **14**, 4364–4370.
85. P. D. Cozzoli, A. Kornowski and H. Weller, Low-temperature synthesis of soluble and processable organic-capped anatase TiO_2 nanorods, *J. Am. Chem. Soc.*, 2003, **125**, 14539–14548.
86. P. D. Cozzoli, R. Comparelli, E. Fanizza, M. L. Curri, A. Agostiano and D. Laub, Photocatalytic synthesis of silver nanoparticles stabilized by TiO_2 nanorods: a semiconductor/metal nanocomposite in homogeneous nonpolar solution, *J. Am. Chem. Soc.*, 2004, **126**, 3868–3879.
87. P. D. Cozzoli, E. Fanizza, R. Comparelli, M. L. Curri, A. Agostiano and D. Laub, Role of metal nanoparticles in TiO_2/Ag nanocomposite-based microheterogeneous photocatalysis, *J. Phys. Chem. B*, 2004, **108**, 9623–9630.
88. P. D. Cozzoli, E. Fanizza, M. L. Curri, D. Laub and A. Agostiano, Low-dimensional chainlike assemblies of TiO_2 nanorod-stabilized Au nanoparticles, *Chem. Commun.*, 2005, 942–944.
89. P. D. Cozzoli, M. L. Curri and A. Agostano, Efficient charge storage in photoexcited TiO_2 nanorod-noble metal nanoparticle composite systems, *Chem. Commun.*, 2005, 3186–3188.
90. R. Buonsanti, V. Grillo, E. Carlino, C. Giannini, M. L. Curri, C. Innocenti, C. Sangregoria, K. Achterhold, F. G. Parak, A. Agostiano and P. D. Cozzoli, Seeded growth of asymmetric binary nanocrystals made of a semiconductor TiO_2 rodlike section and a magnetic γ-Fe_2O_3 spherical domain, *J. Am. Chem. Soc.*, 2006, **128**, 16953–16970.
91. R. Alexandrescu, F. Dumitrache, I. Morjan, I. Sandu, M. Savoiu, I. Voicu, C. Fleaca and R. Piticescu, TiO_2 nanosized powders by $TiCl_4$ laser pyrolysis, *Nanotechnology*, 2004, **15**, 537–545.

92. R. Alexandrescu, I. Morjan, M. Scarisoreanu, R. Birjega, E. Popovici, I. Soare, L. Gavrila-Florescu, I. Voicu, I. Sandu, F. Dumitrache, G. Prodan, E. Vasile and E. Figgemeier, Structural investigations on TiO_2 and Fe-doped TiO_2 nanoparticles synthesized by laser pyrolysis, *Thin Solid Films*, 2007, **515**, 8438–8445.
93. B. Pignon, H. Maskrot, V. Guyot Ferreol, Y. Leconte, S. Coste, M. Gervais, T. Pouget, C. Reynaud, J. F. Tranchant and N. Herlin-Boime, Versatility of laser pyrolysis applied to the synthesis of TiO_2 nanoparticles – application to UV attenuation, *Eur. J. Inorg. Chem*, 2008, **6**, 883–889.
94. G. Ma, X. Zhao and J. Zhu, Microwave hydrothermal synthesis of rutile TiO_2 nanorods, *Int. J. Mod. Phys. B*, 2005, **19**, 2763–2768.
95. X. Wu, Q. Z. Jiang, Z. F. Ma, M. Fu and W. F. Shangguan, Synthesis of titania nanotubes by microwave irradiation, *Solid State Commun.*, 2005, **136**, 513–517.
96. A. B. Corradi, F. Bondioli, B. Focher, A. M. Ferrari, C. Grippo, E. Mariani and C. Villa, Conventional and microwave-hydrothermal synthesis of TiO_2 nanopowders, *J. Am. Ceram. Soc.*, 2005, **88**, 2639–2641.
97. M. Addamo, M. Bellardita, D. Carriazo, A. Di Paola, S. Milioto, L. Palmisano and V. Rives, Inorganic gels as precursors of TiO_2 photocatalysts prepared by low temperature microwave or thermal treatment, *Appl. Catal. B*, 2008, **84**, 742–748.
98. M. P. Soriaga, J. Stickney, L. A. Bottomley and Y.-G. Kim (eds), *Thin Films: Preparation, Characterization, Applications*, Kluwer-Plenum, New York, 2002 (ISBN 0-306-47335-6).
99. J. Fraden, *Handbook of Modern Sensors*, Springer, Heidelberg/New York, 2003.
100. *Engineering Thin Films and Nanostructures with Ion Beams*, ed. E. Knystautas, CRC Press, Taylor & Francis Group, New York, 2005.
101. *Handbook of Advanced Magnetic Materials*, ed. Yi Liu, D. J. Sellmyer and D. Shindo, Volumes 1–4, Springer, Heidelberg/New York, 2006.
102. *Trends in Thin Solid Films Research*, ed. A. R. Jost, Nova Science Publishers, New York, 2007.
103. O. Stern, Zur theorie der elektrolytischen doppelschicht, *Z. Elektrochem.*, 1924, **30**, 508–516.
104. H. Schröder, *Physics of Thin Films*, Academic Press, New York-London, 1969, **5**, 87–141.
105. H. Dislich, New routes to multicomponent oxide glasses, *Angew. Chem. Int. Ed.*, 1971, **6**, 363–370.
106. *Thin films Science and Technology*, ed. G. Siddal, Elsevier, Amsterdam, 1984, vol. 6.
107. M. Addamo, V. Augugliaro, A. Di Paola, E. García-López, V. Loddo, G. Marcì and L. Palmisano, Photocatalytic thin films of TiO_2 formed by a sol-gel process using titanium tetraisopropoxide as the precursor, *Thin Solid Films*, 2008, **516**, 3802–3807.

108. M. Addamo, M. Bellardita, A. Di Paola and L. Palmisano, Preparation and photoactivity of nanostructured anatase, rutile and brookite thin films, *Chem. Commun.*, 2006, 4943–4945.
109. N. Ignjatovíc, Z. Brankovíc, M. Dramićanin, J. M. Nedeljkovíc and D. P. Uskovíc, Preparation of TiO_2 and ZnO thin films by dip-coating method, *Mater. Sci. Forum*, 1998, **282–283**, 147–152.
110. H. Tada and M. Tanaka, Dependence of TiO_2 photocatalytic activity upon its film thickness, *Langmuir*, 1997, **13**, 360–364.
111. Y. Paz and A. Heller, Photooxidatively self-cleaning transparent titanium dioxide films on soda lime glass: the deleterious effect of sodium contamination and its prevention, *J. Mater. Res.*, 1997, **12**, 2759–2760.
112. L. D. Landau and B. G. Levich, Dragging of a liquid by a moving plate, *Acta Physicochim. U.R.S.S.*, 1942, **17**, 42–54.
113. H. Dislich, Sol-gel: science, processes and products, *J. Non-Cryst. Solids*, 1986, **80**, 115–121.
114. D. Meyerhofer, Characteristics of resist films produced by spinning, *J. Appl. Phys.*, 1978, **49**, 3993–3997.
115. B. T. Chen, Investigation of the solvent-evaporation effect on spin coating of thin films, *Polym. Eng. Sci.*, 1983, **23**, 399–403.
116. A. Weill, in *The Physics and Fabrication of Microstructures and Microdevices*, ed. M. J. Kelly and C. Weisbuch, Springer-Verlag, Berlin, 1986, 51.
117. M. H. Habibi, M. Nasr-Esfahani and T. A. Egerton, Preparation, characterization and photocatalytic activity of TiO_2/methylcellulose nanocomposite films derived from nanopowder TiO_2 and modified sol-gel titania, *J. Mater. Sci.*, 2007, **42**, 6027–6035.
118. X. Fan, T. Yu, L. Zhang, X. Chen and Z. Zou, Photocatalytic degradation of acetaldehyde on mesoporous TiO_2: effects of surface area and crystallinity on the photocatalytic activity, *Chin. J. Chem. Phys.*, 2007, **20**, 733–738.
119. E. M. El-Maghraby, Y. Nakamura and S. Rengakuji, Composite TiO_2–SnO_2 nanostructured films prepared by spin-coating with high photocatalytic performance, *Catal. Commun.*, 2008, **9**, 2357–2360.
120. J. Kasanen, M. Suvanto and T. T. Pakkanen, Self-cleaning TiO_2-based multilayer coating fabricated on polymer and glass surfaces, *J. Appl. Polym. Sci.*, 2009, **111**, 2597–2606.
121. M. Uzunova-Bujnova, R. Todorovska, M. Milanova, R. Kralchevska and D. Todorovsky, On the spray-drying deposition of TiO_2 photocatalytic films, *Appl. Surf. Sci.*, 2009, **256**, 830–837.
122. K. Masters, *Spray Drying Handbook*, 4th edn., Longman Scientific & Technical, London, 1985.
123. G. Bertrand, C. Meunier, P. Bertrand and C. Coddet, Dried particle plasma spray in-flight synthesis of spinel coatings, *J. Eur. Ceram. Soc.*, 2002, **22**, 891–902.
124. N. Berger-Keller, G. Bertrand, C. Filiatre, C. Meunier and C. Coddet, Microstructure of plasma-sprayed titania coatings deposited from spray-dried powder, *Surf. Coat. Technol.*, 2003, **168**, 281–290.

125. R. Jaworski, L. Pawlowski, F. Roudet, S. Kozerski and A. Le Maguer, Influence of suspension plasma spraying process parameters on TiO_2 coatings microstructure, *J. Therm. Spray Technol.*, 2008, **17**, 73–81.
126. C. Lee, H. Choi, C. Lee and H. Kim, Photocatalytic properties of nanostructured TiO_2 plasma sprayed coating, *Surf. Coat. Technol.*, 2003, **173**, 192–200.
127. F. L. Toma, D. Sokolov, G. Bertrand, D. Klein, C. Coddet and C. Meunier, Comparison of the photocatalytic behavior of TiO_2 coatings elaborated by different spraying processes, *J. Therm. Spray Technol.*, 2006, **15**, 576–581.
128. H. L. Hartnagel, A. L. Dawar, A. K. Jain and C. Jagadish, *Semiconducting Transparent Thin Films*, Institute of Physics Publishing, Bristol, 1995.
129. P. S. Patil, Versatility of chemical spray pyrolysis technique, *Mater. Chem. Phys.*, 1999, **59**, 185–198.
130. B. Goodbole, N. Badera, S. B. Shrivastav and V. Ganesan, A simple chemical spray pyrolysis apparatus for thin film preparation, *J. Instrum. Soc. India*, 2009, **39**, 42–45.
131. Y. Ohko, S. Saitoh, T. Tatsuma and A. Fujishima, Photoelectrochemical anticorrosion and self-cleaning effects of a TiO_2 coating for type 304 stainless steel, *J. Electrochem. Soc.*, 2001, **148**, B24–B28.
132. M. Okuya, N. A. Prokudina, K. Mushika and S. Kaneko, TiO_2 thin films synthesized by the spray pyrolysis deposition (SPD) technique, *J. Eur. Ceram. Soc.*, 1999, **19**, 903–906.
133. J. K. Park and H. K. Kim, Preparation and characterization of hydrophilic TiO_2 film, *Bull. Korean Chem. Soc.*, 2002, **23**, 745–748.
134. M. O. Abou-Helal and W. T. Seeber, Preparation of TiO_2 thin films by spray pyrolysis to be used as a photocatalyst, *Appl. Surf. Sci.*, 2002, **195**, 53–62.
135. N. Castillo, D. Olguýn and A. Conde-Gallardo, Structural and morphological properties of TiO_2 thin films prepared by spray pyrolysis, *Rev. Mexicana Física*, 2004, **50**, 382–387.
136. A. I. Martínez, D. R. Acosta and A. A. López, Effect of deposition methods on the properties of photocatalytic TiO_2 thin films prepared by spray pyrolysis and magnetron sputtering, *J. Phys.: Condens. Matter*, 2004, **16**, S2335–S2344.
137. M. Shiojiri, Crystallization of amorphous titanium dioxide films prepared by vacuum-evaporation, *J. Phys. Soc. Jpn.*, 1996, **21**, 335.
138. T. Myata, S. Tsukada and T. Minami, Preparation of anatase TiO_2 thin films by vacuum arc plasma evaporation, *Thin Solid Films*, 2006, **496**, 136–140.
139. T. S. Yang, C. B. Shiu and M. S. Wong, Structure and hydrophilicity of titanium oxide films prepared by electron beam evaporation, *Surf. Sci.*, 2004, **548**, 75–82.
140. P. K. Song, Y. Irie and Y. Shigesato, Crystallinity and photocatalytic activity of TiO_2 films deposited by reactive sputtering with radio frequency substrate bias, *Thin Solid Films*, 2006, **496**, 121–125.

141. P. Zeman and S. Takabayashi, Self-cleaning and antifogging effects of TiO_2 films prepared by radio frequency magnetron sputtering, *J. Vac. Sci. Technol. A*, 2002, **20**, 388–393.
142. M. Kitano, M. Takeuchi, M. Matsuoka, J. M. Thomas and M. Anpo, Preparation of visible light-responsive TiO_2 thin film photocatalysts by an RF magnetron sputtering deposition method and their photocatalytic reactivity, *Chem. Lett.*, 2005, **20**, 616–617.
143. H. Wang, T. Wang and P. Xu, Effects of substrate temperature on the microstructure and photocatalytic reactivity of TiO_2 films, *J. Mater. Sci. Mater. Electron.*, 1998, **9**, 327–330.
144. M. Takeuchi, T. Yamasaki, K. Tsujimaru and M. Anpo, Preparation of crystalline TiO_2 thin film photocatalysts on polycarbonate substrates by a RF-magnetron sputtering deposition method, *Chem. Lett.*, 2006, **35**, 904–905.
145. K. Eufinger, E. N. Janssen, H. Poelman, D. Poelman, R. De Gryse and G. B. Marin, The effect of argon pressure on the structural and photocatalytic characteristics of TiO_2 thin films deposited by d.c. magnetron sputtering, *Thin Solid Films*, 2006, **515**, 425–429.
146. M. Kitano, K. Funatsu, M. Matsuoka, M. Ueshima and M. Anpo, Preparation of nitrogen-substituted TiO_2 thin film photocatalysts by the radio frequency magnetron sputtering deposition method and their photocatalytic reactivity under visible light irradiation, *J. Phys. Chem. B*, 2006, **110**, 25266–25272.
147. L. Chen, M. E. Graham, G. Li and K. A. Gray, Fabricating highly active mixed phase TiO_2 photocatalysts by reactive DC magnetron sputter deposition, *Thin Solid Films*, 2006, **515**, 1176–1181.
148. S. Chaiyakun, A. Pokaipisit, P. Limsuwan and B. Ngotawornchai, Growth and characterization of nanostructured anatase phase TiO_2 thin films prepared by DC reactive unbalanced magnetron sputtering, *Appl. Phys. A*, 2009, **95**, 579–587.
149. A. Anders, *Cathodic Arcs: From Fractal Spots to Energetic Condensation*, Springer Series on Atomic, Optical, and Plasma Physics, Springer, New York, 2008.
150. A. Kleiman, A. Márquez and D. G. Lamas, Anatase TiO_2 films obtained by cathodic arc deposition, *Surf. Coat. Technol.*, 2007, **201**, 6358–6362.
151. J. M. Lackner, Industrially-scaled large-area and high-rate tribological coating by pulsed laser deposition, *Surf. Coat. Technol.*, 2005, **200**, 1439–1444.
152. M. Terashima N. Inoue, S. Kashiwabara and R. Fujimoto, Photocatalytic TiO_2 thin-films deposited by a pulsed laser deposition technique, *Appl. Surf. Sci.*, 2001, **169–170**, 535–538.
153. A. P. Caricato, M. G. Manera, M. Martino, R. Rella, F. Romano, J. Spadavecchia, T. Tunno and D. Valerini, Uniform thin films of TiO_2 nanoparticles deposited by matrix-assisted pulsed laser evaporation, *Appl. Surf. Sci.*, 2007, **253**, 6471–6475.

154. M. Walczak, E. L. Papadopoulou, M. Sanz, A. Manousaki, J. F. Marco and M. Castillejo, Structural and morphological characterization of TiO_2 nanostructured films grown by nanosecond pulsed laser deposition, *Appl. Surf. Sci.*, 2009, **255**, 5267–5270.
155. E. György, A. Pérez del Pino, G. Sauthier, A. Figueras, F. Alsina and J. Pascual, Structural, morphological and local electric properties of TiO_2 thin films grown by pulsed laser deposition, *J. Phys. D: Appl. Phys.*, 2007, **40**, 5246–5251.
156. M. F. Brunella, M. V. Diamanti, M. P. Pedeferri, F. Di Fonzo, C. S. Casari and A. Li Bassi, Photocatalytic behavior of different titanium dioxide layers, *Thin Solid Films*, 2007, **515**, 6309–6313.
157. A. Goossens, E.-L. Maloney and J. Schoonman, Gas-phase synthesis of nanostructured anatase TiO_2, *Chem. Vap. Deposition*, 1998, **4**, 109–114.
158. Z. Ding, P. L. Yue, G. Q. Lu and P. F. Greenfield, Synthesis of anatase TiO_2 supported on porous solids by chemical vapor deposition, *Catal. Today*, 2001, **68**, 173–182.
159. A. Watanabe, T. Tsuchiya and Y. Imai, Selective deposition of anatase and rutile films by KrF laser chemical vapor deposition from titanium isopropoxide, *Thin Solid Films*, 2002, **406**, 132–137.
160. N. Kaliwoh, J.-Y. Zhang and I. W. Boyd, Characterisation of TiO_2 deposited by photo-induced chemical vapour deposition, *Appl. Surf. Sci.*, 2002, **186**, 241–245.
161. M. L. Hitchman and F. Tian, Studies of TiO_2 thin films prepared by chemical vapour deposition for photocatalytic and photoelectrocatalytic degradation of 4-chlorophenol, *J. Electroanal. Chem.*, 2002, **538–539**, 165–172.
162. S. C. Jung, B. H. Kim, S. J. Kim, N. Imaishi and Y. I. Cho, Characterization of a TiO_2 photocatalyst film deposited by CVD and its photocatalytic activity, *Chem. Vap. Deposition*, 2005, **11**, 137–141.
163. K. M. K. Srivatsa, M. Bera and A. Basu, Pure brookite titania crystals with large surface area deposited by plasma enhanced chemical vapour deposition technique, *Thin Solid Films*, 2008, **516**, 7443–7446.
164. A. Sobczyk-Guzenda, M. Gazicki-Lipman, H. Szymanowski, J. Kowalski, P. Wojciechowski, T. Halamus and A. Tracz, Characterization of thin TiO_2 films prepared by plasma enhanced chemical vapour deposition for optical and photocatalytic applications, *Thin Solid Films*, 2009, **517**, 5409–5414.
165. Y. Gao, Y. Masuda, Z. Peng, T. Yonezawa and K. Koumoto, Room temperature deposition of a TiO_2 thin film from aqueous peroxotitanate solution, *J. Mater. Chem.*, 2003, **13**, 608–613.
166. Y. F. Gao, M. Nagai, W. S. Seo and K. Koumoto, Thick transparent rutile TiO_2 films crystallized in solution, *Langmuir*, 2007, **23**, 4712–4714.
167. R. C. Jaeger, *Thermal Oxidation of Silicon*, Prentice Hall, Upper Saddle River, NJ, 2001.

168. S. U. M. Khan, M. Al-Shahry and W. B. Ingler Jr., Efficient photochemical water splitting by a chemically modified n-TiO_2, *Science*, 2002, **297**, 2243–2245.
169. A. Fujishima, K. Kohayakawa and K. Honda, Hydrogen production under sunlight with an electrochemical photocell, *J. Electrochem. Soc.*, 1975, **122**, 1487–1489.
170. N. Cabrera and N. F. Mott, Theory of the oxidation of metals, *Rep. Proc. Phys.*, 1948, **12**, 163–184.
171. M. V. Diamanti and M. P. Pedeferri, Effect of anodic oxidation parameters on the titanium oxides formation, *Corrosion Sci.*, 2007, **49**, 939–948.
172. G. K. Mor, O. K. Varghese, M. Paulose, K. Shankar and C. A. Grimes, A review on highly ordered, vertically oriented TiO_2 nanotube arrays: fabrication, material properties, and solar energy applications, *Sol. Energy Mater. Sol. Cells*, 2006, **90**, 2011–2075, and references therein.
173. D. Gong, C. A. Grimes, O. K. Varghese, W. Hu, R. S. Singh, Z. Chen and E. C. Dickey, Titanium oxide nanotube arrays prepared by anodic oxidation, *J. Mater. Res.*, 2001, **16**, 3331–3334.
174. C. Ruan, M. Paolose, O. K. Varghese, G. K. Mor and C. A. Grimes, Fabrication of highly ordered TiO_2 nanotube arrays using an organic electrolyte, *J. Phys. Chem. B*, 2005, **109**, 15754–15759.
175. M. V. Diamanti, M. Ormellese and M. P. Pedeferri, Photoactivity of nitrogen-doped anodic TiO_2 under UV and VIS light irradiation, *Nanotechnology*, 2009, **3**, 198–201.
176. W. F. Pickard, Remarks on the theory of electrophoretic deposition, *J. Electrochem. Soc.*, 1968, **115**, 105C–108C.
177. P. Sarkar and P. S. Nicholson, Electrophoretic deposition (EDP): mechanisms, kinetics, and application to ceramics, *J. Am. Ceram. Soc.*, 1996, **79**, 1987–2002.
178. S. Hayashi, N. Furuhata and Z. Nakagawa, Deposition efficiency in the electrophoretic deposition of ALPHA-Al_2O_3 powder, *J. Ceram. Soc. Jpn.*, 2002, **110**, 135–138.
179. T. Uchikoshi, T. S. Suzuki, H. Okuyama, Y. Sakka and P. S. Nicholson, Control of crystalline texture in polycrystalline TiO_2 (anatase) by electrophoretic deposition in a strong magnetic field, *J. Eur. Ceram. Soc.*, 2004, **24**, 225–229.
180. A. R. Boccaccini, M. Rossetti, J. A. Roether, S. H. S. Zein and M. Ferraris, Development of titania coatings on glass foams, *Constr. Build. Mater.*, 2009, **7**, 2554–2558.
181. A. Fernandez, G. Lassaletta, V. M. Jiménez, A. Justo, A. R. González-Elipe, J. M. Herrmann, H. Tahiri and Y. Ait-Ichou, Preparation and characterisation of TiO_2 photocatalysts supported on various rigid supports (glass, quartz and stainless steel). Comparative studies of photocatalytic activity in water purification, *Appl. Catal. B: Environ.*, 1995, **7**, 49–63.

182. S. Yanagida, A. Nakajima, Y. Kamishima, N. Yoshida, T. Watanabe and K. Okada, Preparation of a crack-free rough titania coating on stainless steel mesh by electrophoretic deposition, *Mater. Res. Bull.*, 2005, **40**, 1335–1344.
183. A. R. Boccaccini, P. Karapappas and J. M. Marijuan, TiO_2 coatings on silicon carbide and carbon fibre substrates by electrophoretic deposition, *J. Mater. Sci.*, 2004, **39**, 851–859.
184. J.-H. Yum, S.-S. Kim, D.-Y. Kim and Y.-E. Sung, Electrophoretically deposited TiO_2 photo-electrodes for use in flexible dye-sensitized solar cells, *J. Photochem. Photobiol. A: Chem.*, 2005, **173**, 1–6.

CHAPTER 3
Unique Properties of Supported TiO_2

3.1 Superhydrophilic *versus* Photocatalytic Character

This chapter will introduce the reader to the most relevant properties of different materials where TiO_2 is supported in the shape of a thin film (for instance, glass) or of a blend exteriorly deposited (such as in concrete and paving blocks). Details about commercial, prototypes and promising products are given in subsequent chapters.

The most popular property of TiO_2-based materials is undoubtedly the ability to degrade pollutants in air and water, affording a cleaning effect for both air or water and the material that TiO_2 is deposited on. This behavior is the result of the semiconducting properties of TiO_2, which, when irradiated with appropriate light, can give rise to redox reactions, as widely discussed in Chapter 1 and in the literature.[1] Applications related to this property are widespread, both in open and in enclosed areas. Thus, indoor car parks have been equipped to test the de-polluting efficiency of a TiO_2-containing paint in an indoor polluted environment; 917 m^3 of enclosed area were monitored by gas chromatography to measure all the gaseous pollutants (NO_x, SO_2, CO) and CO_2, and, in addition, the roof (322 m^2) was covered with an acrylic paint treated with 10% TiO_2 prepared by Millennium Chemicals.[2] The results, taking into account the reduction of pollution due to car emission variation, showed that the photocatalytic removal of NO_x was about 20%, with a photocatalytic rate of the order of 0.1 µg m^{-2} s^{-1}. This study demonstrated that TiO_2-modified materials can work properly even in an enclosed, heavily polluted environment.

Beside the photocatalytic effect, we can regard superhydrophilicity as the other primary characteristic of TiO_2-based materials. This effect was discovered accidentally during experiments carried out at the laboratories of Toto, Inc., in 1995.

When water drops are deposited on a TiO_2-functionalized surface, they show a contact angle of several tens of degrees. Upon irradiation by UV-light (even the small fraction present in solar light) water begins to exhibit a decrease in

---

Clean by Light Irradiation: Practical Applications of Supported TiO_2
By Vincenzo Augugliaro, Vittorio Loddo, Mario Pagliaro, Giovanni Palmisano and Leonardo Palmisano
© V. Augugliaro, V. Loddo, M. Pagliaro, G. Palmisano and L. Palmisano 2010
Published by the Royal Society of Chemistry, www.rsc.org

contact angle, and starts to spread out flat instead of beading up. After a certain time the water contact angle reaches *ca.* 0°, meaning that the surface became superhydrophilic, *i.e.*, totally non-water-repellent.[3] If the material is then kept in the dark, a very low contact angle is retained for a certain time. When the surface is composed of TiO_2 coupled to SiO_2 or Si-containing compounds with siloxane bonding the retention of superhydrophilic behavior in the dark is very high and can last several days. After a given time, however, the angle starts increasing and the surface becomes hydrophobic again. Simple exposure at UV-light, though, can impart again the hydrophilic properties to the surface. This property is permanent and is retained provided that the film is well supported and does not peel off.

Figure 3.1 shows schematically the steps yielding a superhydrophilic surface: upon light irradiation, –OH surface groups are produced with high affinity for water, so that H_2O molecules can occupy a thin layer of space, constituting a uniform sheet rather than forming single drops.

In terms of the chemical mechanism, electrons tend to reduce Ti(IV) cations to the Ti(III) state, and the holes oxidize the O_2^- anions. In the process, oxygen atoms are ejected, creating oxygen vacancies.[4] This process can coexist with the photocatalytic effect, generated by electrons, giving rise to redox reactions involving the target pollutants rather than Ti cations. Depending on the film morphology one phenomenon can prevail over the other. Figure 3.2 shows this mechanism; clearly the hydrophilic behavior depends on the vacancies finally occupied by H_2O molecules. As far as water molecules are concerned, when the TiO_2 surface is irradiated with UV light under ambient conditions, the distribution of hydrogen bonds in the H_2O molecules decreases, along with the surface tension of H_2O clusters. Furthermore, the decrease in the amount of H_2O adsorbed on the TiO_2 surface implies a decrease in the outer surface areas of the H_2O clusters. It can be thus stated that the surface relaxation energies of

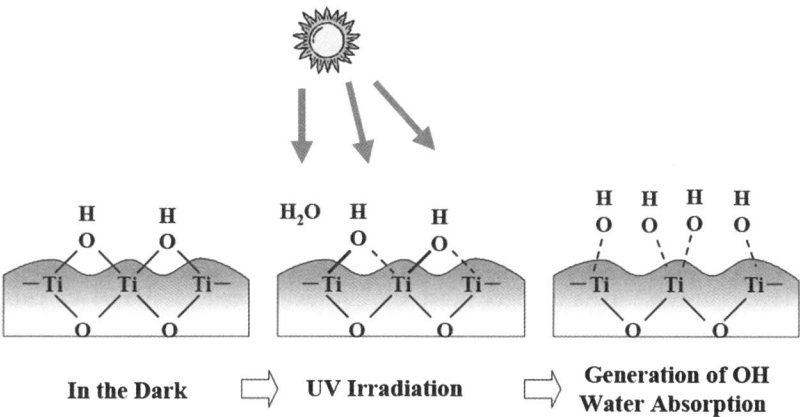

Figure 3.1 Physicochemical principles of light-induced superhydrophilicity. (Reproduced with permission from ref. 3b.)

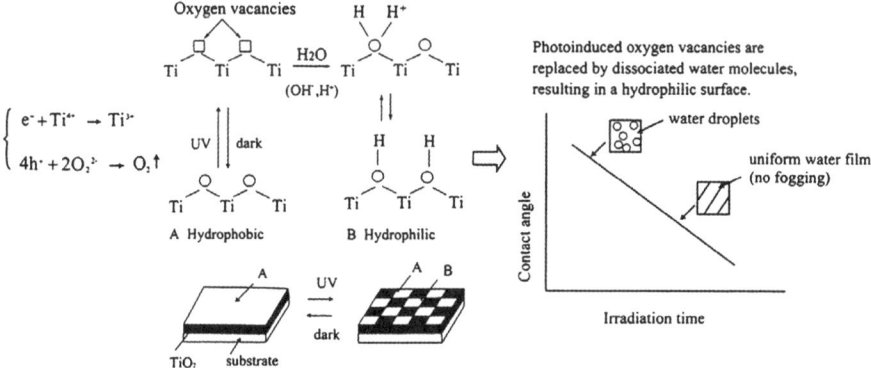

Figure 3.2 Mechanism of light-induced superhydrophilicity. (Reproduced with permission from ref. 4.)

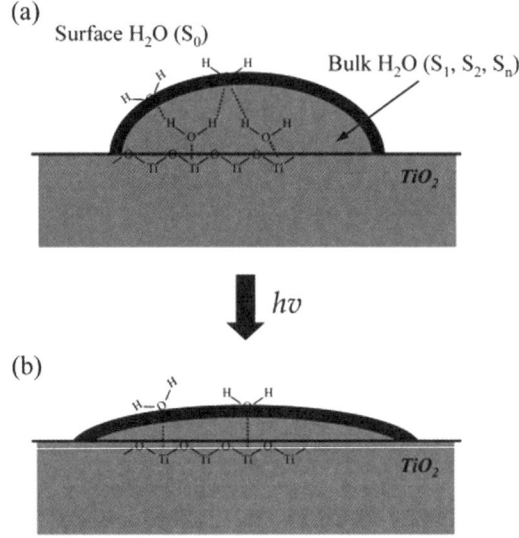

Figure 3.3 Schematic diagrams of shapes of H_2O clusters on TiO_2 surfaces: (a) before and (b) after UV light irradiation. (Adapted with permission from ref. 3a.)

the H_2O clusters are stabilized and the surface areas of the H_2O clusters increase spontaneously. A strict consequence of this behavior is the lower amount of adsorbed H_2O molecules, driven by a force behind the H_2O clusters spreading out, thermodynamically, on the TiO_2 surfaces (Figure 3.3).

The indicated reactions are, notably, completely reversible. Interestingly, the hydrophilicity of a TiO_2 surface hampers the adhesion of fatty compounds, making such surfaces very easy to clean, which is highly desirable for example in kitchens.

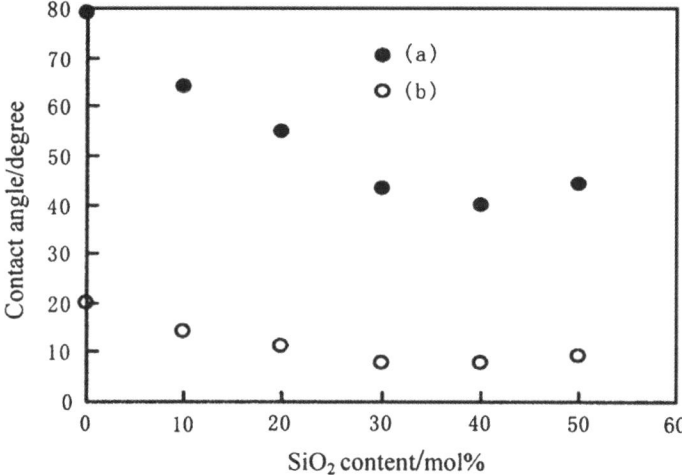

Figure 3.4 Dependence of photo-induced change in water contact angle of TiO_2/SiO_2 composite films after (a) exposure to air for 2 weeks; (b) UV irradiation. (Reproduced with permission from ref. 5.)

Surfaces can have more photocatalytic character and less superhydrophilic character, or *vice versa*, depending upon the composition and the manufacturing process. The presence of SiO_2 greatly enhances the hydrophilic properties of TiO_2 films and in particular allows for long-lasting superhydrophilic behavior even after many days in a dark state.[5] Increasing the SiO_2 amount up to 40% yields an improvement in hydrophilicity (Figure 3.4). Interestingly, the self-cleaning effect has an analogous trend. This synergetic effect can be explained as follows: numerous –OH groups can be adsorbed on the surface due to hydrophilicity, hence the photocatalytic activity is enhanced due to easier formation of HO˙ radicals. Conversely, the film surface can adsorb organic pollutants, which tends to turn the hydrophilic surface into a hydrophobic one. Photocatalysis, however, can decompose the organic compounds on the surface, yielding again a high hydrophilicity. This is how the photocatalytic activity helps in preserving the film hydrophilicity.

The improved performances of SiO_2/TiO_2 binary systems are attributed to the silicon cation, which is still bonded to the same number of oxygen atoms even though the oxygen atoms now have a new coordination; this creates a charge imbalance. Thus, Lewis sites are expected to form due to the positive charge in a TiO_2/SiO_2 binary metal oxide, thereby improving both adsorption of water and photocatalytic reactivity.

Hydrophilicity plays a very important role in maintaining the self-cleaning characteristic of a surface. The vulnerability of an exterior surface to soiling is strictly related to its contact angle with water. For instance, plastic materials, such as fluorocarbons, are more likely to be soiled than sheet glass or tiles. Thus a superhydrophilic material showing a water contact angle of 0° can be very improbably smeared. Nevertheless, a surface that is only superhydrophilic cannot always guarantee a self-cleaning effect. The photocatalytic effect indeed plays an important role in decomposing the soiled compounds and organic compounds to acquire a self-cleaning effect directly, and in decomposing the soiled compounds to recover the hydrophilicity to acquire self-cleaning character indirectly.

The hydrophilic feature plays a primary role in avoiding the undesirable fogging effects typically present in indoor windows, car glass, *etc.* For instance, a TiO_2 coating hampers the water condensation and formation of ice crystals that produce an opaque layer that reduces the transparency of glass or the visibility of traffic mirrors.[6] This point is detailed in Chapters 4–6.

Multicomponent films have been developed to impart particular properties to the resulting coating or to protect the underlying substrate. For instance, a multilayer structure consisting of a polyurethane (PU) protective layer on a substrate covered by two layers of photocatalytic TiO_2 and, finally, by immobilized TiO_2 particles (deposited from a suspension containing PU as binder) has been used with substrates such as high-density polyethylene (HDPE), poly(vinyl chloride) and borosilicate glass (BK7).[7] The fabricated coatings finally underwent an etching treatment by reactive oxygen-plasma, thus exhibiting a photocatalytic activity in the degradation of palmitic acid.

Oxygen-plasma treatment indeed produces hydrophilic groups (such as non-stoichiometric TiO_x) on the surface of TiO_2. Moreover, this treatment can add polar groups, such as hydroxyl groups, to the PU surface. Thus the PU binder between TiO_2 particles can actively contribute to the superhydrophilic effect, once the oxygen-plasma treatment has been performed.

Figure 3.5 shows a sketch of some contact angles for different multilayer films deposited on HDPE substrates. Case B refers to a film with PU binding, thus resulting in a more hydrophobic surface than case A, due to the presence of the organic component. It can be seen that superhydrophilicity is not present unless the oxygen-plasma treatment is performed (case C). The addition of palmitic acid on the film (case D) followed by UV irradiation causes superhydrophilic behavior and photocatalytic degradation of the pollutant, eventually yielding a 0° contact angle (case E).

3.2 Antireflection

Composite films containing TiO_2 are used to obtain antireflective surfaces. In particular, some excellent results have been obtained by coupling TiO_2 and sili*ca*. This property is almost exclusively used in glass or silicon wafer coatings. In the former case the coatings allow for a clear view through the glazing by increasing the light transmittance and avoiding negative effects on visual observation like double image, reflection of light sources, *etc.*[8] The basis of the antireflective behavior of TiO_2-based glass-coatings is widely discussed in Chapter 4. Here we report instead the recent application developed by the German company HEGE Solar, working in the field of photovoltaics.

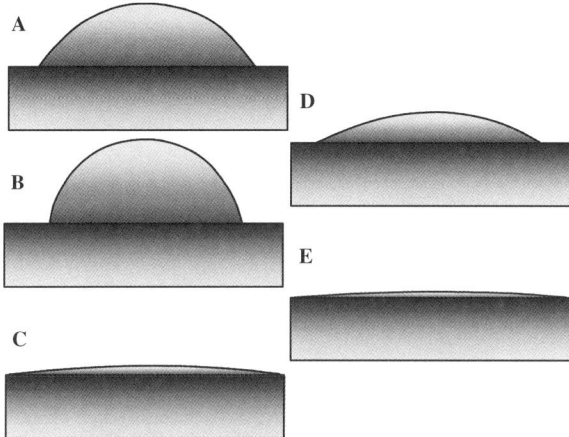

Figure 3.5 Water contact angles on HDPE samples at different stages of fabrication. The basic sample consisted of a substrate (HDPE), a protective PU coating, and two layers of TiO$_2$. The images show a sample (A) without a PU binder, (B) with a PU binder, (C) after a 15 s plasma treatment and 3 h UV illumination, (D) after the addition of palmitic acid to the surface and (E) after 24 h UV illumination of the coating with palmitic acid.

Researchers at this company have realized a new antireflective coating (TOP-NANO-HEGE® Coating) that boosts the efficiency of solar panels and allows sunlight to be absorbed from a very wide range of angles.[9]

Notably, a standard silicon solar cell absorbs just 2/3 the sunlight it receives. By covering the module with the described nanoengineered coating this value rises to more than 95%. Moreover, gains in absorption are consistent across the entire spectrum of sunlight, from ultraviolet to infrared and the absorption even takes place with very low solar rays angles (at dawn and sunset).

The coating is based on seven tiny layers, with a single thickness of 50–100 nm. They are basically made of TiO$_2$ tilted nanorods and silica and can be easily deposited (Figure 3.6).

This series of layers also helps to bend the flow of photons so that they can thus reach the silicon wafers and be absorbed rather than reflected off.

As illustrated in Chapter 2, sol-gel techniques are probably the most used ones to obtain precursors to be deposited in the form of a thin film of *ca.* 100 nm. Thus antireflective coatings can be obtained on silicon substrates by applying a TiO$_2$ sol synthesized through the hydrolysis and polycondensation of a titanium alkoxide (such as titanium tetrabutoxide) with water in anhydrous ethanol such as the solvent.[10]

Figure 3.7 shows the reflectance spectra of the uncoated and coated silicon substrates. The thicknesses of the films were adjusted to give a minimum in reflectance at $\lambda = 600$ nm. However, the reflectance of coated samples is lower for all wavelengths; in particular, a strong decrease from 37% to 1.5% is seen at 600 nm. Owing to a small difference in thickness, samples 2 and 3 show a certain

Figure 3.6 A worker at HEGE Solar applying a nanocoating based on TiO_2 nanorods and silica on a monocrystalline solar module. (Photograph courtesy of HEGE Solar.)

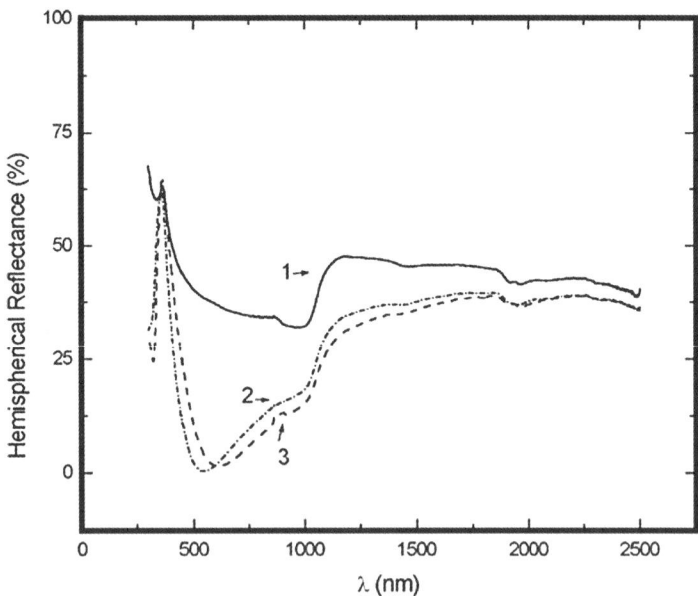

Figure 3.7 Hemispherical reflectance spectra of uncoated silicon (1) and silicon coated with TiO_2 films (2, 3). Curves 2 and 3 refer to sol-gel solutions prepared with different alkoxide/ethanol ratios and dip-coated at different withdrawal rates. (Reproduced with permission from ref. 10.)

shift in wavelength minimum, whereas the shift of the reflectance minimum intensity is determined by the difference in refraction indices.

By calculating the solar averaged reflectance data at the AM1.5 solar photon spectral distribution,[11] it is possible to integrate the values ranging from wavelengths of 320 to 1120 nm. For uncoated silicon a value of 0.387 was obtained while 0.120–0.132 was found for TiO_2-coated silicon, showing a significant improvement in antireflective properties.

3.3 Photoprotection and Anticorrosive Effects

The developed TiO_2 sols (*i.e.* nanoparticles finely dispersed in an aqueous or organic solvent) applicable by means of an easy coating system, such as dip-coating, spray-coating or through a brush, have in some cases become commercialized. TiPE™ for instance is a leading Chinese company in the production of photocatalysts. They use advanced nano-hydrosynthetic technology that has resulted in the constitution of a product line that includes eight series products and more than 30 kinds of different photocatalyst products in these series.[12] They present photocatalyst coating products as suitable devices for high-efficiency and professional application, adding that the line represents the most important one in TiPE series. The possible fields of application cover the environment, healthcare, chemicals, buildings and energy fields. This company provides also Pt-doped TiO_2 sols to improve activity under visible light irradiation.

Beside the role of degrading pollutants, with an important effect in deodorizing the environment by destroying the malodorous species present in air, a superhydrophilic behavior that avoids drops and fog formation on glass favoring its self-cleaning, thus also hampering the deposition of smog on outdoor architectural concrete elements, these kind of coatings can also work as protective layers against UV-irradiation for various materials.

Poly(ethylene terephthalate) (PET) can for instance be effectively photoprotected thus preventing its photo-aging.[13] It is possible indeed to cover the surface of the polymer film by implementing a thin bilayer Al_2O_3/TiO_2 ceramic coating. TiO_2 coating absorbs the UV radiation that could damage the polymer whereas Al_2O_3 provides a barrier against oxygen diffusion.

The films were deposited by means of a sputtering system equipped with an RF generator with bulk Ti and Al_2O_3 targets. They were tested in the photodegradation of Rhodamine B in aqueous solution.

Since PET is a polymer degradable by irradiation at $\lambda < 330$ nm, natural solar light produces stable changes in its structure. Depositing a film of Al_2O_3 does not change the absorption spectra of PET, whereas a TiO_2 layer properly plays this role. The photodegradation of PET produces hydroxylated species such as alcohols, acids and hydroperoxides that can be monitored by IR spectroscopy, at $3470\,cm^{-1}$.

Figure 3.8 shows the change in the absorbance at $3470\,cm^{-1}$ versus irradiation time for PET, Al_2O_3- and TiO_2-coated PET, and PET coated with $Al_2O_3/$

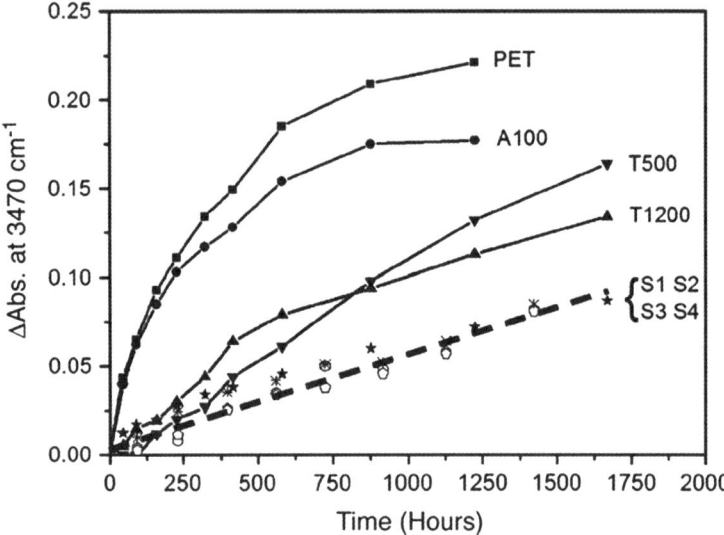

Figure 3.8 Change in absorbance, at 3470 cm^{-1}, determined from IR spectra, *versus* irradiation time. PET: uncoated PET; A100: PET coated by a 100–nm thick Al_2O_3 film; T500, T1200: PET coated by a 500– and 1200-nm thick TiO_2 film, respectively; S1-S4: different samples corresponding to TiO_2/Al_2O_3 coatings. (Reproduced with permission from ref. 13.)

TiO_2 bilayers. The effectiveness of these layers in protecting the PET film from photooxidation is quite evident. While the Al_2O_3-coated PET shows limited improvement over uncoated PET, the TiO_2-coatings exhibit some protection to PET photodegradation. However, the effectiveness of these single coatings was limited with respect to the double-layer coatings.

TiO_2 films can yield significant protection against corrosion for metals such as stainless steel.[14] A two-step treatment can be applied to this material. First, a conversion coating has to be realized, which is obtained by chemical oxidation of the stainless steel substrate in an acidic solution containing additives such as thiosulfate and propargyl alcohol to control the growth of the coating and to obtain a high surface area. Second, a supporting TiO_2 layer is formed by the hydrolysis of titanium(IV) butoxide. A final calcination at 400 °C is required to guarantee good adhesion and a transition from amorphous to anatase phase.

Corrosion tests of such materials have been performed as follows: in a closed chamber a salt solution (3% NaCl, pH 6.8) is sprayed by means of a nozzle at a controlled temperature of 35 °C. The manifestation of the first traces of corrosion or red rust was monitored and the weight was measured at time intervals to quantify the progress of the corrosion process.

Table 3.1 shows the weight measurements carried out during the corrosion tests. Clearly, the protection against corrosion provided by the conversion layer combined with TiO_2 coating on stainless steel is very good. Accordingly, the adhesion of such films was also very good.

Table 3.1 Mass variation measured during corrosion tests performed on stainless steel untreated, treated with a conversion layer and treated with a conversion layer subsequently covered by a TiO_2 film (double coating).[14]

Steel	$t = 168\,h$	$t = 500\,h$
Bare steel	+1.5 g	+5.0 g
With conversion layer	+0.83 g	No measurement (too high)
With double coating	−0.01 g	−0.03 g

3.4 Bactericidal Properties

Upon UV-light irradiation titanium dioxide exhibits a powerful oxidizing effect (fully described in its physicochemical aspect in Chapter 1) that can lead to a proficient destruction of bacteria, viruses and moulds.[15] Since the reaction proceeds through a heterogeneous catalytic reaction, physical contact between the target organism and the catalyst is necessary. This quite obvious fact is indeed important because if our catalyst is supported on a material, such as a ceramic tile, and the target organism is in the surrounding air the reaction occurs only if the ceramic tile works as a highly-adsorbent device. Notably, however, photocatalytic reactions are quite slow so that in the presence of a high concentration of organisms the cleaning effect could be negligible.

The bactericidal effect is relevant since bacteria and viruses commonly present in bathrooms, kitchens and hospitals can grow at an exponential rate unless a tool such as TiO_2 continuously lowers such growth.

The mechanisms of death of *Escherichia coli* cells have been particularly well studied in homogenous media, since *E. coli* is one of the most common pathogenic organisms. The mechanisms are, however, analogous to those occurring on supported TiO_2, such as in the photocatalytic tiles described in Chapter 5. When irradiation of TiO_2 occurs in direct contact with microbes, the microbial surface is the primary target of the initial oxidative attack. Polyunsaturated phospholipids are an integral component of the bacterial cell membrane, and they are very susceptible to be attacked by reactive oxygen species, such as those produced upon irradiation of TiO_2.

Several functions, namely, semipermeability, respiration and oxidative phosphorylation reactions, rely on an intact membrane structure. When lipid peroxidation occurs, thus, all forms of life are negatively affected.[16]

An estimation of membrane damage can be made by monitoring the production of malondialdehyde (MDA), a product of lipid peroxidation. Figure 3.9 shows the results of some experimental studies carried out by incubating *Escherichia coli* cells in suspension with TiO_2 (Degussa P25) and irradiating the suspension with UV light (8 W m^{-2}) for 30 min with continuous stirring. As shown, light and TiO_2 slightly promote the cell destruction, whereas the coupled presence of light and TiO_2 significantly increases cell destruction.

Along with the cell membrane destruction, TiO_2 also influences cellular respiratory activities. The essential components of the respiratory chain are

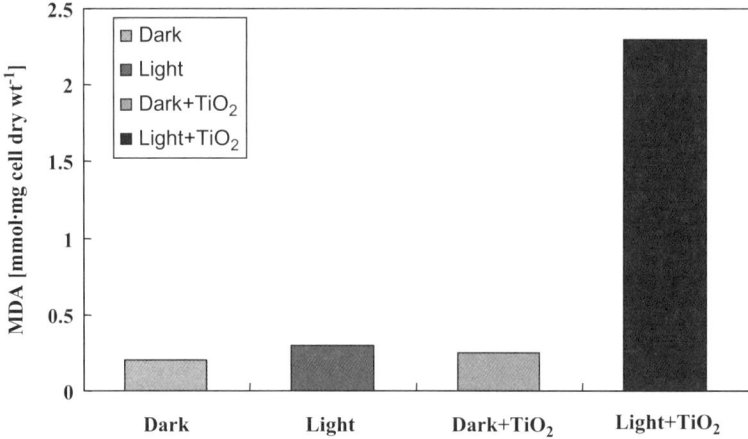

Figure 3.9 Effects of light and TiO_2 on lipid peroxidation of *Escherichia coli*. MDA stands for malondialdehyde, which is the main product generated during lipid peroxidation.

contained in the membrane and damage to them produces irreversible effects on respiration. Cell respiration can be determined by monitoring oxygen uptake in the presence of an electron donor.

Generally, alterations in membrane architecture lead to conformational changes in many membrane-bound proteins and electron mediators and to changes in their orientation across the cell membrane. Functional changes of membrane are thus produced.

Similar studies exhibited analogous results in the case of TiO_2 supported in the form of a thin film on a glass substrate.[17]

Interestingly, the survival change of *Escherichia coli* intact cells significantly differs from that of spheroplasts, *i.e.*, a modified cell lacking the peptidoglycan layer and part of the outer membrane of the cell envelope. Figure 3.10 shows that spheroplasts exhibit a simple single-exponential decay dynamics, with a higher rate constant than that for the intact cells under the same conditions.

This behavior is due to the blocking function of the cell wall (particularly the outer membrane and the peptidoglycan layer) that can oppose reactive species such as holes, $\cdot OH$, O_2^- and H_2O_2 produced by TiO_2 photocatalyst. Following the results shown in Figure 3.10, the corresponding data suggest that the cell wall of the intact cell is damaged during the initial reactive step exhibiting the lower photodegradation rate.

The process of *Escherichia coli* photokilling on a TiO_2 film can be followed by atomic force microscopy (Figure 3.11). Before illumination, the cells are cylindrical (Figure 3.11a). After a short irradiation time, no major morphological change can be observed. However, the outermost layer (indicated in the figure by "A"), disappears after 24 h of illumination (Figure 3.11b). Indeed, partial decomposition of the outer membrane takes place through the reactive

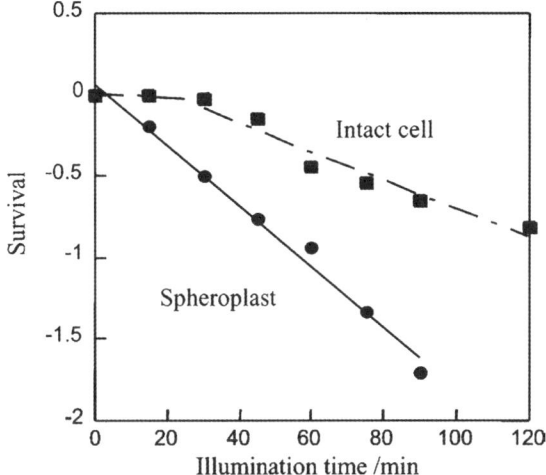

Figure 3.10 Survival of intact cells and spheroplasts of *Escherichia coli* on a TiO_2 film under UV illumination. (Reproduced with permission from ref. 17.)

species produced by TiO_2. However, in the first hours of irradiation (Figure 3.11b), cell validity is not lost very effectively. Notably, the partial decomposition of the membrane changes the permeability to reactive species, thus enabling the reactive species to reach the cytoplasmic membrane of the cell, with its final peroxidation (Figure 3.11c).

Stainless steel and other metals can be functionalized by a titanium dioxide layer, with a self-sterilizing effect. This is related to important applications in medical and cutlery and alimentary accessories in general. For instance, easily implementable sol-gel techniques can be applied to support TiO_2 on metal, as described in Chapter 2. The survival of *Staphylococcus aureus* has been studied on functionalized stainless steel and titanium.[18] The results indicate that the photocatalytic action of TiO_2 is effective against bacteria (Figure 3.12). As shown, titanium is a more effective bactericide than stainless steel. When UV light was screened, negligible degradation occurs. UV light itself degrades bacteria, but the effect is significantly higher in the presence of TiO_2. Complete bacterial inactivation was achieved after 60 min on stainless steel and after 90 min on titanium.

Sterilization of several objects is a primary tool in various fields of everyday life. Thus, TiO_2 can play the double role of antifogging and antibacterial device in dental mirror surfaces routinely used by doctors. Tetraethyl orthotitanate can be hydrolyzed proficiently to support titanium dioxide on a SiO_2-precoated dental mirror.[19] The mirror was immersed in the solution and left to hydrolyze for different times (Figure 3.13). Water contact angles on these glasses were measured to evaluate surface hydrophilicity. The result was a sharp decrease, from 35.5° to less than 1°, when the film was thick enough. Reflectance was also higher than in the case of a plain mirror for any film

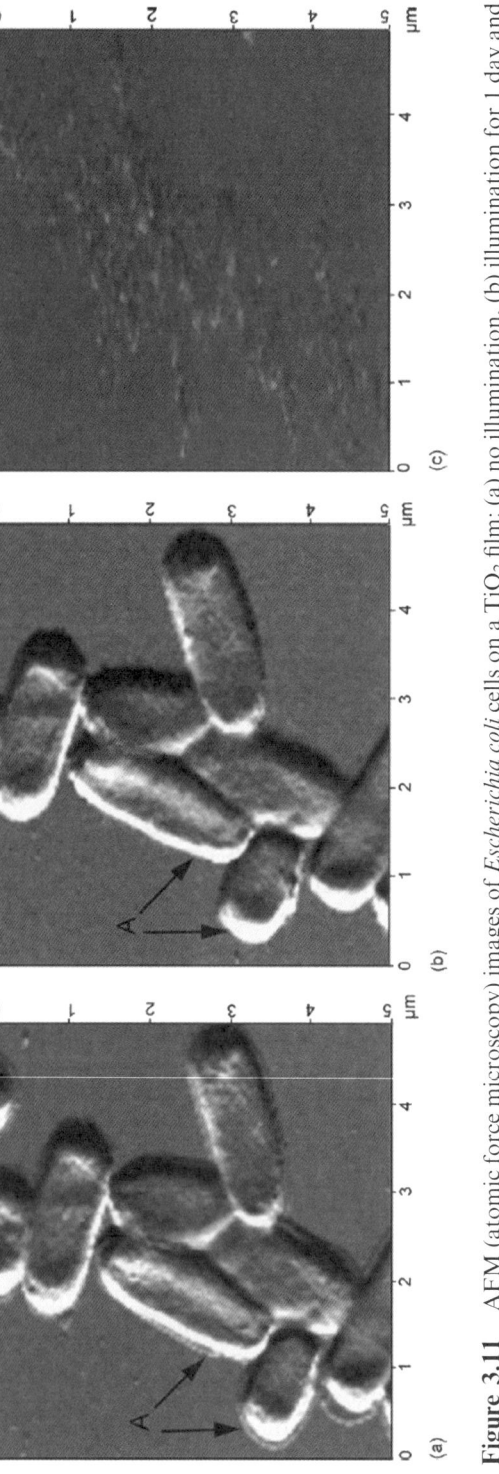

Figure 3.11 AFM (atomic force microscopy) images of *Escherichia coli* cells on a TiO$_2$ film: (a) no illumination, (b) illumination for 1 day and (c) illumination for 6 days. Radiation flux was 1.0 mW cm^{-2}. (Reproduced with permission from ref. 17.)

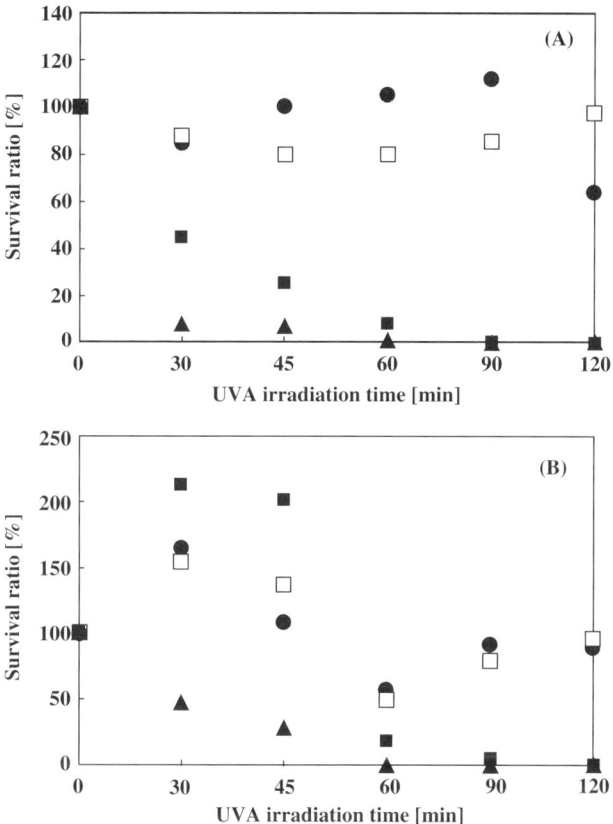

Figure 3.12 Photocatalytic inactivation of *Staphylococcus aureus* on (A) titanium and (B) stainless steel; anatase TiO_2 with UV illumination (▲), anatase TiO_2 screened from UV light (●), irradiated naked disks (■) and naked disks screened from UV light (□).

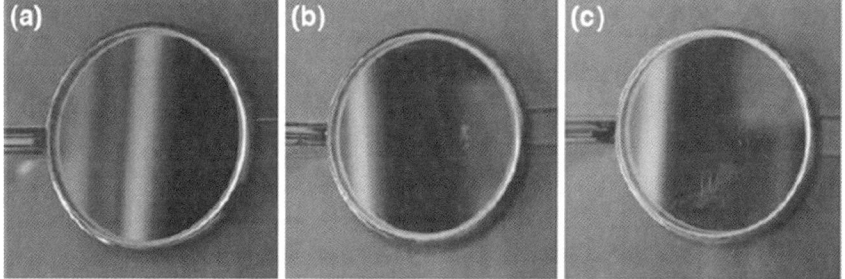

Figure 3.13 Appearance of a dental mirror surface with different tetraethyl orthotitanate hydrolysis times: (a) 10, (b) 30 and (c) 60 min. (Reproduced with permission from ref. 19.)

thickness, for both visible and UV wavelengths. Even the transparency of films was very high (Figure 3.13).

An antibacterial effect is often needed in areas that are poorly irradiated; for example, the areas in a room that are screened by furniture. In these cases it is possible to load TiO_2 with a small amount of doping cations that are *per se* able to impart a sterilizing effect, such as Ag^+ and Cu^{2+}. After an initial irradiation with UV light these cations are reduced to the metal state in the form of nanopowders finely adherent to TiO_2 nanoparticles. This approach is extremely efficient, especially in ensuring a permanent action, even when light does not reach the photocatalytic material for different reasons. Some experimental results have shown that photocatalytic tiles loaded with Ag or Cu can degrade significant amounts of bacteria for several hours even in the dark.[15]

Figure 3.14, thus, shows the results obtained with the installation of photocatalytic tiles in an operating room: it is possible to see that the bacteria present in air were reduced to negligible amounts and that the sterilizing effect is fully retained for several months. Practically, it is possible to obtain an effect that lasts for 24 h, which is ideal for instance in a hospital.

Following the same principles it is possible to sterilize toilets, reducing the bad smell resulting from the ammonia and ammonia-derivatives arising mainly from the decomposition of urine carried out by different bacteria.

Accordingly, algae and moulds growth is prevented in TiO_2-covered tiles, even under natural conditions favoring their development. It is important to stress once again that TiO_2 not only kills bacteria, but also completely

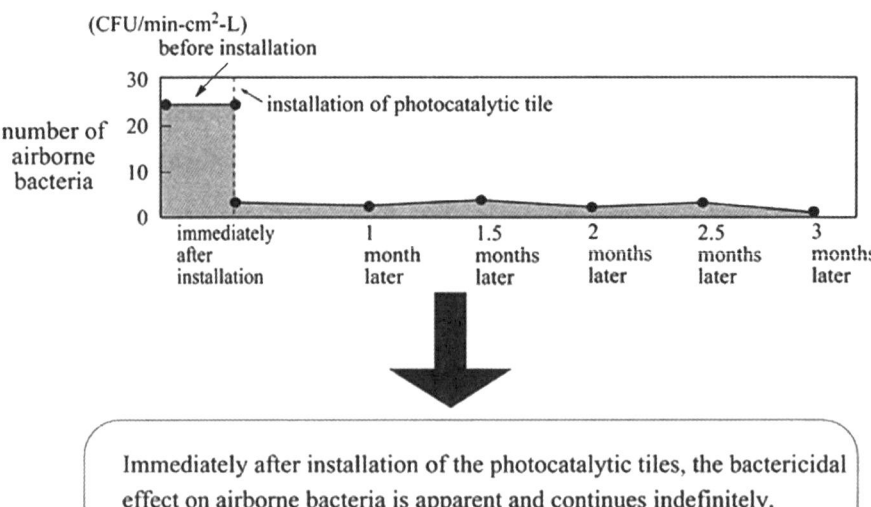

Figure 3.14 Sterilizing effect on airborne bacteria present in a hospital operating room. (Reproduced with permission from ref. 15.)

Unique Properties of Supported TiO$_2$ 113

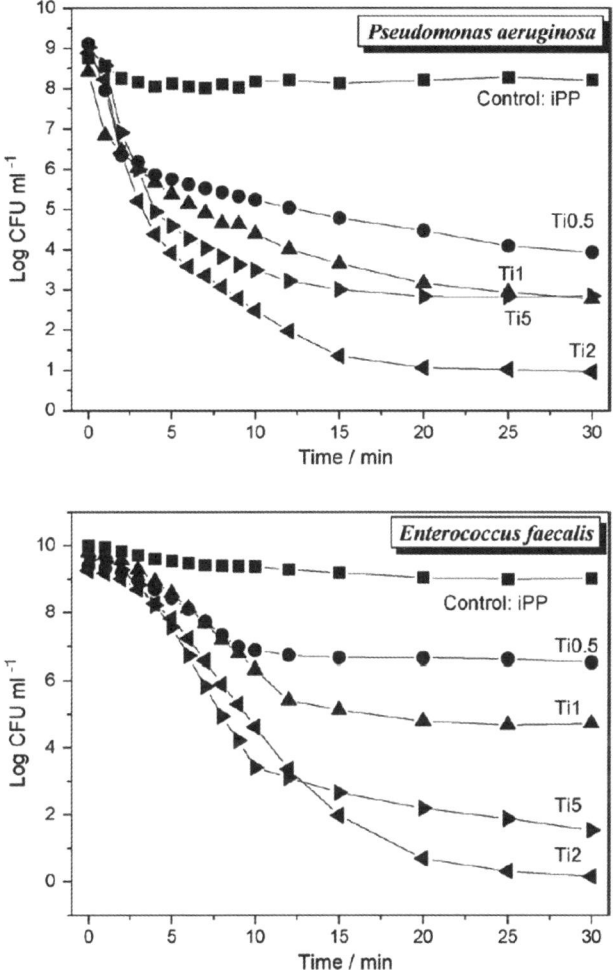

Figure 3.15 Logarithmic reduction of microorganism population as a function of irradiation time. (TiX indicates a polypropylene/TiO$_2$ composite with X wt% TiO$_2$.) (Reproduced with permission from ref. 20.)

decomposes cells, unlike typical bactericidal agents such as silver that leave the cell intact. This is clearly evident in Figure 3.11.

Antimicrobial polymers for food packaging applications can be prepared by implementing a certain amount of TiO$_2$ in an isotactic polypropylene matrix.[20] A polypropylene partially modified with maleic anhydride is added as a compatibilizing agent to improve interfacial adhesion between the isotactic polypropylene and TiO$_2$. These composite materials display maximum activity for samples containing 2 wt% of anatase-TiO$_2$ irrespective of the microorganism. Two bacteria with different features were tested: a Gram negative (*Pseudomonas aeruginosa*) and a Gram positive one (*Enterococcus faecalis*). Interestingly, both bacteria are antibiotic resistant. In both cases, the antibacterial activity decreased with increasing TiO_2 amount up to 2 wt% (Figure 3.15). Tests performed with plain irradiation with UV-light exhibited a negligible degradation ("Control: iPP" in Figure 3.15).

Such functionalized polymers can open the way to a new class of materials, *i.e.*, photoactive food packaging, that could help in avoiding bad smells typically deriving from fish deterioration (mainly caused by trimethylamine) or from fruits and vegetables.

References

1. (a) M. R. Hoffmann, S. T. Martin, W. Choi and D. W. Bahnemann, Environmental applications of semiconductor photocatalysis, *Chem. Rev.*, 1995, **95**, 69–96; (b) X. Chen and S. Mao, Titanium dioxide nanomaterials: synthesis, properties, modifications, and applications, *Chem. Rev.*, 2007, **107**, 2891–2959; (c) I. P. Parkin and R. G. Palgrave, Self-cleaning coatings, *J. Mater. Chem.*, 2005, **15**, 1689–1695.
2. T. Maggosa, J. G. Bartzis, M. Liakou and C. Gobin, Photocatalytic degradation of NO_x gases using TiO_2-containing paint: a real scale study, *J. Hazard. Mater.*, 2007, **146**, 668–673.
3. (a) M. Takeuchi, K. Sakamoto, G. Martra, S. Coluccia and M. Anpo, Mechanism of photoinduced superhydrophilicity on the TiO_2 photocatalyst surface, *J. Phys. Chem. B*, 2005, **109**, 15422–15428; (b) M. Shimohigoshi and Y. Saeki, Research and applications of photocatalyst tiles, *Photocatalysis, Environment and Construction Materials*, ed. P. Baglioni and L. Cassar, RLEM, pp. 291–297, ISBN: 978-2-35158-056-1.
4. A. Fujishima, T. N. Rao and D. A. Tryk, Titanium dioxide photocatalysis, *J. Photochem. Photobiol. C: Photochem. Rev.*, 2000, **1**, 1–21.
5. K. Guan, Relationship between photocatalytic activity, hydrophilicity and self-cleaning effect of TiO_2/SiO_2 films, *Surf. Coat. Technol.*, 2005, **191**, 155–160.
6. See, for example, Hydro-Klean Mirror by Sekisui Jushi Corporation online at http://www.sekisuijushi.co.jp.
7. J. Kasanen, M. Suvanto and T. T. Pakkanen, Self-cleaning, titanium dioxide based, multilayer coating fabricated on polymer and glass surfaces, *J. Appl. Polym. Sci.*, 2009, **111**, 2597–2606.
8. C. Battaglin, F. Caccavale, A. Menelle, M. Montecchi, E. Nichelatti, F. Nicoletti and P. Polato, Characterisation of antireflective TiO_2/SiO_2 coatings by complementary techniques, *Thin Solid Films*, 1999, **351**, 176–179.
9. Details on TOP-NANO-HEGE® Coating are available online at http://www.hege-solar.de/inovation2009.html.
10. G. San Vicente, A. Morales and M. T. Gutierrez, Preparation and characterization of sol-gel TiO_2 antireflective coatings for silicon, *Thin Solid Films*, 2001, **391**, 133–137.

11. The air mass (AM) coefficient characterizes the solar spectrum after the solar radiation has traveled through the atmosphere. For a full explanation see: http://rredc.nrel.gov/solar/spectra/am1.5/#about.
12. TiPE™: http://www.tipe.com.cn.
13. K. O. Awitor, A. Rivaton, J.-L. Gardette, A. J. Down and M.B. Johnson, Photo-protection and photo-catalytic activity of crystalline anatase titanium dioxide sputter-coated on polymer films, *Thin Solid Films*, 2008, **516**, 2286–2291.
14. L. Bamoulid, M.-T. Maurette, D. De Caro, A. Guenbour, A. Ben Bachir, L. Aries, S. El Hajjaji, F. Benoît-Marquié and F. Ansart, An efficient protection of stainless steel against corrosion: combination of a conversion layer and titanium dioxide deposit, *Surf. Coat. Technol.*, 2008, **202**, 5020–5026.
15. A. Fujishima, K. Hashimoto and T. Watanabe, *TiO₂ Photocatalysis, Fundamentals and Applications*, BKC, Tokyo, 1999.
16. P.-C. Maness, S. Smolinski, D. Blake, Z. Huang, E. J. Wolfrum and W. Jacoby, Bactericidal activity of photocatalytic TiO_2 reaction: toward an understanding of its killing mechanism, *Appl. Environ. Microbiol.*, 1999, **65**, 4094–4098.
17. K. Sunada, T. Watanabe and K. Hashimoto, Studies on photokilling of bacteria on TiO_2 thin film, *J. Photochem. Photobiol. A: Chem.*, 2003, **156**, 227–233.
18. K. Shiraishi, H. Koseki, T. Tsurumoto, K. Baba, M. Naito, K. Nakayama and H. Shindo, Antibacterial metal implant with a TiO_2-conferred photocatalytic bactericidal effect against *Staphylococcus aureus*, *Surf. Interface Anal.*, 2009, **41**, 17–22.
19. K. Funakoshi and T. Nonami, Photocatalytic treatments on dental mirror surfaces using hydrolysis of titanium alkoxide, *J. Coat. Technol. Res.*, 2007, **4**, 327–333.
20. A. Kubacka, M. Ferrer, M. L. Cerrada, C. Serrano, M. Sánchez-Chaves, M. Fernández-García, A. de Andrés, R. J. Jiménez Riobóo, F. Fernández-Martín and M. Fernández-García, Boosting TiO_2-anatase antimicrobial activity: polymer-oxide thin films, *Appl. Catal. B: Environ.*, 2009, **89**, 441–447.

CHAPTER 4
Photocatalytic Glass

4.1 Improving Glass Performance by Functionalization with TiO_2

The development of transparent thin coatings that impart various properties to different kinds of glass is one of the main goals of applied research in the glass industry at present. Typical examples are car glass and building glasses for windows, facades, roofs, *etc.*

In this context, titanium dioxide is a material with outstanding properties, such as its high optical index, refraction and transparency across the visible wavelength range, cut-off effect towards highly damaging UV light, intrinsic semiconducting properties, compatibility with the environment and complete atoxicity, along with photocatalytic, anticlouding and antifogging properties.[1]

TiO_2 thin films can also adsorb species and, hence, degrade organic and inorganic (chemically present in a reduced form) pollutants, eventually discharging their oxidation products by easy washing with plain water (even through naturally occurring rain), thus preventing glass from fouling. In addition, a superhydrophilic surface cannot adsorb organic oily liquids but, instead, strongly expel them. Since TiO_2 films are both photocatalytically active and superhydrophilic, they can work in both ways (photocatalytic degradation of pollutants on one hand and rejection of organic liquid species caused by the superhydrophilic characteristic on one other hand). These properties arise due to light-induced redox reactions and superhydrophilicity, the latter avoiding the formation of rain-droplets or water vapor (Figures 4.1 and 4.2) and favoring the free flowing of water, which washes away both dirt (such as fine dusts) and oxidation residuals of polluting environmental agents. Environmental malodorous species in air, both indoors and outdoors, are typically present in very low concentrations, even when human sensitivity notices a very strong smell. This is why TiO_2 is an ideal species for deodorization of air; photocatalysis works indeed unselectively by degrading almost all organics present in the environment if adsorbed on its surface, although reaction rates are typically very low, so that large amounts of pollutants cannot be degraded

Photocatalytic Glass 117

Figure 4.1 The glass shown on the left-hand side is prevented from fogging even in a water vapor saturated environment such as the represented room. (Photograph courtesy of Toto Ltd.)

quickly by using light and a photocatalyst. Conversely, very small amounts of organic species can be degraded easily. If we think that common malodorous species, such as methyl mercaptan, must be kept in concentrations lower than 0.01 ppm, it becomes clear how powerful photocatalytic devices for air decontamination can be.[2]

Figure 4.1 depicts the common case of a room saturated by a high amount of water vapor, such as in an indoor kitchen or in a particularly crowded room. The typical fogging effect appearing on the inner side of the window glass is absent on the left-hand glass, functionalized by a TiO_2 film.

Similarly, Figure 4.2 shows (a) a car side-view mirrors (produced by Tokai Rika) and (b) curved road mirrors under rainy conditions (manufactured by Sekisui Jushi Corporation) when droplets of different sizes form on the surface and remain there unless a TiO_2 film activated even by a low environmental light ensures superhydrophilicity to the surface and, thereby, prevents even fogging effects. The result is a perfect rear-view for the driver, whose security increases even in environmentally difficult conditions. Figure 4.2(c) shows a lamp (by Toshiba Lighting & Technology Corporation) typically used for lighting outdoor areas at airports: its glass cover is self-cleaning and is activated by the light emitted by the lamp.

The glass in Figures 4.1 and 4.2 has been modified through Hydrotect, a technical brand name of Toto Ltd., by implementing a surface treatment as part of the manufacturing process.[3] The layer is perfectly adherent and transparent, so that a colored surface remains esthetically unaffected by deposition.

Figure 4.2 The right-hand part of a side-view mirror (a) and the curvet mirror on the right in (b) are covered with Hydrotect and, hence, clouding and the formation of water droplets are prevented. (Photograph courtesy of Toto Ltd.) (c) The glass cover of a lamp (produced by Toshiba Lighting & Technology Corporation) is covered with a TiO_2 film. (Reproduced with permission from ref. 1.)

It is ideal for application in areas were hygiene matters are relevant, such as lavatories, swimming pools and hospitals. Hydrotect has been implemented by several companies for many ceramic and polymeric products, such as in polycarbonate-made soundproof walls for the sides of roads.

Standard soda-lime glass, mainly made of SiO_2, Na_2O, CaO, Al_2O_3 (Table 4.1),[4] amounting to more than 90% of the globally manufactured glass, can be warmed to 500 °C, without any modification in structure. The glass transition temperature is in fact always higher than 540 °C, depending on the treatments undertaken. Hence supporting a TiO_2 thin transparent film on glass is a rather easy matter, because one can achieve both crystallization to the photocatalytically most active crystalline phase (anatase) and mechanical adherence through particle sintering and agglomeration at only 350 °C.

Films as thin as a few hundred nanometers are enough to ensure a significant improvement in self-cleaning performance, without forgetting TiO_2 intrinsic bactericidal effect. This is particularly relevant where a sterilized environment must be guaranteed, such as in hospital, especially in surgery rooms, laboratory clean rooms, *etc*. Moreover, the antibacterial effect can be increased greatly by doping TiO_2 with noble metals.

Table 4.1 Soda-lime glass: typical composition and approximate limits (mol.%). (Reproduced with permission from ref. 4.)

	Typical container glass	Typical float glass	Approximate limits
SiO_2	74.42	71.86	63–81
Al_2O_3	0.75	0.08	0–2
MgO	0.30	5.64	0–6
CaO	11.27	9.23	7–14
Li_2O	0.00	0.00	0–2
Na_2O	12.9	13.13	9–15
K_2O	0.19	0.02	0–1.5
Fe_2O_3	0.01	0.04	0–0.6
Cr_2O_3	0.00	0.00	0–0.2
MnO_2	0.00	0.00	0–0.2
Co_3O_4	0.00	0.00	0–0.1
TiO_2	0.01	0.01	0–0.8
SO_3	0.16	0.00	0–0.2
Se	0.00	0.00	0–0.1

4.2 TiO₂ on Glass: More Tasks or Benefits?

Supporting a photocatalytic film on a glass substrate presents two main issues that have to be addressed and which are not very significant for other common ceramic materials, such as tiles: (i) the glass optical properties should not be negatively affected by the TiO_2 film; (ii) the very smooth surface of glass does not support good adhesion, which is indeed necessary to manufacture a commercial product that can be washed for years without modification of its quality. Pretreatments to modify glass surface are possible to increase roughness before depositing TiO_2.

As far as deposition technologies are concerned, sol-gel precursors are typically used in industrial production, due to the high activity of the resulting films and good general performances in terms of mechanical resistance, light transmittance and durability.[5] The process is realized by dipping, spinning or spraying a liquid sol on the glass, followed by a thermal treatment. The inorganic or organic TiO_2 precursor thus hydrolyzes and the resultant product crystallizes in an active crystalline phase. The latter can be controlled by changing the preparation parameters. Many other techniques are available and present some advantages over sol-gel, but usually at much higher costs. For instance chemical vapor deposition,[6] evaporation,[7] various sputtering depositions,[8] and ion beam-assisted processes[9] have been used successfully.

Electron beam physical vapor deposition, for example, gives very regular films, with a tunable thickness, whereas film uniformity is less accurate when employing sol-gel preparation. Photoactivity, conversely, is always higher in the case of sol-gel deposition.

The use of a common glass, *i.e.*, made not only of silica but containing significant amounts of other oxides such as Na_2O, can give limited reactivity. Thus, supporting films on soda-lime glass can give rise to diffusion of Na^+ ions during calcination at 400–500 °C. Warming the film at high temperature is often needed to promote particle crystallization and aggregation, eventually resulting in fine adhesion and photoactivity. A low Na content could act as a recombination centre whereas higher amounts favor the formation of the brookite allotropic phase or of sodium titanate, which is much less photoactive than TiO_2 in its anatase form.[10]

To prevent the negative effect of Na^+ ions on the photocatalytic activity, many methods have been developed, such as precoating soda-lime glasses with silica-based layers, thus creating a barrier layer, or implementing a glass pre- or post-treatment with an acidic aqueous solution aimed at exchanging Na^+ ions.[11] Minimizing the rate of temperature increase can also reduce the diffusion phenomenon, since the glass structure is less subject to thermal stress and modification is limited. A rapid increase of temperature from 25 to 400–500 °C can give rise to a partial softening of glass structure at the interface with TiO_2, producing sodium-titanium-oxygen compounds without any photocatalytic activity. Figure 4.3 shows that supporting TiO_2 on glass in the absence of a

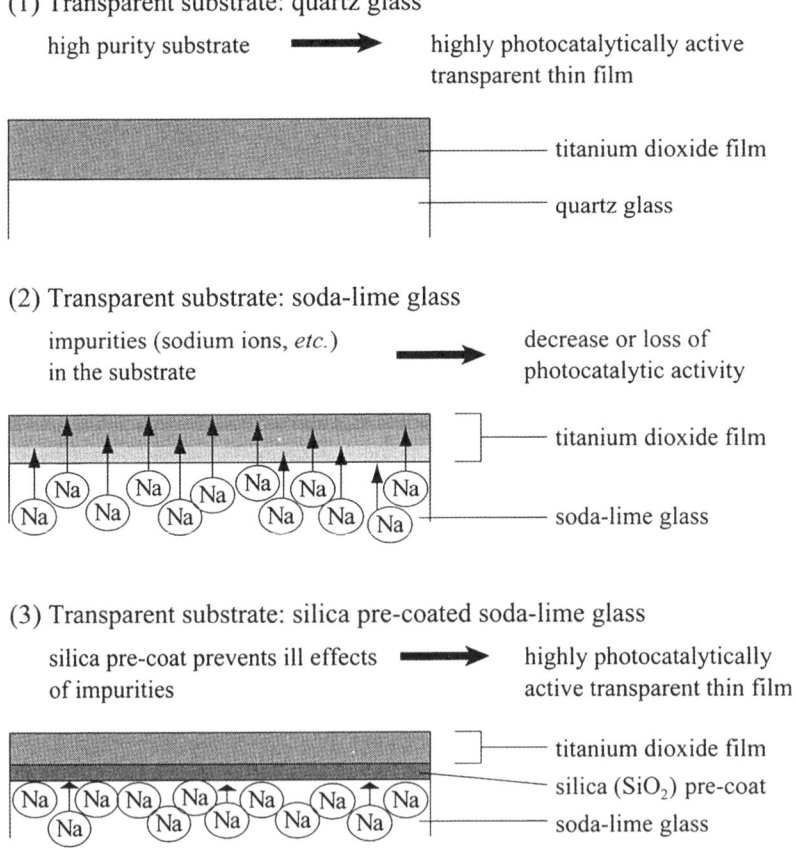

Figure 4.3 Preparation processes for TiO_2 films on glass substrates. (Reproduced with permission from ref. 1.)

barrier film, and without negatively altering the photocatalytic activity, is possible only when a highly pure glass (such as quartz) is used. Conversely, diffusion of Na^+ into TiO_2 has to be prevented in the case of common glass.

Clearly, moving to high TiO_2 film thickness or performing a low-temperature preparation can assure the absence of Na^+ diffusion but usually the quality (in terms of adherence, transparence and activity) of the resulting film is not comparable with a high-temperature preparation.

Silicon nitride has been used recently as a better barrier film than silica. For example, a three-fold improvement in photocatalytic film activity can be obtained by interposing a 100 nm thick SiN_x film between a soda-lime glass substrate and an anatase film. The amount of sodium present in the cross-sectional profile of a TiO_2 film is 7–8 at.% when no barrier film is present, 4 at.% when interposing a SiO_2 film, and decreases to under 2 at.% when using a SiN_x film.[12] Figure 4.4 shows scanning electron micrographs of the top TiO_2 films before and after calcination. The TiO_2 film was deposited by direct current sputtering of a metallic titanium target in argon–oxygen reactive mixtures, whereas the barrier films (both SiN_x and SiO_2) were deposited by reactive magnetron sputtering, using a conductive boron-doped silicon target. Nitrogen or oxygen is added to Ar for the deposition of the nitride and oxide barriers, respectively. Calcination at 450 °C resulted in crystallization to anatase phase and particle aggregation (Figure 4.4b and c).

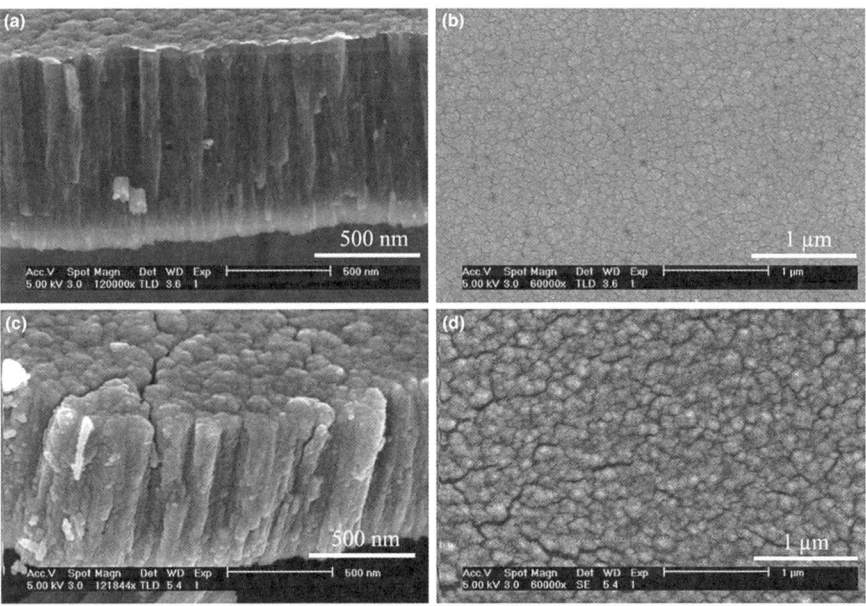

Figure 4.4 SEM micrographs of cross-section (a) and top surface (b) of an amorphous TiO_2 coating, and the cross-section (c) and top surface (d) of the coating after annealing at 450 °C. (Reproduced with permission from ref. 12.)

A TiO_2-modified glass has to meet different criteria to be suitable for industrial manufacturing and commercial access to the market. Some important attributes are needed for both application in cars and also for building facades and windows, or for indoor applications, as previously mentioned.

The main concerns regarding a glass to be used in everyday life are listed below:

- *Optical properties*: the light transmittance has to be checked, because TiO_2 absorbs, on the one hand, the UV fraction of solar light (a desirable property yielding an important UV cut-off effect) but can, on the other hand, give rise to light scattering effects with films that are either partially inhomogeneous or too thick. An ideal film has to absorb as much as possible UV light at $\lambda < 380$ nm and transmit visible light as a normal glass. Electron microscopy and UV transmission spectra (Figure 4.5) can give useful information about this point. Measurement of film thickness and refractive index is also necessary and can be carried out by ellipsometry. Refractive index in particular is a relevant parameter: soda-lime glass shows values of *ca*. 1.5, whereas that of TiO_2 is about 2.5 under solar irradiation. This is why a larger portion of the incident light can be reflected by TiO_2 with respect to glass, if the surface composition is not well controlled. The air refractive index is indeed much lower, with a value of *ca*. 1.0, and light reflection in external areas can yield undesirable optical effects. However, nanoporous TiO_2–SiO_2 coatings have been prepared recently with a very low refractive index, enhancing the transmission

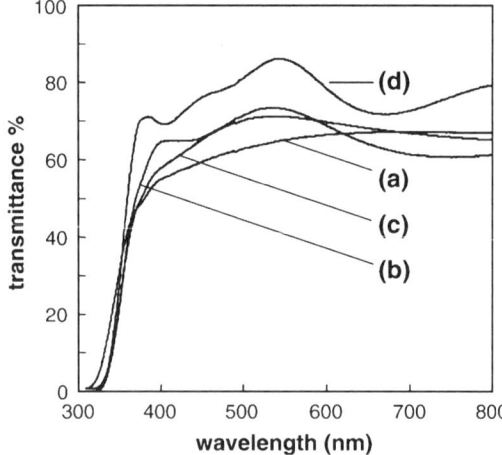

Figure 4.5 UV–Visible transmittance spectra of TiO_2 films obtained by dip-coating of SiO_2-precoated glass into sols derived from titanium tetraisopropoxide: (a) 8, (b) 15, (c) 20 layers calcined layer-by-layer at 400 °C; (d) 15 layers with only one final calcination at 400 °C. (Reproduced with permission from ref. 14.)

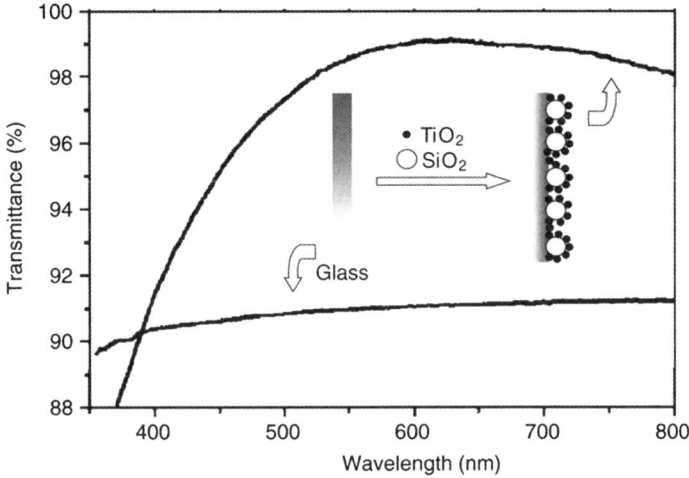

Figure 4.6 Transmittance of bare glass and a glass coated with a double-layer TiO_2–SiO_2 film. The maximum transmittance for the coated glass is over 99%. (Reproduced with permission from ref. 15.)

of the glass to more than 97% for visible light (>90–92% is usual for glass) (Figure 4.6).[13–15]

- *Photocatalytic activity*: photocatalytic sites have to be able to both adsorb and oxidize inorganic and organic pollutants, washed out by natural water or human clean up. Glass stains under outdoor conditions usually derive from salts, silicates, carbonates, soots (inorganic species) and polycyclic aromatic hydrocarbons, fatty acids, and long-chain alkanes. Pollutant degradation becomes possible when TiO_2 is present in an active crystalline phase and if it is accessible to pollutants present in the gas phase. X-Ray diffraction, along with Raman spectroscopy, can be used to identify and quantify the crystalline phases. Biological tests can yield specific information on the bactericidal effects of the active coating. Specific tests under conditions simulating real ones can be carried out to study photocatalytic activity in gas-phase reactors by degrading target compounds. Solar (direct or diffuse) light in outdoor application is always sufficient to activate the redox reactions, whereas a study of indoor lighting is needed to understand if natural light is enough for activation or a specific lamp is required. The study of photocatalytic activity is somewhat complex, since it is influenced by many other factors, such as relative humidity, temperature, kind of catalyst, irradiation type and intensity, *etc.*
- *Superhydrophilicity*: this is probably the most important property for glasses, since it allows the avoidance of fogging and droplet formation, as previously explained, hence imparting self-cleaning features to glass. Superhydrophilic surfaces also have another (often forgotten) property, one that is particularly relevant for surfaces where organic liquids of

various types can accidentally spill or be left by recondensation from air: they are highly averse towards absorption of organic liquids, thus being easily cleanable by plain water. This is highly desirable in kitchens or lavatories, and also in very environmentally polluted areas, such as industrial sites. Measurement of the water contact angle during irradiation is important as this parameter is needed to understand the hydrophilic properties of a film. As noted above, the study of irradiation field is important in understanding whether superhydrophilicity is induced in a specific environment and in glasses with a certain inclination angle (from 0 to 90°).

- *Retention of self-cleaning properties*: in a typical environment where photocatalytic glasses are used, their ability to keep their self-cleaning feature should be carefully checked. Cleaning has to be performed without aggressive detergent and avoiding strong mechanical friction that could damage the surface. Conversely, a particularly polluted or dirty environment could result in coverage of the functionalized surface, thus hindering physical contact between TiO_2 and external agents. These phenomena are usually easily reversible by plain cleaning, although prolonged deposition of fine organic/inorganic particulates over the glass, left unwashed for a very long time, could cause a temporary deactivation that is solvable only by a significant warm-up.

As already mentioned, glass superhydrophilicity is the most important property and, hence, several methods have been implemented to enhance performances in this direction. Controlled porosity on photocatalytic films deposited on glass can, for example, improve significantly both film hydrophilicity and photoactivity. Adding PEG [poly(ethylene glycol)] to a sol-gel TiO_2 precursor, deposited over glass, gives rise to higher reactivity and superhydrophilicity performance. Film calcination at 520 °C affords mineralization of all the carbon present, in place of which a tunable porosity is left. The much higher number of hydroxyl surface groups present when adding PEG can influence greatly the performance in terms of contact angle decrease.[16]

Figure 4.7 shows that, while there is a plateau of the hydroxyl content by increasing the amount of added PEG (Figure 4.7a), the optimal hydrophilic behavior is at the beginning of this plateau (Figure 4.7b). The decrease of water contact angle from a value in the 20–40° range to 0° is indeed much faster for sample c, *i.e.*, the one prepared by using 0.5 g PEG. Moreover, when the light is turned off, the contact angle value of the sample prepared in the absence of polymer reaches its initial value fastest among all samples. This results in better self-cleaning properties of PEG-templated samples; in fact, glass typically has water contact angles of *ca.* 30°, not influenced by light, whereas a much lower angle is retained even in dark conditions for many days only by the PEG-templated films.

Doping with transition metals is a widely investigated field, since the band gap can be reduced and, hence, the fraction of visible light absorbed can be

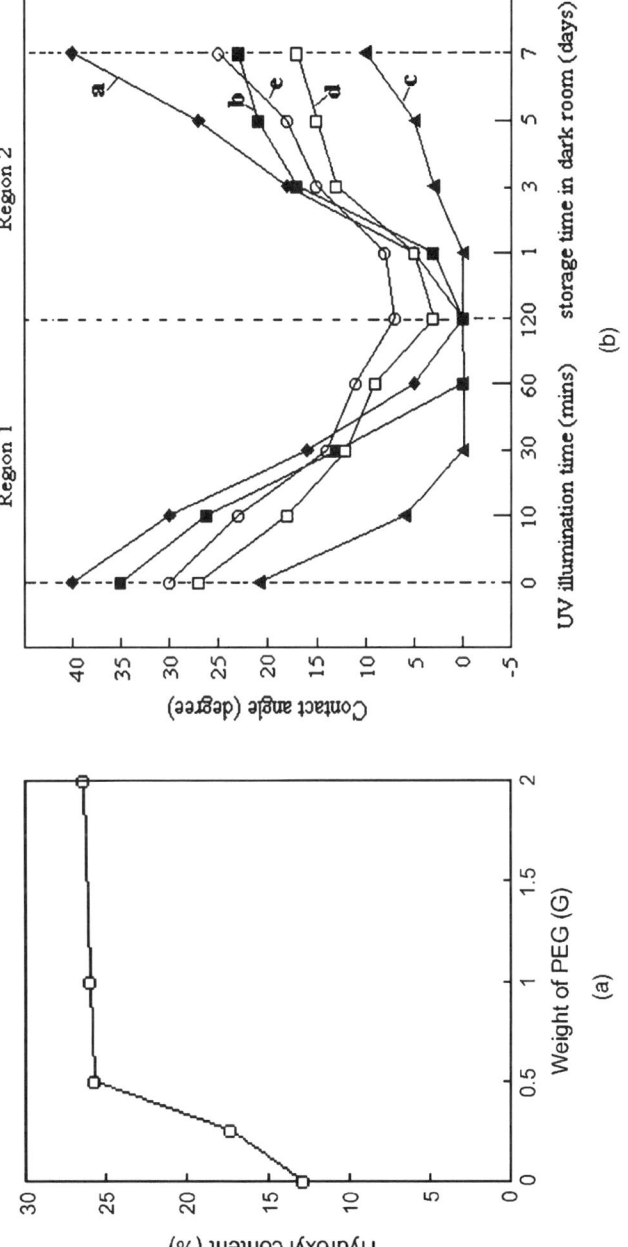

Figure 4.7 Hydroxyl content (a) and contact angle variation in light and dark conditions (b) upon introducing various amounts of PEG in a sol-gel preparation. Samples a–e were prepared by ten coating cycles from precursor solutions containing 0, 0.25, 0.5, 1.0 and 2.0 g PEG, respectively. (Reproduced with permission from ref. 16.)

enhanced. In particular, very limited Ag-doping can give bactericidal properties to a TiO_2 film even in the absence of light.

Iron-doped TiO_2 films can also yield better superhydrophilic attributes.[17] Both films shown in Figure 4.8, deposited through spray pyrolysis, present a smooth surface, although the 5 at.% Fe-doped surface is less smooth than the undoped one. Interestingly, contact angles, even under dark conditions, are quite small ($<5°$) both for doped and undoped samples, due to the deposition

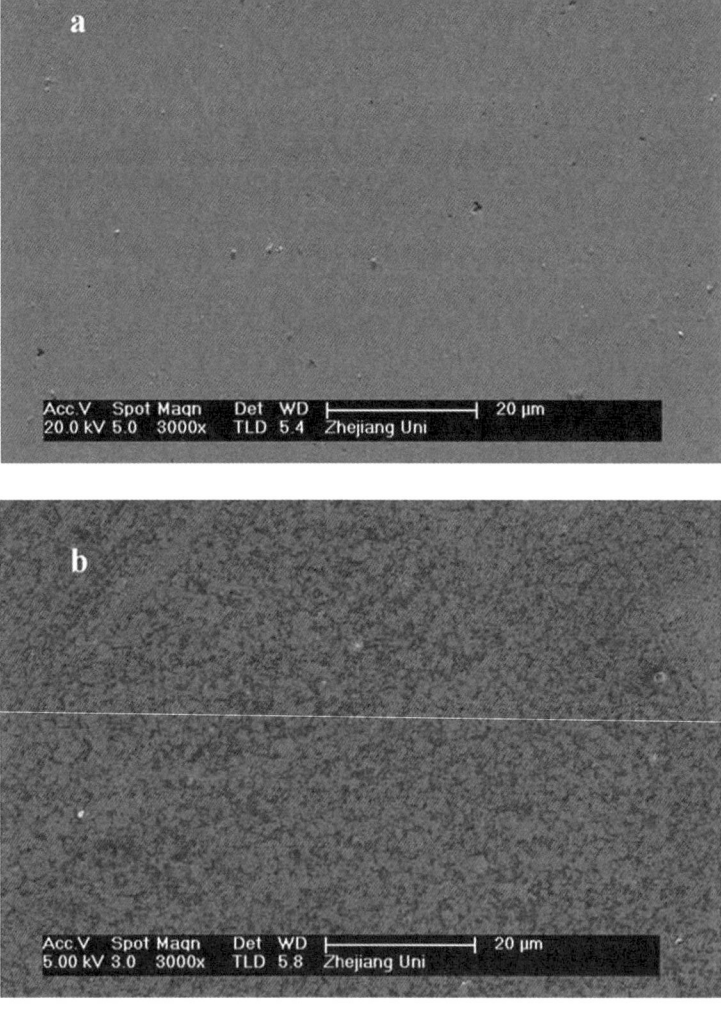

Figure 4.8 SEM images of the surface morphology of (a) undoped and (b) 5 at.% Fe-doped TiO_2 thin films deposited by spray pyrolysis. (Reproduced with permission from ref. 17.)

method, while the 5 at.% Fe-doped sample reaches 0° upon irradiation much faster than the undoped TiO$_2$ film.

4.3 Antireflection and Composite Multilayer Films for Advanced Applications

Composite films have been developed to improve the transmittance of visible light of photocatalytic glasses, which is highly desirable in many applications such as solar collectors, greenhouses and windows. Dense TiO$_2$ films transmit, typically, 75–85% of visible light (see for instance Figure 4.5), but work as cut-off devices in the UV region (for $\lambda < 380$ nm). While the latter parameter is very attractive since UV light is harmful for human skin, eye and immune system, and also damages indoor objects, through photodegradation generated by radical reactions, the low transmittance of visible light can be enhanced by integrating antireflective properties in TiO$_2$ films.

The high refractive index under visible radiation is responsible for this property (Table 4.2) but attaining antireflection features is possible, for instance, by using double layered films and tailoring their porosity, in most cases employing TiO$_2$ and SiO$_2$.[18]

The glass reflection coefficient, or reflectance, can be calculated by the Fresnel equation and, assuming that light is perpendicularly incident, it can be expressed by:[19]

$$R = \frac{(n_0 - n_G)^2}{(n_0 + n_G)^2} \quad (4.1)$$

where n_0 is the refractive index of the medium the light is coming from (usually air, and water in some cases) and n_G is the glass refractive index. Moreover, since light is reflected not only on the front of the glass but also on its back, and can travel back and forth several times as well, the total reflectance through a window is expressed as:

$$R_{TOT} = \frac{2R}{1 + R} \quad (4.2)$$

Table 4.2 Rough refractive index of some common substances under visible irradiation. (Accurate values can be found in ref. 18.)

Substance	Refractive index
Air	1.0
Water	1.3
Soda-lime glass	1.5
Pure silica	1.4
Titanium dioxide (anatase)	2.3

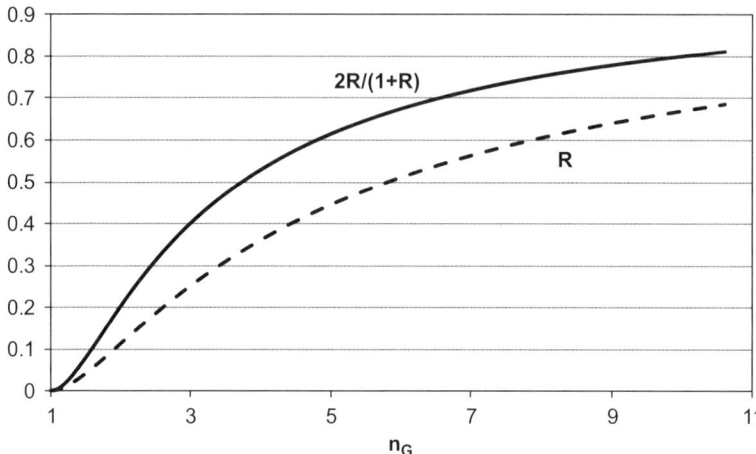

Figure 4.9 Reflectance and total reflectance *versus* refractive index [see Equations (4.1) and (4.2)].

Figure 4.9 plots R and R_{TOT} values against n_G, assuming that the medium is air with $n_0 = 1$. It can be seen that for $n_G = 1.5$, the characteristic value of common windows glass, the total reflectance, is *ca.* 8%. If we neglect absorption, non-reflected light is transmitted, *i.e.*, $T = 1 - R$, and soda-lime glass reflectance is about 92%, assuming that light rays are perpendicular to the glass.

Sol-gel methods can easily yield a low refractive index, resulting in a maximal transmittance of 99%, which is much more than plain glass (characteristic values are *ca.* 90–92%). SiO_2–TiO_2 films can be obtained by depositing first a silica film from a colloidal solution and afterwards doing the same with TiO_2.[20] The resulting transmission spectra are reported in Figure 4.10. Notably, silica itself plays the role of antireflection layer with respect to nude glass (Figure 4.10b *versus* a). In contrast, TiO_2 alone strongly increases the glass reflection, especially in the 380–600 nm visible range, whereas for higher wavelength the difference in performance is quite low. The presence of a double-layered film, instead, strongly enhances the transmittance of visible light, and ensures that all the other properties typical of TiO_2 films are present, resulting in self-cleaning surfaces, which are not available with monolayered silica films.

The coupling of SiO_2 with TiO_2 films, however, is not the only way to attain good antireflective properties. Extensively used as substitutes for diamonds in their cubic phase single-crystal form, zirconium dioxide can also be coupled to TiO_2 to yield films with a low refractive index.[21] Films can be prepared by a sol-gel method, allowing the optical properties to be tailored over a wide range by simple control of the sol composition. Figure 4.11 shows a clear decrease, from *ca.* 2.3 to *ca.* 1.9 (at 630 nm), for films with different percentages of ZrO. In the weak and medium absorption region, the refractive indices of the composite ZrO_2–TiO_2 thin films were calculated through some relationships employing parameters mainly derived from transmission spectra.

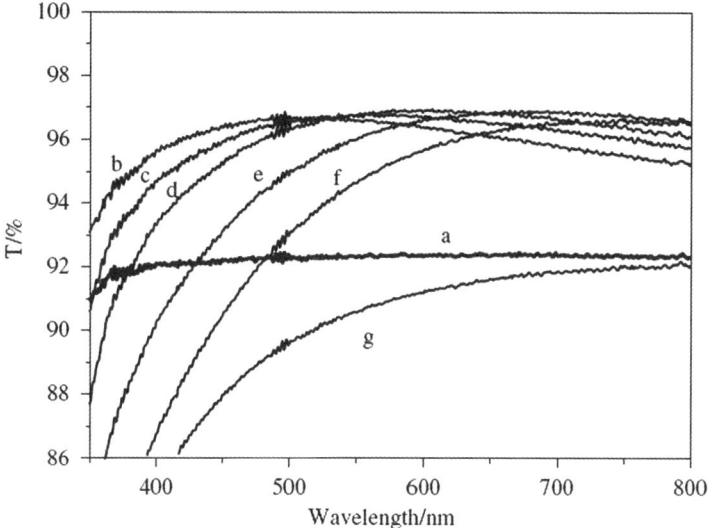

Figure 4.10 Transmission spectra of a bare glass substrate (a), SiO_2 film (b), SiO_2–TiO_2 layered films with an increasing amount of TiO_2 (c)–(f) and TiO_2 films on a glass substrate (g). (Reproduced with permission from ref. 20.)

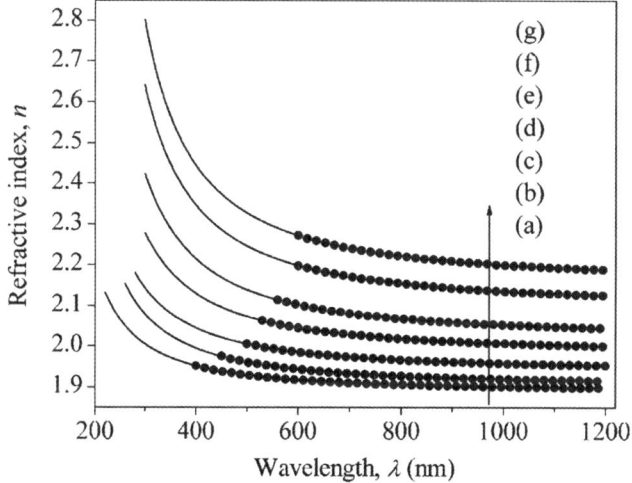

Figure 4.11 Refractive index *versus* wavelength for composite ZrO_2–TiO_2 film samples: (a) 100% ZrO_2; (b) 90% ZrO, 10% TiO_2; (c) 70% ZrO_2, 30% TiO_2; (d) 50% ZrO_2, 50% TiO_2; (e) 30% ZrO_2, 70% TiO_2, (f) 10% ZrO_2, 90% TiO_2; and (g) 100% TiO_2. (Reproduced with permission from ref. 21.)

Figure 4.12 The Fioravanti Hidra is the first car that will not have windscreen wipers (a), thanks to its windshield (b), which is functionalized by a four layer film that prevents the formation of droplets and fogging, thus ensuring perfect visibility for the driver. (Photograph courtesy of Fioravanti S.r.l., ref. 23.)

Multilayered films can be used for several applications by integrating the exceptional properties of TiO_2 with other technologies and materials. Hence, the engineer Leonardo Fioravanti, previously a designer of Ferrari cars, has designed Hidra, a hydrogen-powered car without windscreen wipers (Figure 4.12a), that was presented at the Geneva Motor Show 2008.[22,23]

A four-layer film imparts self-cleaning properties to the windshield under any environmental conditions and, notably, at any rate: it does not matter if the car is parked or is travelling rapidly along a highway. The windshield is moreover equipped with a hot air fan system to avoid snow deposits and hot water is sprayed when necessary to avoid stain deposits on the glass. Moving from outside to inside, the four thin films are constituted of:

1. a superhydrophilic layer of TiO_2, to avoid the formation of droplets and degrade volatile pollutants;
2. nanosized spherical particles to clean dirt deposits by mechanical (nano)friction;
3. a sensor layer, detecting the presence of water over the windshield;
4. an electrical net, enabling current to reach every area of the glass (Figure 4.12b).

The amount of current employed is very low and, indeed, is negligible compared with the energy consumed by a car. Moreover, hot water is available to clean the most persistent dirt when the car is used after a long parking time. Besides the already cited function, the TiO_2 layer also provides an excellent shield against the UV fraction of solar light. Fioravanti is working on a partnership for serial manufacturing. The car could be on the market in 2–3 years.

4.4 Industrial Overview and Commercial Products

The glass industry, along with cementitious and ceramic construction materials industries, has probably been the most active in looking for new solutions in the field of self-cleaning materials. Glass, in particular, is an old though still very attractive material but all its appeal is strictly linked to its optical properties, which decline dramatically when any kind of dust deposits on it. This explains the strong interest of industry in developing low maintenance products in this field.

The main sources of dust are meteorological events, essentially rain and wind, and in urban areas also smog. Powders of different kinds can deposit on glass, while rain absorbs all the gaseous pollutants present in air. Droplet formation, due to the intrinsic low wettability of glass, moreover, does not favor the leaching of soluble and particulate species. Glass forms in fact a contact angle with water in 20–30° range, but uncleaned glass can give rise to much higher contact angles, behaving as a truly hydrophobic material.

Taking a look at the glass industry, 62% of the international flat glass market is, interestingly, shared among only four big companies (Table 4.3) and Europe, in particular, shows a prominent market size of over 10 million ton (*i.e.* more than 20% of the global production). The European market is largely directed by high value-added products such as coated and laminated products, and per capita glass consumption is around 17 kg, which is the biggest value in the world (the United States and Japan have, for example, a per capita consumption of 10 and 8 kg, respectively).[24]

Nowadays the market offers a wide variety of self-cleaning glasses functionalized with TiO_2 on sale worldwide.

Pilkington, a 200-year-old company, and member of the NSG Group from June 2006, developed in 2001 a dedicated line of TiO_2-modified glasses. Pilkington Activ™ is indeed the world' first dual-action self-cleaning glass, and is sold in major markets worldwide. The line includes three products, covered by a thin transparent layer of TiO_2: Activ™ Clear, Activ™ Blue and Activ™ Neutral, available in 4, 6 and 10 mm thickness.[25]

Activ™ Clear can be used for conservatories, roofs, skylights, covered swimming pools and interior or exterior glass doors (Figures 4.13 and 4.14). After the first stage of coating activation (5–7 days are typically needed), after installation and also after each mechanical cleaning of the surface, the glass

Table 4.3 Global market share of flat glass industry in 2008. (Source: Pilkington, ref. 24.)

Company	Country	Percentage of world capacity
AGC	Japan	17.5
NSG Group	Japan	17.0
Saint-Gobain	France	15.5
Guardian	United States	12.0
Others		38.0

Figure 4.13 Application of Pilkington Activ™ Clear in a house (left, photograph courtesy of Pilkington) and in a conservatory (right, photograph courtesy of Oakland Group Limited).

Figure 4.14 Matsushita Denso building covered with Pilkington self-cleaning glass. (Photograph courtesy of Nippon Sheet Glass.[26])

self-cleaning properties are retained for a long time even in the presence of cloudy weather and during the night. The second stage takes place when rainwater hits the glass. The coatings react with ultraviolet rays present in natural daylight to break down and disintegrate organic dirt. During the third stage Activ™ Clear works as a superhydrophilic coating, thus impeding droplet

Photocatalytic Glass 133

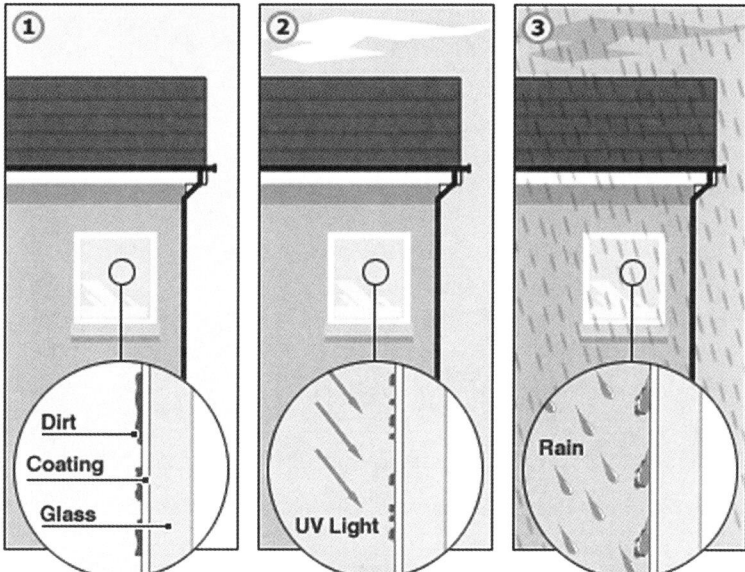

Figure 4.15 The three working stages of Pilkington Activ™: (1) the coating is activated by solar light (typically, 5–7 days required); (2) light destroys organic and many inorganic pollutants, reducing their adherence to the glass thanks to photocatalytic effect; (3) rain washes dirt away by forming a uniform water "sheet". (Reproduced with permission from ref. 25.)

formation, so that water can spread evenly over the surface and, while running off, take the dirt with it (Figure 4.15). Moreover, water dries off very quickly, with respect to conventional glass, without leaving unsightly drying spots. The reduction of glass cleaning maintenance is also a significant economic advantage, especially in buildings where glass is a major component of the structure, as in the case of skylights or conservatories. A light mechanical cleaning with a soft cloth is sufficient and no abrasive cloth or aggressive detergents are needed.

The coating of Pilkington Activ™ glass, deposited during the installation phase, is designed to last as long as the glass and re-application is not needed. Ideal working conditions start with sloping angles of at least 10°, and optimal inclination values are around 30°, but Pilkington Activ™ can be also used in vertical façades and with slopes lower than 10°.

Activ™ Blue is an attractive self-cleaning glass that evolved from Activ™ Clear. It offers attractive aesthetics with its blue color and allows for solar control and heat transmission. Compared with Activ Clear it has a significant shading coefficient, thanks to its color, which permits indoor light regulation also for long-wavelength light. The UV cut-off effect is comparable in both cases. Activ Blue is, for example, ideal for conservatory roofs (Figure 4.16).

Activ™ Neutral is a newly developed product that is ideal even for roofs and vertical façades. Its neutral, though slightly darker, appearance, which

Figure 4.16 Pilkington Activ™ Blue allows for solar control performance and light transmission even at long wavelengths. Its slightly blue color keeps the inside environment cooler. (Photograph courtesy of Pilkington.[25])

helps to reduce heat and light transmission, gives this product superior performances to Activ™ Clear, especially in well irradiated areas.

Activ™ Blue and Neutral can be combined with a thermally insulating glass, thus functioning as both UV and IR cut-off (Figure 4.17). This configuration yields a comfortable temperature throughout the year in highly exposed closed environments. In this way high energy savings for thermal conditioning can be readily achieved. Figure 4.18 reports the performance of composite glasses (with argon-filled cavities) of this kind compared with normal double glazing and polycarbonate sheets. The former present much more limited light transmittances, cutting the UV and IR radiation. Accordingly radiant heat transmittance is low too, meaning that the warming effects caused by solar rays are not so significant. Moreover, by considering the heat transmittance coefficient, U, the superior performance of double glazing employing Pilkington self-cleaning glass together with Optitherm™ and K Glass™, which are two thermally insulating glasses, is obvious.

Finally, the self-cleaning features are present neither in polycarbonate nor in normal double glazing configurations.

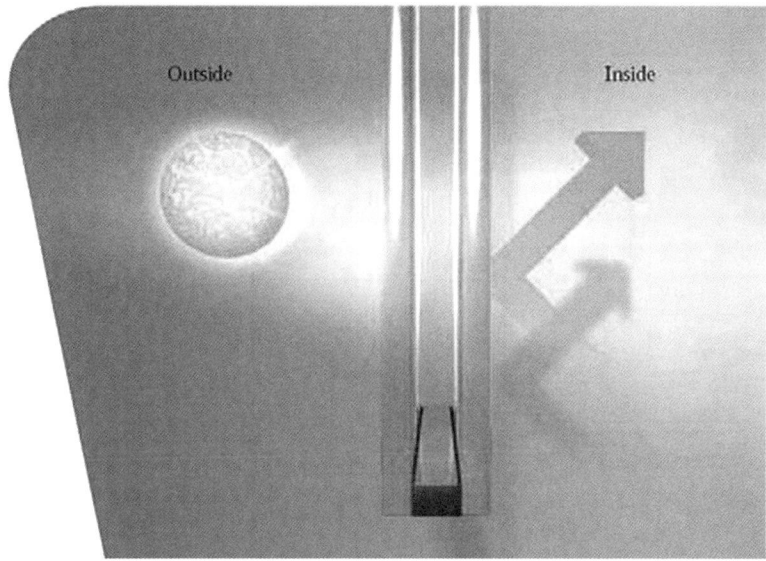

Figure 4.17 Pilkington Activ™ Blue or Neutral in double glazing configuration with a thermally insulating glass. (Reproduced with permission from ref. 25.)

Soon after the introduction of self-cleaning glass by Pilkington (2001), the 125-year old Pittsburgh Plate Glass Company (PPG) launched a new product, named SunClean®, implementing a durable, transparent coating of TiO_2 applied during the manufacturing process.[27] The transparent coating is applied to hot glass during the forming process to produce a strong, long-lasting bond, which makes the coating an integral part of the outer glass surface, through an application process patented by PPG.

The product exhibits properties similar to other self-cleaning glasses: easier maintenance is allowed by photocatalytic and superhydrophilic properties of the coating (Figures 4.19 and 4.20). Since UV light is abundant even on cloudy days or in shaded areas, this process works continuously throughout the day and, after an initial activation period, even at night. The hydrophilicity makes water droplets spread out, or sheet, across the surface of the glass. This sheeting action also helps the window dry quickly with minimal spotting or streaking.

SunClean® is distributed to makers of wood, vinyl and aluminium windows that can integrate this glass in their frames. Coupling self-cleaning properties together with energy saving is also possible by employing integrated solutions where SunClean® and thermally insulating glasses complement each other. PPG can also provide such products, like Solarban® 70XL glass or Sungate® Low-E Glass.

The features of PPG SunClean® can be summarized as follows:

1. The glass is photocatalytic and the self-cleaning coating is activated by UV solar rays.

	Normal double glazing	25mm Polycarbonate	4mm Pilkington Activ™ Blue 16mm Argon filled cavity* 4mm Pilkington K Glass™	4mm Pilkington Activ™ Blue 16mm Argon filled cavity* 4mm Pilkington Optitherm™ S4	4mm Pilkington Activ™ Neutral 16mm Argon filled cavity* 4mm Pilkington K Glass™	4mm Pilkington Activ™ Neutral 16mm Argon filled cavity* 4mm Pilkington Optitherm™ S4
Light	81%	68%	49%	52%	42%	44%
Total Solar Radiant Heat Transmittance, g	0.76	0.55	0.41	0.39	0.43	0.39
U-Value	2.7	1.75	1.5*	1.2*	1.5*	1.2*
Self-cleaning plus Solar Control on exterior surfaces	✗	✗	✓	✓	✓	✓

*Based on 90% gas fill.

Figure 4.18 Pilkington Activ™ *versus* normal double glazing and polycarbonate performances in terms of light and heat transmittance. (Reproduced with permission from ref. 25.)

Photocatalytic Glass 137

Figure 4.19 PPG SunClean® compared with regular windows during raining. Water sheeting is well apparent here. (Photograph courtesy of PPG.)

2. The glass is hydrophilic, so that the coating allows water sheeting action.
3. It offers improved thermal/optical qualities compared to regular clear glass.
4. The self-cleaning coating is applied using PPG' patented process.

Other benefits are:

1. UV rays work to slowly break down and loosen organic dirt on the glass surface;
2. the glass surface dries more quickly with minimal spotting and streaking;
3. 40% reduction in UV transmittance;
4. outward appearance of the glass is slightly brighter than regular glasses;
5. solar heat gain coefficient is improved by about 0.05 points;
6. the coating is durable and long-lasting.

An overview of commercial products shows that a primary position is occupied by Saint Gobain Glass (SGG), which is one of the three biggest glass manufacturers in the world. The company' proprietary product line is SGG Bioclean,[28] which is presented as the most neutral self-cleaning glass on the market. The glass is not characterized by unsightly tints or reflective surfaces and basically looks just like normal glass. Other properties are analogous to those described for other products, thus superhydrophilic and photocatalytic properties, durability of performance and reduced cleaning maintenance are guaranteed.

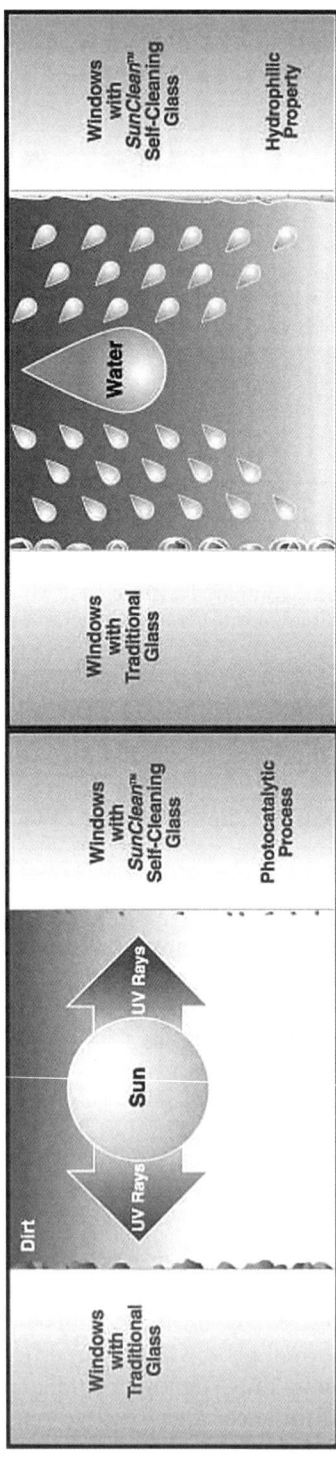

Figure 4.20 Scheme of PPG SunClean® working steps and comparison with regular glass. (Reproduced with permission from ref. 27.)

Photocatalytic Glass 139

This glass can similarly be integrated together with an appropriate inner glass, yielding SGG Bioclean Cool-Lite ST, ideal for contemporary solar radiation and heat transfer control. This outstanding product can hamper heat radiation transfer from the sun to the inside of a conservatory, keeping it cooler in warmer months (Figure 4.21). The advantages of applying such a glass can be listed as follows:

1. neutral appearance of the roof glass matches naturally with the rest of the conservatory glazing without compromising performance;
2. permanent dual-action SGG BIOCLEAN® self-cleaning coating stays on the glass for the lifetime of the window;
3. much easier cleaning as less dirt and grime adhere to the glass;

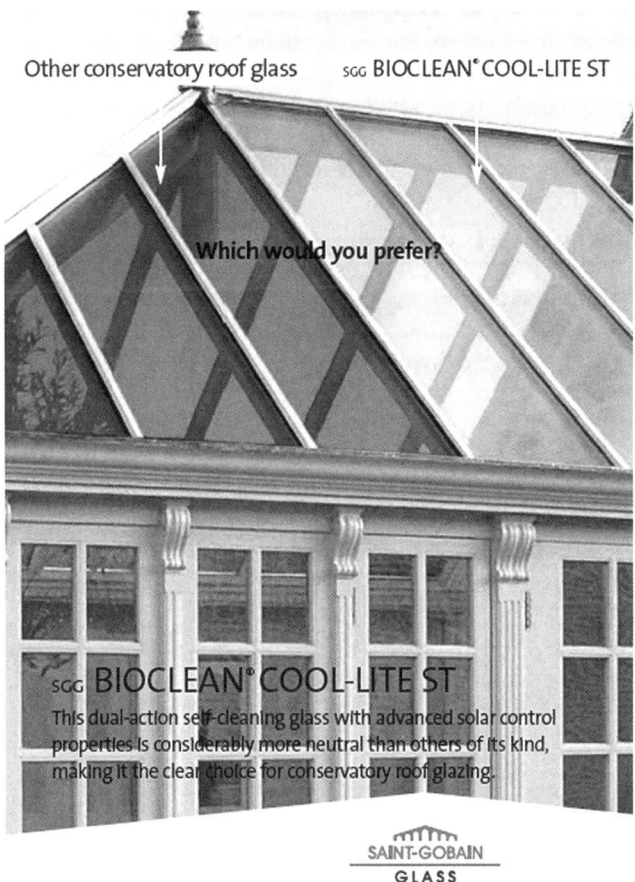

Figure 4.21 SGG Bioclean® Cool-Lite ST is ideal for conservatory roofs. (Reproduced with permission from ref. 28.)

4. ideal for hard to reach areas such as conservatory roofs and overhead glazing;
5. saves on the cost of window cleaning;
6. up to $\frac{2}{3}$ of heat from the sun is reflected by the advanced solar control coating;
7. keeps conservatories cooler during sunny periods without costly air conditioning or blinds;
8. reduces uncomfortable glare from the sun by over 50%, whilst letting in lots of natural light;
9. 100% noise reduction compared to polycarbonate roofs;
10. provides year round comfort when combined with SGG PLANI-THERM, a glass ensuring heat gain from the sun during cold seasons.

Competitive products are produced by Cardinal Glass Industries, a company working on the development of residential glass for windows and doors that has grown rapidly to more than 5500 employees located at 27 manufacturing locations around the United States.[29] Their Neat™ glass is a technologically advanced product and a commercial example of TiO_2–SiO_2 composite glass. The proprietary company products compare well with some of those of the present worldwide giants (Figure 4.22). The visible transmittance and the visible reflectance are in fact very high and low, respectively, by comparison with the other products. Remember that the presence of a TiO_2 film can alter the optical glass properties drastically if adequate preventative measures are not taken.

Neat™ glass is exceptionally smooth thanks to a thin SiO_2 layer that improves hydrophilicity, too, by reducing significantly light reflectance. This behavior is due to the much smoother surface of this film, compared to different solutions (Figure 4.23). Moreover, when Neat™ glass is fabricated with low-emissivity coatings it provides ideal temperature indoors. Hence year-round comfort can be appreciated inside the home, with an average annual energy saving of 25%. Double-pane, tempered or laminated configurations are available in customized shapes and size.[30]

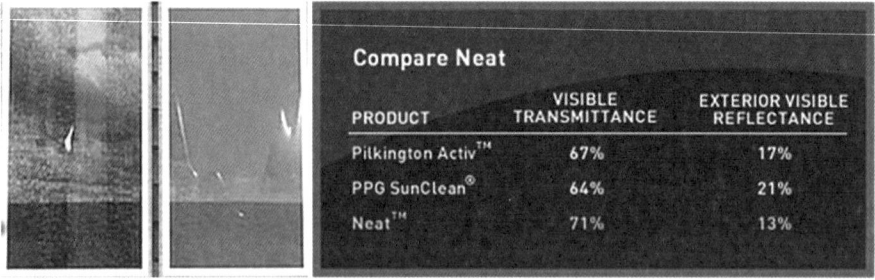

Figure 4.22 Wet plain glass compared with Neat™ glass (left-hand panels); visible transmittance and reflectances of Neat™ glass compared to other commercial glasses (right-hand panel). (Reproduced with permission from ref. 29.)

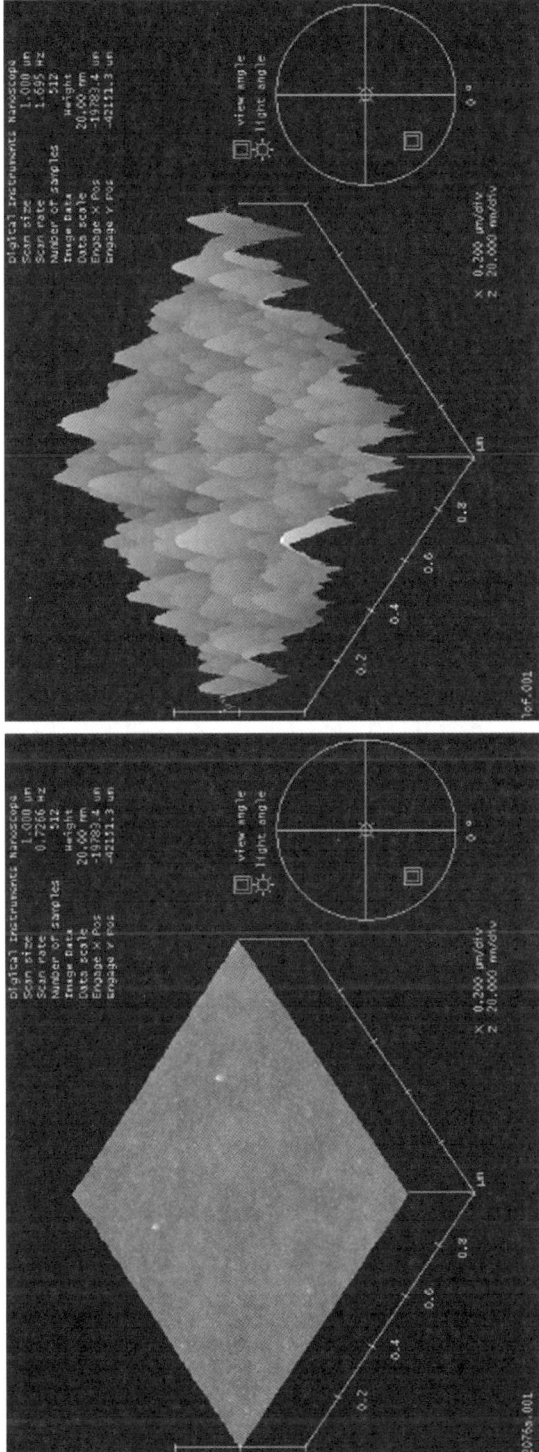

Figure 4.23 These AFM images show that Neat™ coatings (left) are much smoother than pyrolytic photoactive coatings typically used glass surface (right). (Reproduced from ref. 30.)

References

1. A. Fujishima, K. Hashimoto and T. Watanabe, *TiO$_2$ Photocatalysis. Fundamentals and Applications*, BKC, Tokyo, 1999.
2. (a) M. Schiavello, *Heterogeneous Photocatalysis*, John Wiley & Sons, Chichester, 1997; (b) I. P. Parkin and R. G. Palgrave, Self-cleaning coatings, *J. Mater. Chem.*, 2005, **15**, 1689–1695.
3. Toto' Hydrotect at http://www.toto.co.jp/products/hydro/genri_en.htm
4. Statistical Calculation and Development of Glass Properties, online at http://www.glassproperties.com
5. M. Addamo, M. Bellardita, A. Di Paola and L. Palmisano, Preparation and photoactivity of nanostructured anatase, rutile and brookite TiO$_2$ thin films, *Chem. Commun.*, 2006, 4943–4945.
6. H. M. Yates, M. G. Nolan, D. W. Sheel and M. E. Pemble, The role of nitrogen doping on the development of visible light-induced photocatalytic activity in thin TiO$_2$ films grown on glass by chemical vapour deposition, *J. Photochem. Photobiol. A*, 2006, **179**, 213–223.
7. T. Miyata, S. Tsukada and T. Minami, Preparation of anatase TiO$_2$ thin films by vacuum arc plasma evaporation, *Thin Solid Films*, 2006, **496**, 136–140.
8. O. Treichel and V. Kirchhoff, The influence of pulsed magnetron sputtering on topography and crystallinity of TiO$_2$ films on glass, *Surf. Coat. Technol.*, 2000, **123**, 268–272.
9. X. Cheng, S. Hu, P. Zeng, T. Kuang, G. Xie and F. Gao, Structure and properties of TiO$_2$ films prepared by ion beam assisted deposition, *Surf. Coat. Technol.*, 2007, **201**, 5552–5555.
10. (a) J. Yu and X. Zhao, Effect of substrates on the photocatalytic activity of nanometer TiO$_2$ thin films, *Mater. Res. Bull.*, 2000, **35**, 1293–1301; (b) Y. Paz and H. Heller, Photooxidatively self-cleaning transparent titanium dioxide films on soda lime glass: the deleterious effect of sodium contamination and its prevention, *J. Mater. Res.*, 1997, **12**, 2759–2766.
11. (a) J. M. Herrmann, H. Tahiri, C. Guillard and P. Pichat, Photocatalytic degradation of aqueous hydroxy-butandioic acid (malic acid) in contact with powdered and supported titania in water, *Catal. Today*, 1999, **54**, 131–141; (b) H. Yu, J. Yu and B. Cheng, Facile preparation of Na-free anatase TiO$_2$ film with highly photocatalytic activity on soda-lime glass, *Catal. Commun.*, 2006, **7**, 1000–1004.
12. E. Aubry, M. N. Ghazzal, V. Demange, N. Chaoui, D. Robert and A. Billard, Poisoning prevention of TiO$_2$ photocatalyst coatings sputtered on soda-lime glass by intercalation of SiNx diffusion barriers, *Surf. Coat. Technol.*, 2007, **201**, 7706–7712.
13. X. Zhang, A. Fujishima, M. Jin, A. V. Emeline and T. Murakami, Double-layered TiO$_2$-SiO$_2$ nanostructured films with self-cleaning and antireflective properties, *J. Phys. Chem. B*, 2006, **110**, 25142–25148.
14. M. Addamo, V. Augugliaro, A. Di Paola, E. García-López, V. Loddo, G. Marcì and L. Palmisano, Photocatalytic thin films of TiO$_2$ formed by a sol–

gel process using titanium tetraisopropoxide as the precursor, *Thin Solid Films*, 2008, **516**, 3802–3807.
15. A. Fujishima, X. Zhang and D. A Tryk, TiO_2 photocatalysis and related surface phenomena, *Surf. Sci. Rep.*, 2008, **63**, 515–582.
16. X. Zhao, Q. Zhao, J. Yu and B. Liu, Development of multifunctional photoactive self-cleaning glasses, *J. Non-Cryst. Solids*, 2008, **354**, 1424–1430.
17. W. Weng, M. Ma, P. Du, G. Zhao, G. Shen, J. Wang and G. Han, Superhydrophilic Fe doped titanium dioxide thin films prepared by a spray pyrolysis deposition, *Surf. Coat. Technol.*, 2005, **198**, 340–344.
18. E. D. Palik (ed.), *Handbook of Optical Constants of Solids*, Academic Press, Orlando, 1985.
19. E. Hecht and A. Zajac, *Optics*, 4th edn, Addison Wesley, 2002.
20. Z. Liu, X. Zhang, T. Murakami and A. Fujishima, Sol–gel SiO_2/TiO_2 bilayer films with self-cleaning and antireflection properties, *Sol. Energy Mater. Sol. Cells*, 2008, **92**, 1434–1438.
21. L. Liang, Y. Sheng, Y. Xu, D. Wu and Y. Sun, Optical properties of sol–gel derived ZrO_2–TiO_2 composite films, *Thin Solid Films*, 2007, **515**, 7765–7771.
22. Geneva Motor Show, online at http://www.salon-auto.ch/en
23. Fioravanti S.r.l. online at http://www.fioravanti.it
24. Pilkington and the Flat Glass Industry, 2008 report, available online at http://www.pilkington.com/resources/pfgi2008final.pdf.
25. Pilkington Activ[TM] range is presented online at http://www.pilkingtonselfcleaningglass.co.uk
26. Nippon Sheet Glass is distributor of Pilkington Active in Japan and South East Asia (http://www.nsggroup.net).
27. PPG URL: http://corporateportal.ppg.com/ppg
28. SGG Bioclean is presented online at http://www.selfcleaningglass.com
29. Cardinal Glass Industries: http://www.cardinalcorp.com
30. Cardinal Glass Industries Neat Glass, description online at http://www.cardinalcorp.com/data/tsb/cg/CG05.pdf

CHAPTER 5
TiO_2-modified Cement and Ceramics

5.1 Keeping Structures and Air Clean Indoors and Outdoors

Urban buildings are often fouled by several volatile pollutants and fine dusts of any kind present in our crowded towns. When used in a blend with a concrete structure, TiO_2 decomposes such organic fouling species, which include not only soot, grime, oil and particulates but also biological organisms, such as mold, algae, bacteria and allergens, and air-borne pollutants, *i.e.* VOCs, including, for instance, benzene, tobacco smoke, and all the nitrous oxides (NO_x) and sulfuric oxides (SO_x) that are significant factors in smog. Nitric and sulfuric acids, formed by oxidation of NO_x and SO_x, respectively, are eventually washed away as nitrate and sulfate salts by rainwater. Chemicals causing strong odors should not be forgotten, especially in residential areas with several restaurants or in industrial districts.

Photocatalytic pollutant degradation can be also exploited to significantly reduce air pollution *via* application of TiO_2 on pavers. These modified concrete materials are already widely used (Figure 5.1); some hundreds of square meters have been installed in Japan, and a preliminary calculation in Harris County in Texas estimates that a paving area costing *ca.* $200 is enough to destroy 1 ton of NO_x, assuming a quite limited paver duration (5 years).[1]

Fabrication of ceramic tiles has been another major field of application in construction materials, both for external walls and for indoor applications (Figure 5.2).[2] In the latter case, properties such as antimicrobial and deodorizing ones are valuable when paving, for instance, lavatories. It has been clearly demonstrated indeed that compounds from urine, and in particular urea, can be degraded, and bacteria propagation inhibited. In addition, kitchens are typical rooms where photocatalytic tiles can play an important role, by facilitating removal of oily compounds.

Clean by Light Irradiation: Practical Applications of Supported TiO_2
By Vincenzo Augugliaro, Vittorio Loddo, Mario Pagliaro, Giovanni Palmisano and Leonardo Palmisano
© V. Augugliaro, V. Loddo, M. Pagliaro, G. Palmisano and L. Palmisano 2010
Published by the Royal Society of Chemistry, www.rsc.org

Figure 5.1 On-site application of Noxer paving blocks. (Photograph courtesy of Marshalls and Mitsubishi Materials Corporation.)

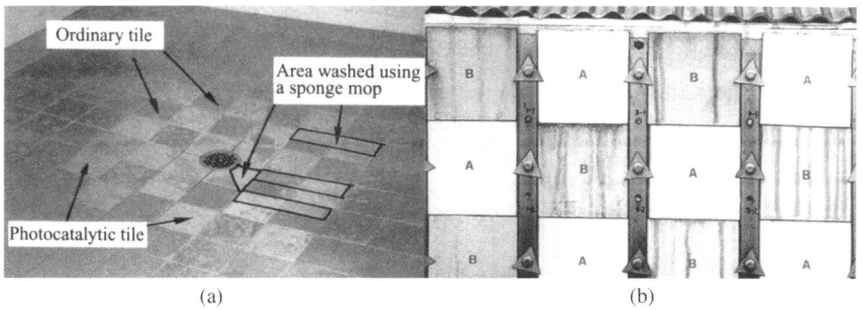

Figure 5.2 Photocatalytic ceramic tiles (a) in a dormitory shower room and (b) comparison between outdoor photocatalytic/superhydrophilic (A) and ordinary painted (B) wall tiles. (Reproduced with permission from ref. 2.)

Many other applications in construction materials are possible, such as in tunnel walls, which quickly become dark since air flow is very difficult and smog mounts up easily.

The use of TiO_2 in construction and building materials relies on the advantages distinctive of this semiconductor – superhydrophilicity, photocatalytic activity and self-cleaning properties are guaranteed. Moreover, the concentration of characteristic pollutants in an urban environment or in tunnels (ranging from 0.01 to 10 ppm) is compatible with photocatalytic degradation even with low ultraviolet sun radiation.

Starting in the early 1990s with the first attempts at preparing photocatalytic concretes, the use of photocatalysts together with building materials has been favored by the versatile features of TiO_2, which, besides playing the role of

Table 5.1 Main applications of TiO$_2$ in construction materials.

Horizontal applications	Concrete pavements
	Paving blocks and plates
	Roofing tiles
	Roofing panels
	Cement-based tiles
Vertical applications	Indoor and outdoor paints
	Finishing coatings
	Covering precast panels
	Permanent formworks
	Masonry blocks
	Sound-absorbing elements
	Traffic dividers elements
	Street furniture
Tunnels	Paints and renderings
	Concrete panels
	Concrete pavements
	Ultra-thin whitetopping

light-activated semiconductor material, is also a good structural mineral that in addition matches well and is easily implemented in existing industrial manufacture of all the most common (inorganic) construction materials. These are basically the reasons why, already in 2003, sales of photocatalytic building materials (including glass products, tents, plastic and wallpapers, soundproof walls – topics not discussed in this chapter) accounted for *ca.* 60% of the whole photocatalytic market share in Japan.[3] Table 5.1 reports the main application of TiO$_2$ in construction materials, divided into horizontal and vertical applications and tunnel materials.[4]

5.2 Merging TiO$_2$ and Cementitious Materials

Concrete materials with TiO$_2$ can be modified simply by dry adding a certain amount of semiconductor. TiO$_2$ weight percentages are in the 0.1–50% range with respect to the binder. The resulting material is commonly used only in an exterior layer with a thickness of *ca.* 1 cm and not in the entire structure, as Figure 5.3 shows in the case of paving blocks used in a pilot (10 000 m^2 large) area in downtown Antwerp (Belgium).[5] The environmental abatement of NO$_x$ was measured in standard laboratory experiments and *in situ*.

Photocatalytic concrete paving blocks have been contemporaneously patent-protected by Murata *et al.* (Mitsubishi Materials Corporation)[6] and Cassar *et al.* (Italcementi S.p.A.).[7] The first includes the application of TiO$_2$ in a functional surface layer of a double-layer paving block with enhanced NO$_x$-cleaning. Parameters such as thickness, porosity or surface texture layer are all claimed by the patent. The concentration of TiO$_2$, use of appropriate aggregates having a high NO$_x$ adsorption rate or high light transmission properties are specified, too. Mitsubishi's commercial product, called Noxer, recalling

TiO₂-modified Cement and Ceramics 147

Figure 5.3 Paving blocks used in Antwerp city centre (Belgium). (Reproduced with permission from ref. 5.)

something like "NO_x killer," has been specifically designed to degrade by photooxidation exhaust automobile gases and is available in different sizes and colors, with strength and slipping resistance equal to those of standard interlocking blocks. The paving works by setting up a photocatalytic reaction between titanium dioxide contained in the top 15 mm of the block and ultraviolet radiation from the sun. Air oxygen oxidizes NO_x in the air to nitric acid. The latter is eventually washed away and highly diluted by rain to realize active sites for more reactions, which can take place both in dry and wet conditions.

The patent by Italcementi covers the application of appropriate photocatalytic particles (not necessarily TiO_2) in paving tiles able to oxidize polluting substances present in the environment. The patent describes of course the application of the semiconductor material in terms of mass with respect to the amount of cement or binder used. Moreover, the composition of a dry premix containing a hydraulic binder and a TiO_2-based photocatalyst is also described in order to preserve the resulting material brilliance and color. In all cases, anatase is indicated as the preferable crystalline phase, although Italcementi specifies that rutile/anatase mixtures also work well, provided that a minimum of 25% anatase is present.

Table 5.2 compares these two most relevant patents.[8]

Primary information when adding TiO_2 powder to a concrete has to be given about the mechanical performances of the resulting blend. Parameters such as compressive strength, tensile strength, flexural strength and modulus of elasticity of the mortars are all needed before applying a photocatalytic concrete in buildings. Figure 5.4 shows that by increasing the semiconductor amount in

Table 5.2 Comparison between Mitsubishi and Italcementi patents on photocatalytic concrete materials. (Adapted from ref. 8.)

	Mitsubishi Materials Corporation	*Italcementi S.p.A.*
Patent title	NO_x-cleaning paving block.	Paving tile consisting of a hydraulic binder and photocatalyst particles.
Field of the invention	NO_x-cleaning paving block with enhanced efficiency of fixing NO_x from the air and increased pluvial NO_x cleaning efficiency.	Cement composition made of a hydraulic binder dry-premixed with an additive, having improved property to maintain the brilliance and color quantity, to prevent aesthetic degradation and to degrade environmental pollutants.
Working principles	• NO_x-cleaning paving block consisting of a surface layer that contains TiO_2 and a concrete made base layer; • NO_x-cleaning paving block with or without adsorbing material in the surface layer; • replacement of the sand used with 10–50% of glass grains or silica sand having a particle size of 1–6 mm; surface layer having a void fraction of 10–40% and water permeability of 0.01 cm s^{-1}; • NO_x-cleaning paving block roughened with a surface roughening tool.	• Use of a photocatalyst, which can oxidize in the presence of light and air environmental polluting substances, for the preparation of a hydraulic binder for manufacturing paving tiles that maintain after installation, for a longer time, brilliance and color quantity; • use of a dry premix containing a hydraulic binder and a photocatalyst that can oxidize in the presence of light and air environmental polluting substances, for manufacturing paving tiles that maintain after installation, for a longer time, brilliance and color quantity.

Binder	Cement	Hydraulic binder; white, grey or pigmented cement; cement used for debris dams; hydraulic lime.
Photocatalyst	TiO$_2$ without any further requirement	• TiO$_2$ or a precursor thereof, mainly in the form of anatase; • TiO$_2$ with anatase structure for at least 25%, 50% and 70%; • blend of anatase and rutile TiO$_2$ (70 : 30); • TiO$_2$ doped with one or more atoms other than Ti • TiO$_2$ doped with one or more atoms selected from Fe(III), Mo(V), Ru(III), Os(III), Re(V), V(V), Rh(III); • photocatalyst selected from the group consisting of tungstic oxide (WO$_3$), strontium titanate (SrTiO$_3$) and calcium titanate (CaTiO$_3$).
Amount of photocatalyst	• 0.6–20% by weight; • 5–50% by weight with respect to the binder.	• 0.01–10% by weight; • 0.1% by weight with respect to the binder; • 0.5% by weight with respect to the binder.

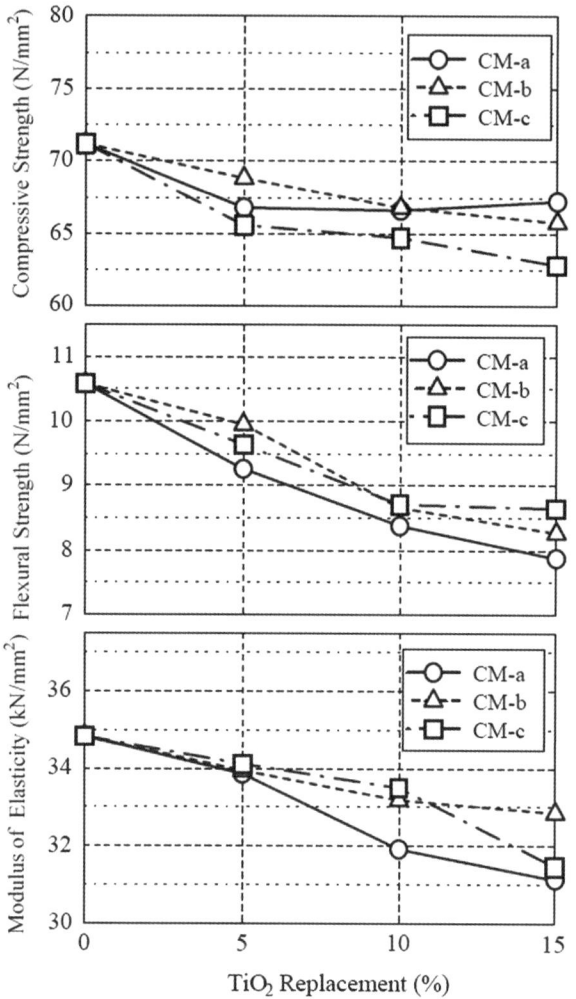

Figure 5.4 Compressive strength, flexural strength and modulus of elasticity *versus* TiO$_2$ percentage in cement mortars. CM-a, CM-b and CM-c refer to different TiO$_2$ types (see text). (Reproduced with permission from ref. 9.)

a cement mortar all the mechanical properties gradually worsen, but are still adequate for practical application.[9] Conversely, the capability of NO$_x$ adsorption shows an opposite trend. The three different curves refer to anatase TiO$_2$ in dry powdered form of 7 nm (CM-a) and 21 nm (CM-b) average particle diameters, and in a colloidal suspension (CM-c) with a particle diameter of 20 nm and a density of 40%. CM-a showed the best performance in terms of NO$_x$ adsorption, probably because of its higher specific surface area. A commercial Portland was used as cement with a density of 3.13 g cm^{-3}.

Combining the traditional bituminous pavement with the proficient photocatalytic action of the discussed cement mortars is also possible, as recently shown in a demonstration test.[10] Common asphalt, characterized by granulometry, has a degree of compactness giving rise to *ca.* 20% void content, with a distribution compatible with the penetration of photocatalytic cement mortar in the upper part of the asphalt layer; 1–1.5 cm has been found as the ideal thickness of photocatalytic concrete over asphalt. Both manual and automatic deposition are possible – the mortar can be induced to penetrate into the asphalt sub-layers through scrapers working on the surface, which also ensure a uniform distribution of the paste (Figure 5.5). The temperature has to be lowered to 30 °C to avoid rapid evaporation that would prevent mortar penetration into asphalt voids.

Compression tests made on these materials suggested that accelerants were needed, since the mechanical performances increased slowly throughout tens of days. Hundred cycles of water freezing/melting leave the mortar in the original form, and no difference in performance is noticeable.

Finally, the product shown in Figure 5.5 is characterized by similar properties to that of a common asphalt (provided that an accelerant is added before deposition), guaranteeing efficient degradation of pollutants, too.

Photocatalytic concrete has the big advantage of maintaining its clear color for a very long time, even if the corresponding building is situated in a very crowded and polluted town. Thus, a white cement containing TiO_2 (TX Active), commercialized by Italcementi at *ca.* 1 € kg^{-1}, has already been used in the construction of the "Cité de la Musique et des Beaux Arts" in Chambéry, France (2001), the "Dives in Misericordia" church in Rome (2003) (Figure 5.6) and for other applications in Italy and France such as the Roissy airport in Paris and the Saint John's Court Hotel in Monte Carlo. The imposing structure and complex construction features of the "Dives in Misericordia" church built in Rome's Tor Tre Teste district were designed by US architect Richard Meier. The structure resembles three great sails apparently

Figure 5.5 Application of a photocatalytic mortar on a common bituminous asphalt. (Reproduced with permission from ref. 10.)

Figure 5.6 The Dives in Misericordia church (Rome) was built using photocatalytic cement. (Photograph courtesy of Italcementi.)

swollen by an easterly wind, and consists of the actual church and the community center.

TX Active is presented by Italcementi as the perfect means to optimize the aesthetic durability of top quality, cement-based elements.[11] Once the white cement has hardened in the form of paste, mortar or concrete, the photocatalytic particles contained inside can oxidize organic and inorganic air pollutants in the presence of air and light. Pollutants that can easily come into contact with the cement, *i.e.*, exhaust fumes, emissions from residential heating systems, industrial emissions of aromatic chemical substances, and pesticides, do not adsorb stably onto TX Active since they are oxidized to carbon dioxide. Moreover, liquid organic pollutants do not adhere to the building materials but are washed away by rain thanks to the light-induced superhydrophilicity of TiO_2.

The use of this photocatalytic cement does not require particular caution or specific procedures: the mortar and/or concrete preparation requires a suitable water/cement ratio to enhance strength and prevent shrinkage; correct binder content and adequate mixing to ensure uniform color; correct granulometric curve; use of clean aggregates since dust particles adhering to the aggregates can cause their removal from the binding matrix; suitable mixing time and correct curing of casts. The concrete mix design, along with a continuous assistance provided by Italcementi to all the customers using its photocatalytic cement, is suitable for a wide range of applications, such as restoration projects, building projects with complex static and aesthetic specifications, high-quality flooring, elements produced with or without steam curing, architectural concrete casts, stuccos and sealings.

Recently, the two main buildings realized with TX Active have been monitored to check the prolonged stability of the original color. The results for both the Italian church and the French construction are excellent. The colorimetric measurements were performed through a CIE $L^*a^*b^*$ method.[12] This method allows us to determine the colorimetric parameters for quantitative evaluation of surface color changes. In particular, the L^* parameter refers to the lightness of luminance (included in the 0–100 range) whereas a^* and b^* indicate the red-magenta and blue-yellow color-opponent dimensions, respectively. In the case of white cement the main reference parameter is L^*, which is equal to 0 for black and 100 for diffuse white.[13]

Measurements were carried out during the building of Dives in Misericordia church (in 2000) until 2006 on different points of the three sails and of the northern wall (Figure 5.6). The differences in lightness of luminance were very low in all cases: Table 5.3 reports in particular the values corresponding to the external sail. The difference in b^* is due to inorganic inert materials mainly caused by sand deposition carried from the Sahara through sirocco wind. This is confirmed by the absence of b^* variation in the northern wall protected from sirocco.

Finally, the white color of the building was preserved for 6 years with less than 1% variation in lightness. Analogous results were recorded in Chambéry. Table 5.4 reports the L^* values along with the standard variations for the ground floor of the main building façades. Similar results were obtained in the other buildings and on first floor walls. Finally, a^* and b^* values did not change and the sand phenomenon recorded in Rome was indeed absent.

Italcementi offers two products containing TX Active:[14] the first is the binder TX Aria, which is ideal for the preparation of paints, mortars, plasters and concretes and can be applied in horizontal or vertical structures and in tunnels to promote the abatement of air pollutants and improve safety. It is especially suitable in urban and highly polluted areas. TX Arca cements are, instead, ideal for prestigious architectural buildings, allowing preservation of both top

Table 5.3 Mean values of L^*, a^* and b^* values corresponding to the external sail of Dives in Misericordia church. (Taken with permission from ref. 13.)

	Year 2000 (building period)	September 2003 (end of works)	February 2005		June 2006	
			External	Internal	External	Internal
L^*	91.75	90.81	89.37	90.44	89.68	90.31
a^*	−0.41	−0.11	0.30	0.06	0.44	0.05
b^*	2.53	4.51	7.75	3.10	7.83	3.35

Table 5.4 Mean values of L^* and standard deviations (s.d.) corresponding to the main building (ground floor) of Cité de la Musique et des Beaux Arts in Chambéry. (Taken with permission from ref. 13.)

Service life (months)	Facade orientation							
	West		North		East		South	
	No. of measurements							
	26		30		26		30	
	L^*	s.d.	L^*	s.d.	L^*	s.d.	L^*	s.d.
$t_0 = 0$	71.5	1.8	72.0	1.6	71.6	2.0	71.3	1.6
$t_1 = 7$	71.5	1.7	71.7	1.7	71.0	1.8	71.3	1.4
$t_2 = 15$	71.7	1.7	71.6	1.1	71.8	1.4	72.0	1.4
$t_3 = 19$	71.6	1.3	71.9	1.5	71.5	1.6	71.2	2.4
$t_4 = 26$	71.3	1.6	71.4	2.0	70.8	1.5	70.4	2.0
$t_5 = 40$	71.4	2.1	71.1	1.6	70.8	1.7	70.7	1.8
$t_6 = 56$	71.3	2.1	71.5	1.9	71.3	1.5	71.1	2.0

physico-mechanical properties and their early esthetical appearance for many years (it was used in the Dives in Misericordia).

TX Active products are commercialized also in North America by ESSROC Cement,[15] a leading North American cement producer with over 5.5 million metric tons of annual capacity and part of the Italcementi Group.

TiO_2-modified cementitious materials are not used as supporting structural elements, since the inner non-irradiated areas would not work photocatalytically. Thus only an external wearing thin layer should be used.

To impart photocatalytic properties to cementitious materials it is also possible to deposit a TiO_2 sol over it by simple spray-coating. This method was proved by the Kon Corporation (Figure 5.7), a Japanese company founded at the beginning of the twenty-first century, producing a brand of TiO_2 aqueous sol, based on peroxide titanium photocatalysts, developed in the Saga Prefecture Ceramic Research Laboratory. The Kon Corporation produces various kinds of titanium dioxide coating products, the choice of which depends only on the property of main interest for a specific application. Table 5.5 describes the characteristics and weight content of different kinds of sol.[16] PTA is an amorphous sol in water medium and hence does not have any photocatalytic function at room temperature unless it is first heated to more than 250 °C. A very thin TiO_2 film can be deposited on the surface of a substrate at room temperature. After it is dried completely the coated film containing TiO_2 can no longer be dissolved in water. PTA-85 is suitable for spray coating whereas the more concentrated PTA-170 is used for dip coating.

The TO series differs from PTA since it is already crystallized and shows a strong photocatalytic activity when it is coated and dried. On a flat, smooth surface the TiO_2 coated to the substrate comes off as a powder when rubbed with a finger because TO cannot form a hard film at room temperature. However, the hardness of the film coated with TO can be improved when it is heated to a certain temperature.

Figure 5.7 Aesthetics salon with exterior and indoor walls covered with a TiO_2 sol. (Photograph courtesy of Kon Corporation.)

Table 5.5 Kon Corporation TiO_2 sol series. (Reproduced with permission from ref. 16.)

Name	Description	Weight content (%)
PTA-85	Amorphous condition aqueous sol	0.85
PTA-170	Amorphous condition aqueous sol	1.70
TO-85	Photocatalytic active sol	0.85
TO-240	Photocatalytic active sol	2.40
TPX-85	Photocatalytic active sol	0.85
TPX-220	Photocatalytic active sol	2.20
TPX-HL	Visible activated photocatalytic active sol	0.85
TPX-VB	Hybrid type sol for sterilizing	0.85
TPX-AD	Hybrid type sol for odor reduction	0.85
TPX-HP	Hybrid type sol for superhydrophilicity	0.85

TPX is a mixed solution of PTA and TO. A film coated by TPX achieves a photocatalytic effect and hardness of film after drying at room temperature. TPX is a neutral solution that can be allied to various materials including metal and resin. The photocatalytic effect and hardness of the film can be further increased by heating the film. The resulting film is very stable over time.

The other solutions are visible-light activated (TPX-HL), hence they exploit not only the solar UV fraction but also the visible one, which is ideal for sterilization even in dark conditions (TPX-VB), or for destroying high amounts of volatile organic compounds thanks to their strong adsorption on this kind of film (TPX-AD), and, finally, the ideal film for window glass, with the highest hydrophilicity (TPX-HP).

Many patents have been registered that describe the preparation of TiO_2 colloidal solutions, even metal-doped ones to ensure a good absorption of visible light. For instance,[17] colorless colloidal solutions can preserve the original appearance of a material's surface, without altering the characteristics, and can be applied on cementitious, marble or stone supports. The doping agent represents an atomic percentage ranging from 0.1 to 1% with respect to titanium atoms. Many precursors can be used, such as titanium tetrachloride, oxysulfate or alkoxides. The hydrolysis of TiO_2 precursors can take place directly in the presence of salts containing the doping metal by co-precipitation or mixing. The resulting liquid can be deposited by spraying or by using a brush. Repeated depositions can be performed to reach the desired thickness.

Similarly, in 2008 A*STAR (Singapore) developed a TiO_2 sol to be used as coating for building exteriors, which has been patented and licensed to Haruna.[18]

In Europe, the Italian company Global Engineering S.p.A.[19] is the leader in the development, production, certification and application of materials with photocatalytic properties. Founded by the Milanese entrepreneur Claudio Terruzzi, Global Engineering has launched the Ecorivestimento® trademark into the Italian and international markets as ideal for use as a cement photocatalytic covering. These photocatalytic coverings have been developed with the exclusive PPS® technology (Proactive Photocatalytic System), which was derived from research activity carried out by the scientific partner of Global Engineering, *i.e.*, Millennium Inorganic Chemical (Group Cristal Global – a worldwide leading company in the treatment of titanium). This collaboration between the two companies has led to the development of photocatalytic water-based paints, internationally marketed under the trademarks of Ecopittura® and Ecopaint®.

5.3 Photocatalytic Ceramic Tiles

With a difference in chemical composition, but similar in employment with respect to self-cleaning cementitious materials, TiO_2-based self-cleaning tiles have been widely commercialized and applied. The Japanese company Toto Ltd., in close collaboration with Professor Akira Fujishima, has registered *ca.* 270 patents in the photocatalytic technology domain.[20] Its Hydrotect series – aqueous suspensions containing TiO_2 in powder or gel form – can be sprayed over the surface of ceramic tiles that are in turn heated at 600–800 °C. The resulting product shows superior superhydrophilic properties and provides degradation of air pollutants both indoors and outdoors (see the white tiles shown in Figure 5.2). The heat treatment is essential to make TiO_2 particles sinter and strongly adhere to the tile surface. Finally, a thin film is formed with a thickness ranging from fractions to a few micrometers.

The big success of photocatalytic tiles, since their appearance in 1998, has given rise to more than 5000 projects and their introduction in the European market. They are indeed not only produced and sold in Japan. The leading European ceramic manufacturer Deutsche Steinzeug Cremer & Breuer AG introduced this technology from Toto and started industrial manufacturing in 2000.[21] Toto indeed transferred a technical package, including the license, some production facilities and a supply of the coating fluid to be deposited over tiles. The advantages of such products, guaranteed for the entire life of the tile, can be summarized as follows:

- decomposition of bacteria, fungi, algae, moss and germs,
- elimination of odors,
- improvement of room climate,
- retention of the original characteristics of the tile such as, for example, resistance to abrasion, resistance to chemicals, *etc.*,
- easy cleaning,
- low frequency of cleaning,
- free of irritating substances, non-toxic,
- environmentally-friendly,
- long-term guarantee,
- wide variety of products.

Areas where the tiles can be applied with great benefits (Figure 5.8) are:

- catering trade and food processing,
- medical areas,
- swimming pools and health areas,
- façades,
- representative areas,
- sanitary rooms.

Figure 5.9 shows how dirt is washed away by natural water (rain) or wiped off easily by manual cleaning since it does not adhere to the ceramic. This is also true for oil and grease, which show a natural inclination to detach from the tiles.

Self-cleaning and stain-free performance were confirmed by many experiments on site, outdoors.[22] Likewise dirt and stains of oily pollutants deposited on interior tiles used in lavatories or kitchens are always a problem. For example, the fatty acids from soap can form chemical bonds with calcium and magnesium in hard water and adhere to the tile surface, which are difficult to clean after the accumulation of dirt.[23] Hence the role of the TiO_2 film interposed between the ceramics and the organic dirt surface can break their binding, rendering the washing process much easier (Figure 5.10).

Photocatalytic activity has to be checked with care since (i) the catalyst properties have to be preserved during the manufacture and (ii) its accessibility to both the organic compounds and to sufficient UV-light have to be verified.[24]

Superhydrophilicity, on one other hand, is promptly promoted when exposing the surfaces to light, but can be highly prolonged even in dark conditions by combining TiO_2 with SiO_2 or other silicon compounds with siloxane

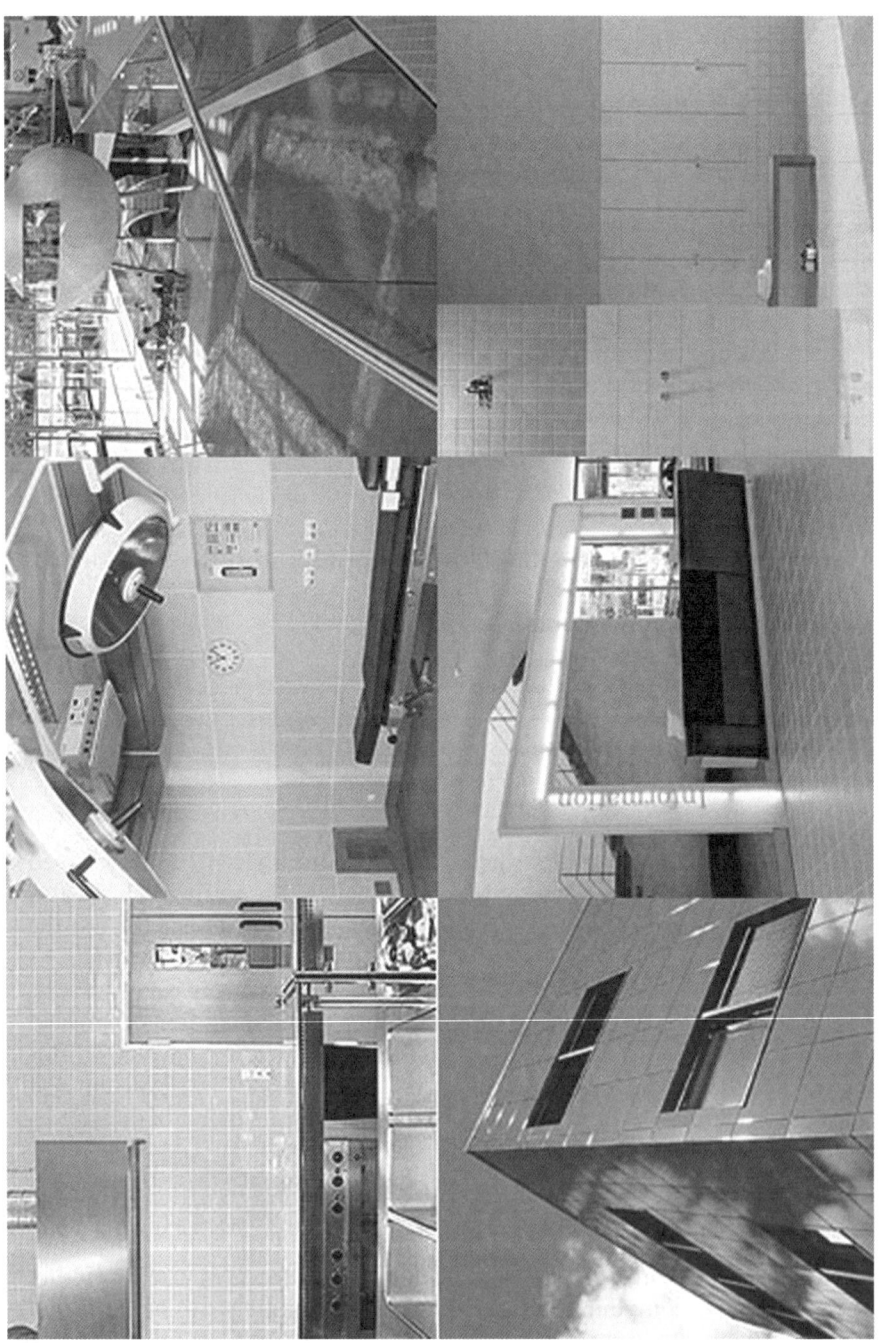

Figure 5.8 Application areas of ceramic tiles. (Photograph courtesy of Deutsche Steinzeug/brand Agrob Buchtal.)

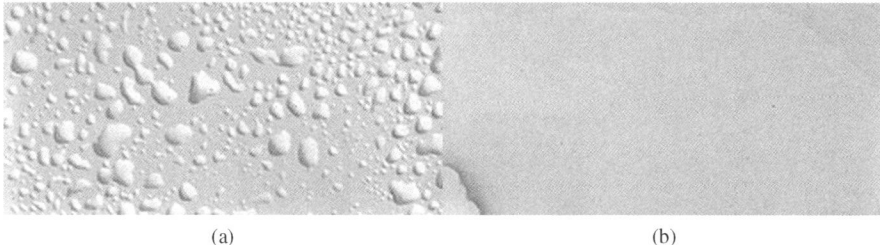

Figure 5.9 Functionalization of simple ceramic tiles (a) with TiO_2 (b) ensures superior performances by avoiding drop formation and promoting the abatement of air pollutants. (Photograph courtesy of Toto Ltd.)

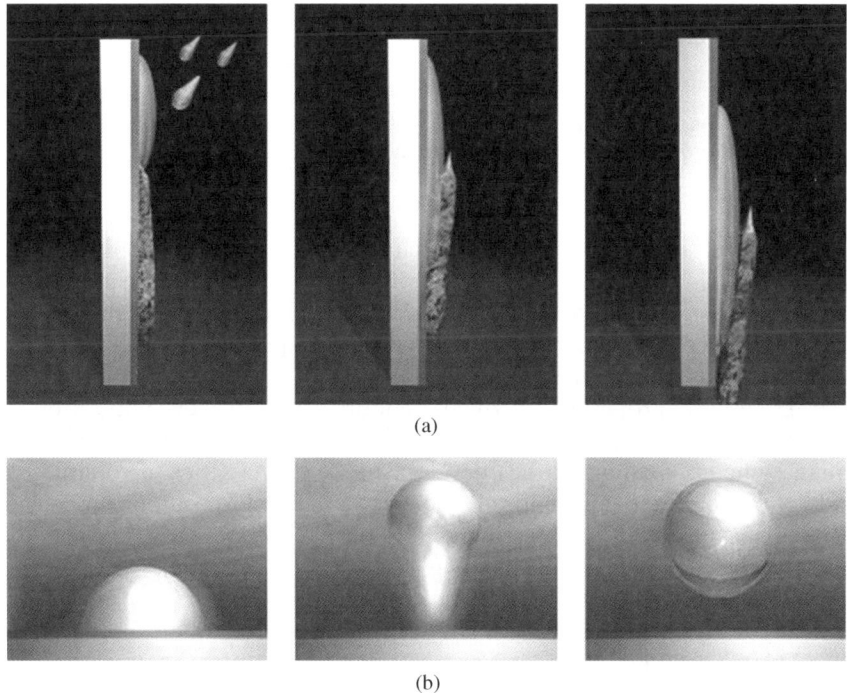

Figure 5.10 Dirt does not adhere to the photocatalytic tiles and is easily wiped off (a); oily liquids that do not have any affinity with the TiO_2 film readily detach (b). (Photograph courtesy of Deutsche Steinzeug/brand Agrob Buchtal.)

bonding.[25] Consequently, a right balance between photocatalytic oxidative activity (decreased by silica) and superhydrophilic effect should be found.

Application of photocatalytic ceramic tiles can ensure a relevant sterilization, thanks to their antimicrobial features. Figure 5.11 shows how a photocatalytic tile can decompose bacteria such as *Escherichia coli*, once they adsorb

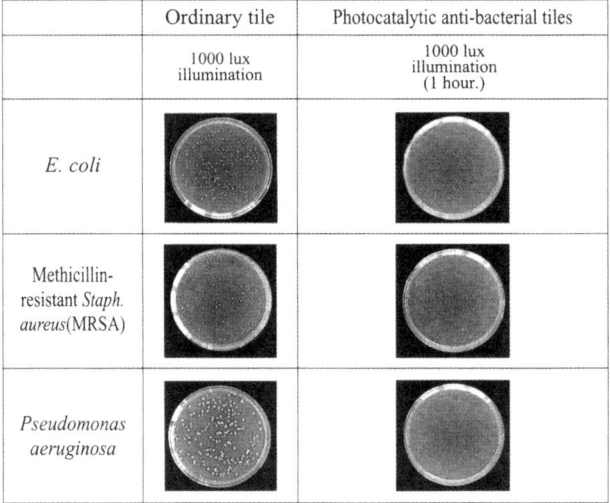

Figure 5.11 Bactericidal activity of photocatalytic tiles is clear for all three bacteria shown. (Reproduced with permission from ref. 2a.)

on its surface. This effect becomes much stronger when TiO_2 is doped with silver or copper nanoparticles. Doping with silver and copper ensures, moreover, a modest antimicrobial activity even in the absence of light, so it is strongly advised when the bactericidal effect is important even in shadowed surfaces such as in hospitals.

Other useful properties, verified indoors, are the decomposition of cigarette smoke and of organic compounds present in urine. On this basis, the Italian company Gambarelli has developed a complete series of ceramic tiles made of porcelain stoneware for all rooms of a flat and for outdoor use. Their brand name Oxygena[26] refers to the role played by the active oxygen that promotes the bactericidal and antismog effects. Oxygena ceramics are protected by an international patent deposited in 2003, where the major innovation lies in the firing system. Indeed, while TiO_2 loses its photocatalytic properties at 900 °C Gambarelli has succeeded in creating a firing system that prevents the loss of anti-pollution properties, enabling the tiles to be fired at 1130 °C, thus ensuring the best qualities of mechanical strength along with those of anti-pollution.

A test protocol has been drawn up that demonstrates that $1\,m^2$ of Oxygena tiles exposed to daylight eliminates in eight hours pollutants such as NO_x, purifying a volume of air as high as $72\,m^3$.

The only by-products are very small quantities of salts that are easily removed by wind and rain outside or a simple washing indoors. Several hotels, touristic villages, fitness centers, airports and also privates have already chosen Oxygena.

TiO$_2$-modified Cement and Ceramics

Before entering the market, these ceramics were tested by simulating the air quality of towns such as Rome or Milan in the laboratory. Gas containing nitrogen monoxide and dioxide was fed into glass domes placed over a titanium dioxide tile. The concentrations were in the 1–10 ppm range, corresponding to medium-high and very high pollution levels. Lamps reproducing the solar spectrum were used to simulate radiation by sunlight. The clear result has been that the quantity of pollution left was by far lower than the gas pumped in at the beginning of the experiment. This gives a direct and clear relation between the effect of the tiles and the reduction of pollution. Regarding indoor application, the activating effect lies in daylight and works when windows are present in a room. It is, however, possible to install a lamp emitting near-UV light in the case of rooms without windows. When turned on for a few hours during the night it manages to clean the tiles and air through activation of photocatalytic reactions. Even ordinary lamps used for lighting activate the reaction, even if in a limited way.

Figure 5.12 shows some of the porcelain stoneware tiles proposed by Gambarelli. Both the glazed version and the more traditional full body type are produced and sold under the Oxygena brand.

Figure 5.12 Porcelain tiles of the Oxygena series range from traditional style to modern and parquet-like. (Photograph courtesy of Gambarelli.)

Figure 5.13 Dahlia Design® ceramic tiles are ideal in kitchens and bathrooms. (Photograph courtesy of Dahlia.)

Dahlia produces and sells in Singapore ceramic and glass tiles functionalized with nanosized anatase TiO_2 with particular emphasis towards application in baths and kitchens.[27] After spending many years on research, Hongsheng Ceramics (Suzhou Industrial Park) Co. Ltd. has successfully developed titanium dioxide coated ceramic tiles (Dahlia Design®), without compromising the exquisite finish of the ceramic tiles. They claim that the photocatalytic coating gave rise to a new generation of self-cleaning, antibacterial, antifungal crystal glass tiles and ceramic tiles, meeting expectations of a high standard of living, environmental quality of the surrounding space and everyone's wish to live in a comfortable, clean and healthy environment. These tiles, which show major anti-abrasion properties, also proved to be bactericidal and to degrade volatile benzene under solar irradiation. Figure 5.13 shows an application of Dahlia tiles in a cooking area. Dahlia has also recently launched a new series of products with highly vibrant colors in the shape of crystal glass mosaics, and sparkling glass.

Tiles indoor and on façades are not the only applications of photocatalytic ceramic materials. Roof-tiles have indeed been commercially developed by the German company Erlus, in a vast variety, different in shape, size and color (red, brown and black) (Figure 5.14).[28] The product name is Erlus Lotus, and it is the first self-cleaning clay roof, destroying organic dirt particles such as fat deposits, grime, moss and algae using sunlight. Rain washes away the residuals produced from reaction. In 2004 this product line won the *Materialica* Design Award in the category of "materials."

TiO$_2$-modified Cement and Ceramics 163

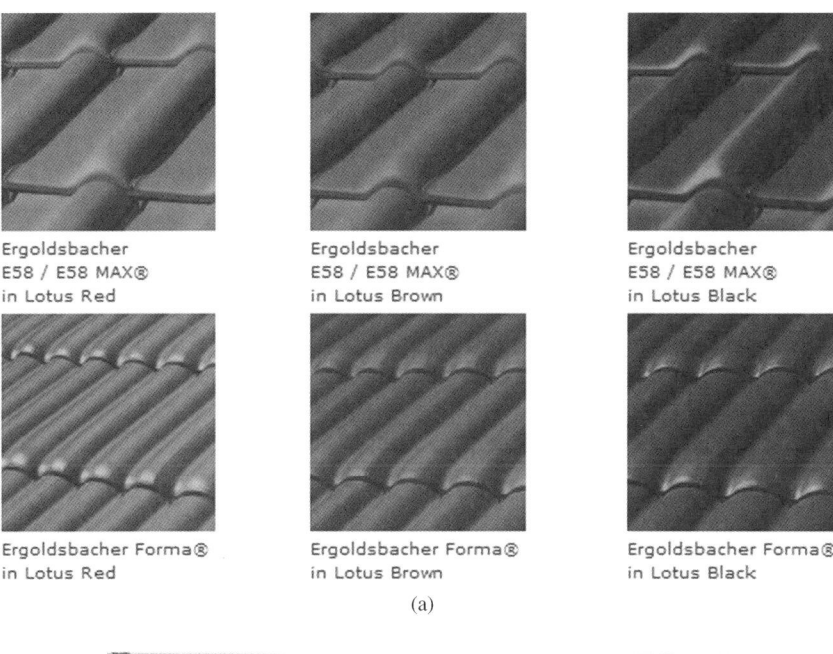

Figure 5.14 Erlus available roof-tiles (a) and an example of a roof with Erlus Lotus self-cleaning tiles (b). (Photograph courtesy of Erlus.)

5.4 New Concepts

Two relevant inconveniences when depositing a layer (commonly 0.3–1.5 cm thick) of photocatalytic concrete over a substrate are that (i) reactions and diffusion of products can take place from this layer to the substrate material,

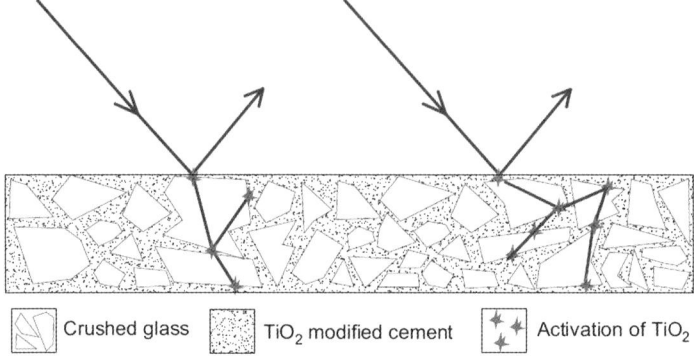

Figure 5.15 Pathways of light and activation of TiO_2 in a concrete surface layer using glass as aggregate. (Reproduced with permission from ref. 30.)

which can undergo damage, and (ii) only a small fraction of the employed TiO_2 truly works photocatalytically, since most of it is not reached by irradiation. The former problem can be solved by using a thin layer of inert silica that can work as a proper barrier against ionic and molecular compound diffusion.[29] To enlarge the active semiconductor surface and hence the amount of pollutants destroyed in a given area, we should work to make solar rays reach as much catalyst as possible.

The replacement of the high amount of sand used in the preparation of concrete with recycled glass cullets is surely an interesting way to achieve this aim.[30] Thanks to the transparency of glass, light rays can indeed penetrate through a higher thickness with respect to normal concrete, where SiO_2 particles give rise to significant light scattering. Light is thus reflected by glass and the pathway of rays is longer than in the presence of sand (Figure 5.15). Moreover, the use of recycled material is of sound added value to this composite material.

The positive influence of transparence has been confirmed by comparing performances of colored and transparent glass of similar sizes. Hence, lightly/ mildly colored glass or clear glass performs better than strongly colored ones and brown glass performs even worse than sand (Figure 5.16).

Finally, glass particle size is not an important factor, probably because compaction is necessary in concrete preparation.

TiO_2 can be efficiently supported by impregnation on porous supports such as pumice stone.[31] Such techniques allow one to adsorb as much as $50\,\mathrm{g\,m^{-2}}$ of catalyst on pumice stone pellets. The latter, being soft materials, have to be fixed on a hard matrix such as cement before its hardening. The resulting materials can easily constitute construction materials and degrade air pollutants. The porosity of the described pellets is a smart means of enhancing photoreactivity.

On one other hand, deposition of TiO_2 over glazed ceramic tiles can be achieved through screen-printing, a very cheap technique that is well established in different manufacturing fields, from textile to glass and ceramics industry. TiO_2 nanoparticles can be firstly suspended in an organic solvent and

TiO$_2$-modified Cement and Ceramics 165

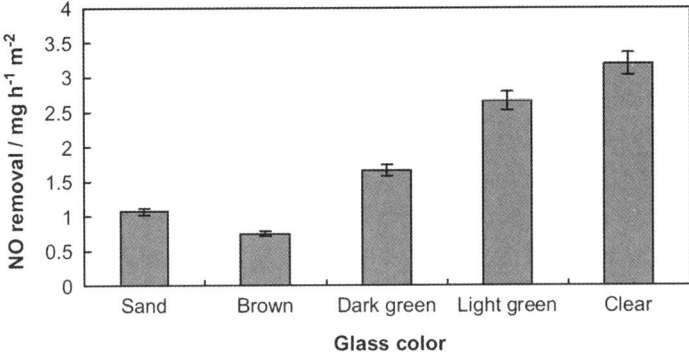

Figure 5.16 Comparison of NO removal by samples containing different colored glass and sand under the same experimental conditions. (Reproduced with permission from ref. 30.)

Figure 5.17 Eco-life-type houses using self-cleaning photocatalytic materials. (Photograph courtesy of PanaHome.)

then deposited. A final calcination at high temperatures ($>300\,°C$) imparts strong adhesion and all the properties characteristic of TiO$_2$ films. The results in terms of structure, agglomeration size, and crystalline phase strongly depend on calcination temperature.[32]

The integrated application of photocatalytic tiles together with cement and glass yields exceptional buildings, such as the Eco-life homes shown in Figure 5.17, realized by PanaHome, a housing company within the Panasonic Group.[33] These houses are designed with the following features:

- *safety and security*: highly durable structures ensuring safety with their high earthquake resistance, obtained by using special kinds of strong steel-frame structures;
- *health and comfort*: by using wind power and natural materials, the quality of air is improved, combining mechanical and natural ventilation, whereas humidity is controlled by means of natural materials absorbing and emitting moisture;

- *energy generation and conservation*: solar power generation systems are integrated in roofs or façades, and a self-cleaning effect brought about by the photocatalytic technology keeps the outer wall of the house clean, at the same time degrading air pollutants.

References

1. D. H. Chen, K. Li and R. Yuan, Photocatalytic Coating on Road Pavements/Structures for NO_x Abatement, presented to Houston Advanced Research Center, available online at http://files.harc.edu/Projects/JointCenter/Meetings/RR200512/ProjectPhotocatalyticCoating.pdf.
2. (a) A. Fujishima, K. Hashimoto and T. Watanabe, *TiO_2 Photocatalysis. Fundamentals and Applications*, BKC, Tokyo, 1999; (b) A. Fujishima, X. Zhang and D. A. Tryk, TiO_2 photocatalysis and related surface phenomena, *Surf. Sci. Rep.*, 2008, **63**, 515–582.
3. A. Fujishima and X. T. Zhang, Titanium dioxide photocatalysis: present situation and future approaches, *C. R. Chim.*, 2006, **9**, 750–760.
4. L. Cassar, A. Beeldens, N. Pimpinelli and G. L. Guerrini, *Photocatalysis, Environment and Construction Materials*, ed. P. Baglioni and L. Cassar, RILEM, 2007, p. 131, ISBN: 978-2-35158-056-1.
5. A. Beeldens, ed., *Photocatalysis, Environment and Construction Materials*, P. Baglioni and L. Cassar, RILEM, 2007, p. 187.
6. Y. Murata, H. Tawara, H. Obata and K. Murata, NO_x-cleaning paving block, *EP-patent 0 786 283 A1*, Mitsubishi Materials Corporation, Japan, 1997.
7. L. Cassar and C. Pepe, Paving tile comprising an hydraulic binder and photocatalyst particles, *EP-patent 1 600 430 A1*, Italcementi S.p.A., Italy, 1997.
8. G. Hüsken, M. Hunger and H. J. H. Brouwers, Experimental study of photocatalytic concrete products for air purification, *Build. Environ.*, 2009, **44**, 2463–2474.
9. M. Kawakami, T. Furumura and H. Tokushige, *Photocatalysis, Environment and Construction Materials*, ed. P. Baglioni and L. Cassar, RILEM, 2007, p. 163.
10. M. Crispino and S. Lambrugo, *Photocatalysis, Environment and Construction Materials*, ed. P. Baglioni and L. Cassar, RILEM, 2007, p. 211.
11. Details of TX Active cement are presented (in Italian) online at http://www.italcementigroup.com/NR/rdonlyres/5A1BE31D-49DE-4C7C-9B95-B49C8DE9F1A7/0/Comunicato_TXActive_EN.pdf
12. Report of Committee on Colorimetry, *J. Opt. Soc. Am.*, 1944, **34**, 633.
13. G. L. Guerrini, A. Plassais, C. Pepe and L. Cassar, *Photocatalysis, Environment and Construction Materials*, ed. P. Baglioni and L. Cassar, RILEM, 2007, p. 219.
14. TX Aria® and TX Arca® are presented at http://txactive.us/product.html
15. ESSROC Cement: http://www.essroc.com/
16. Kon Corporation products are described online at http://www.saga-kon.co.jp/english/product.html

17. R. Amadelli, L. Cassar and C. Pepe, Use of photocatalytic preparations of colloidal titanium dioxide for preserving the original appearance of cementitious, stone, or marble products, *US Patent 6824826*, Italcementi S.p.A., Italy, 2004.
18. A*STAR press release at http://www.nanotech-now.com/news.cgi?story_id = 31500
19. Information about Global Engineering products is available online at http://www.globalengineering.info/index.php?idx = prodotti
20. Toto Ltd. patents: http://www.toto.co.jp/docs/hyd_patent_en/case_001.htm
21. A products brochure is available online at http://www.deutsche-steinzeug.de/en/hydrotect_new/grafik/Hydrotect_GB.pdf
22. See, for example, R. Wang, K. Hashimoto, A. Fujishima, M. Chikuni, E. Kojima, A. Kitamura, M. Shimohigoshi and T. Watanabe, Photogeneration of highly amphiphilic TiO_2 surfaces, *Adv. Mater.*, 1998, **10**, 135.
23. J. Chen and C. Poon, Photocatalytic construction and building materials: from fundamentals to applications, *Build. Environ.*, 2009, **44**, 1899–1906.
24. D. M. Tobaldi, A. Tucci, G. Camera-Roda, G. Baldi and L. Esposito, Photocatalytic activity for exposed building materials, *J. Eur. Ceram. Soc.*, 2008, **28**, 2645–2652.
25. M. Shimohigoshi and Y. Saeki, *Photocatalysis, Environment and Construction Materials*, ed. P. Baglioni and L. Cassar, RILEM, 2007, p. 291.
26. Information about the Oxygena brand can be found online at http://www.gambarelli.it/catalogue/pdf/Catalogo_Tecnico_Oxygena.pdf
27. Information about the Dahlia Design® brand can be found online at http://www.asiadynasty-dahlianano.com/en/index.php?option = com_content& view = article&id = 128:photocatalyticproperty&catid = 58:setsco-test-report&Itemid = 182
28. Information about the Erlus Lotus brand can be found online at http://www.erlus.de/rooftilemodels/lotus/
29. T. Yuranova, V. Sarria, W. Jardim, J. Rengifo, C. Pulgarin, G. Trabesinger and J. Kiwi, Photocatalytic discoloration of organic compounds on outdoor building cement panels modified by photoactive coatings, *J. Photochem. Photobiol. A: Chem.*, 2007, **188**, 334–341.
30. J. Chen and C.-S. Poon, Photocatalytic activity of titanium dioxide modified concrete materials – influence of utilizing recycled glass cullets as aggregates, *J. Environ. Manage.*, 2009, **90**, 3436–3442.
31. K. V. Subba Rao, A. Rachel, M. Subrahmanyam and P. Boule, Immobilization of TiO_2 on pumice stone for the photocatalytic degradation of dyes and dye industry pollutants, *Appl. Catal. B: Environ.*, 2003, **46**, 77–85.
32. P. S. Marcos, J. Marto, T. Trindade and J. A. Labrincha, Screen-printing of TiO_2 photocatalytic layers on glazed ceramic tiles, *J. Photochem. Photobiol. A: Chem.*, 2008, **197**, 125–131.
33. Information about Eco-life homes can be found online at http://www.panahome.jp/english/ecolife/index.html

CHAPTER 6
TiO_2 on Plastic, Textile, Metal and Paper

6.1 TiO_2 Supported on Plastic Materials

The preparation of photocatalytic polymers is an attractive field, in which some results have already been obtained, with a few commercial materials available on the market. The application of such light, sometimes flexible and cheap products is very wide and ranges from household appliances, to the automobile industry, soundproof road barriers, tents for outdoor applications (*e.g.*, gazebos) to various indoor functions. Figure 6.1 shows a sample present in the "Photocatalyst Museum" at the Kanagawa Academy of Science and Technology. The tent material, produced by Taiyo Kogyo, is made of a PVC film coated with TiO_2 photocatalyst on the left-hand half. The pictures were taken in 2004 and 2007 and it is clear how outdoor exposure to smog has dirtied the right-hand uncovered area.[1]

The methods used to obtain such materials are several and varied.[2] The first point that should be stressed is the huge chemical difference between titanium dioxide and polymeric materials. The former is an inorganic mineral, organized in a crystalline structure, whereas polymers are mainly made of organic monomer units with a complex organization and vary one from another. Needless to say, the physicochemical affinity between TiO_2 and (generally speaking) plastics is usually very low. This is why supporting TiO_2 on polymeric materials is challenging; another primary reason resides in the impracticality of warming at high temperature to afford titanium dioxide crystallization (if the deposited film is an amorphous film) or particle sintering (if powdered TiO_2 is supported) and thus a good adhesion to the substrate. Polymers are thermolabile and cannot be treated at temperatures above 200 °C, otherwise their structure undergoes irreversible modifications giving rise to poor mechanical and physicochemical properties.

Introducing TiO_2 crystalline particles during a polymer synthesis is not a good way of obtaining a photocatalytic material, for different reasons: on one hand we should use significant percentages of inorganic material, thus

---

Clean by Light Irradiation: Practical Applications of Supported TiO_2
By Vincenzo Augugliaro, Vittorio Loddo, Mario Pagliaro, Giovanni Palmisano and Leonardo Palmisano
© V. Augugliaro, V. Loddo, M. Pagliaro, G. Palmisano and L. Palmisano 2010
Published by the Royal Society of Chemistry, www.rsc.org

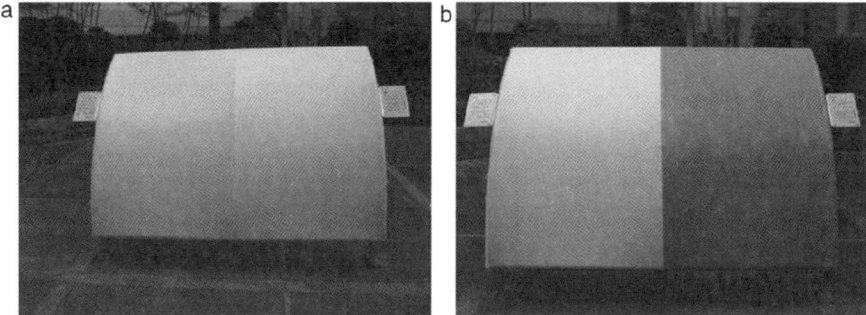

Figure 6.1 PVC material manufactured by Taiyo Kogyo and kept outdoors. The left-hand half part of the tent material was coated with TiO_2. Picture taken: (a) 22 July 2004; (b) 23 April 2007. (Reproduced with permission from ref. 1.)

drastically affecting the final polymer characteristics; on one other hand, TiO_2 particles, even if surrounded by a thin polymeric films, do not come into contact with target pollutants present in air or water and no redox reaction occurs. Another effect that should not be neglected is the thermal catalytic activity of TiO_2, which can radically affect the reaction of monomers if the polymerization is carried out at certain temperatures.

It is, however, possible to spray-coat room temperature curable resins with TiO_2 particles to obtain a photocatalytic external layer. For instance, Smooth-On Crystal Clear® 202, a low viscosity urethane resin that cures at room temperature with negligible shrinkage, can be functionalized by spraying with Degussa P25 particles during resin curing.[3] Samples had to be washed because a significant amount of excess powdered TiO_2 remained on the resin surface. SEM images revealed that the TiO_2 layer was quite thick, up to 70 µm, and it depended on the application moment with respect to curing development. The described coating method works mechanically rather than through chemical bonding, and thus requires a certain degree of TiO_2 particle encapsulation. By tailoring experimental conditions it should be possible to avoid encapsulating all of the TiO_2, which would prevent it participating in reactions with pollutants in the reaction medium (either air or liquid phases). These materials have been tested for propene gas phase oxidation under UV irradiation, performed with a medium-pressure mercury lamp. Acetone, CO, H_2O and CO_2 were detected as oxidation products. Even though this process has been applied to room curable polymer resins, it could be similarly extended to other polymeric materials. A key factor influencing the properties of the final material is the time at which the coating is applied during the resin curing process, which eventually determines the thickness (Figure 6.2) and quality of the TiO_2 layer.

If we choose to support a film by, for example, means of a sol-gel method, this implies that the deposited titanium dioxide film has to be either crystalline or warmed at 400 °C to obtain a crystalline photoactive structure, and the same temperature should be reached if we use titanium dioxide crystalline particles

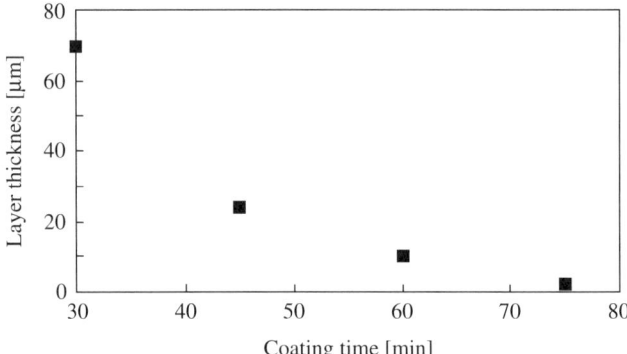

Figure 6.2 Thickness of TiO_2 film spray-coated on Smooth-On Crystal Clear® 202 resin with respect to coating time. Time zero is represented by the beginning of resin curing. (Reproduced with permission from ref. 3.)

along with an organic gluing template, in order to get rid of the organic species and to obtain, mainly, particle sintering. Nevertheless, the hydrolysis of titanium alkoxides, carried out at moderate temperatures (40–100 °C), can yield a crystalline anatase structure without the need for calcination. For instance, acrylonitrile–butadiene–styrene polymer (ABS) and polystyrene (PS) substrates have been covered by means of a TiO_2 transparent nano-sol.[4] The latter was obtained by hydrolysis of titanium tetraisopropoxide in isopropanol, acetylacetone and water. The eventual addition of nitric acid and warming at 80 °C made the hydrolysis complete, modifying the complexing power of the inhibiting ligand (acetylacetone) and providing some thermal energy favoring the structural reorganizations needed for crystallization of TiO_2 nanoparticles. The final transparent sol (applied by dip-coating) showed different transmittance spectra depending on TiO_2 amount (ranging between 0.3 and 3 wt%) and acetylacetone/Ti ratio (Figure 6.3).

Film quality is strongly influenced by some parameters, such as the number of coating cycles. By increasing this number from 5 to 10, the surface starts to crack and deep pores appear with increasing film thickness. This behavior can be ascribed to the large mass change during volatilization of solvent and acetylacetone. Similarly, the increase of TiO_2 content produces severe cracks and pores on the surface of the films. The adhesion of films was tested in various ways and found to be satisfactory. Indeed, even treatments with organic solvents did not give rise to film peeling. The photocatalytic activity of these films was evaluated by decomposing Methylene Blue.

Notably, methods involving the hydrolysis of a titanium alkoxide to yield a transparent sol with a certain degree of crystallization hardly ever permit TiO_2 amounts higher than 10%.

Polycarbonate and poly(methyl methacrylate) substrates have also been functionalized by spin-coating of a sol derived from titanium tetraisopropoxide

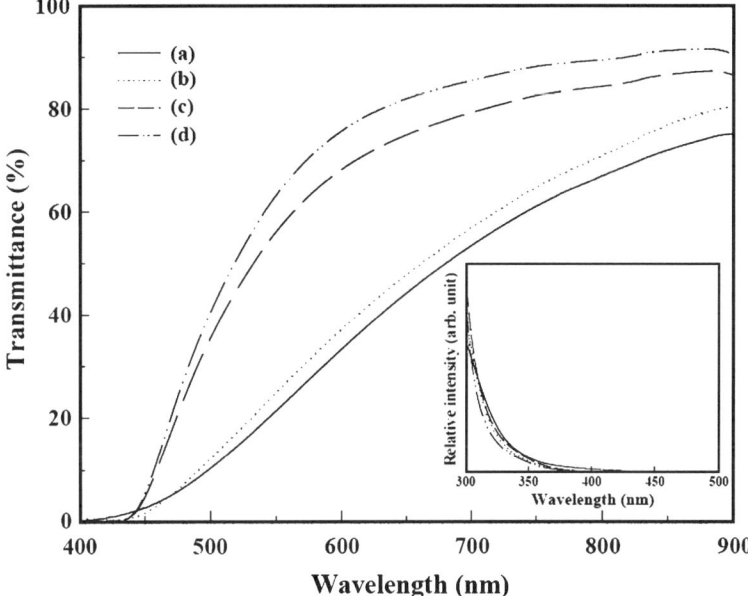

Figure 6.3 Variation of UV–Vis spectra of TiO_2 hybrid sols with complexing ratio (acetylacetone/Ti = x); $x =$ (a) 0, (b) 1, (c) 2 and (d) 3. Inset: corresponding optical absorption spectra. (Reproduced with permission from ref. 4.)

hydrolysis in ethanol, water and hydrochloric acid.[5] Once deposited, the films were heat-treated in an autoclave placed inside a tubular furnace at 90–140 °C. This treatment was carried out in the presence of ethanol–water vapors. Water/ethanol ratios (r_{we}) between 0 and 100 were tested. Anatase crystals (*ca.* 10 nm) were obtained along with traces of brookite, detected by means of selected area electron diffraction patterns, and, notably, autoclaving time strongly influences both refractive index and thickness (Figure 6.4). As crystallization proceeds, the refractive index was observed to decrease. This behavior is related to a continuous increase in porosity. Moreover, it should be underlined that the refractive index value of autoclaved films was very low compared to the value measured on films heat-treated in air.

The photocatalytic activity of these films was studied through malic acid photodecomposition tests. All the autoclaved films showed a constant rate of malic acid disappearance over a period of 3 h, thus indicating apparent zero-order kinetics. To obtain a proper indication of photoactivity, films autoclaved for 5 h in the range 90–140 °C were compared with others of similar thickness that had been heat-treated in air for 1 h at 300 and 500 °C by using a silicon substrate (instead of a polymeric one). The anatase crystallized film heat-treated at 500 °C transformed about 30% of the malic acid after 3 h UV exposure. With respect to this film, the photocatalytic yield of autoclaved films was about 50%.

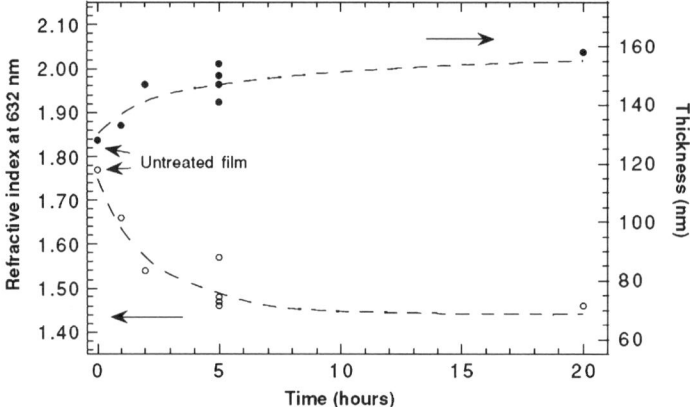

Figure 6.4 Thickness and refractive index *versus* autoclaving time for TiO$_2$ films spin-coated as described in the text. Untreated film accounts for a film dried only at room temperature for 24 h. The drawn lines simply show the trends and do not represent any studied model. (Reproduced with permission from ref. 5.)

The low affinity between the polymeric substrate and TiO$_2$ can be moderated by implementing chemical treatments to perform a surface functionalization of the polymeric interface or by choosing a polymer with appropriate functional moieties. For instance, sulfonic groups of a Nafion film were reported to enhance TiO$_2$ anchoring thanks to the mutual electrostatic interaction.[6] Thus, the resulting thin film photocatalyst showed stable performance during long-term operation, without TiO$_2$ leaching from the polymer surface, providing acceptable kinetics during the photocatalytic degradation of an azo dye.

TiO$_2$ can also be bound on inert thin polymer films without charged groups (such as sulfonic, carboxylic or phosphonic) able to interact electrostatically with the TiO$_2$.[7] Thus, heating Tedlar®, Parylene or low density polyethylene films, inside an alcoholic suspension of TiO$_2$ particles, introduces oxidative binding sites on the polymer surface that properly anchor TiO$_2$. The resulting materials are shown in Figure 6.5 in which Degussa P25 TiO$_2$ nanoparticles are clearly visible. Notably, the layer of TiO$_2$ particles on the polymeric film after *ca.* 6 h reaction was found to be quite stable, thus explaining the long-term activity of the materials under irradiation. Photocatalytic degradation of two dyes (Methyl Orange and Orange II) was tested in aqueous solution irradiated by simulated solar light. Significant differences were found, depending on the polymeric substrate used; the kinetics of the Tedlar/TiO$_2$ and Parylene/TiO$_2$ photocatalyst for azo-dyes degradation were about twice that of polyethylene/TiO$_2$. These materials were able to resist highly oxidizing hydroxyl radicals and to give rise to photocatalytic reactions. However, many aspects of either physical or chemical binding have still to be investigated.

Polycarbonate substrates can be functionalized with TiO$_2$ by direct current reactive magnetron sputtering. This method has several benefits, such as

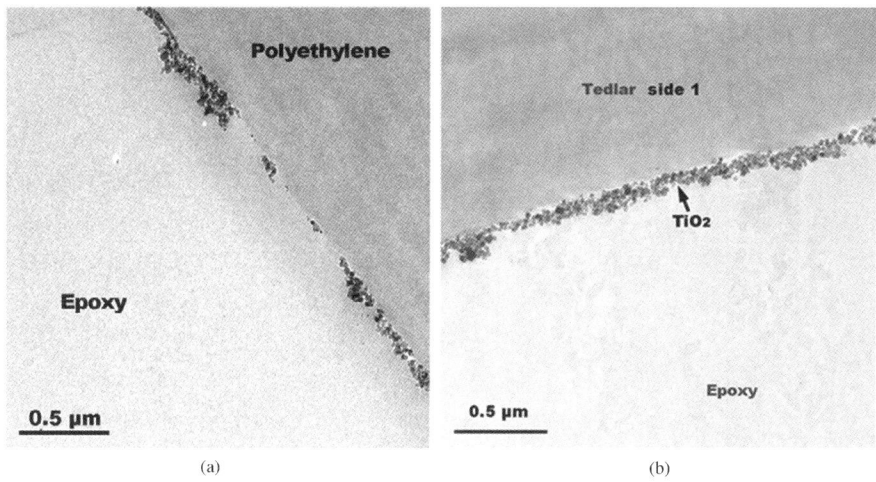

Figure 6.5 Transmission electron microscopy (TEM) of (a) a TiO_2/low density polyethylene film and (b) a TiO_2/Tedlar film before use in the photocatalytic reaction. "Epoxy" indicates a resin polymerized at 60 °C over the film before TEM analysis. (Reproduced with permission from ref. 7.)

applicability to large-areas, even in industrial productions, and can be applied at room temperature. The resulting films are poorly crystalline, but they are photocatalytically active. Polycarbonate is one of the most used thermoplastics, with applications mainly in the construction and automotive industry due to its excellent impact resistance, low weight and transparency. Examples of objects where polycarbonate can be found are building roofs and windows, car roof elements, automotive headlamps and ophthalmic lenses. As far as the application method is concerned, total sputtering pressure has a prominent effect on the film microstructure, light absorption and photocatalytic activity of the resulting TiO_2.[8] The doping effect of iron has been studied, along with the influence of substrate (polycarbonate *versus* glass). UV–Vis transmittance (Figure 6.6) gives useful information when comparing different doping amounts and substrates. In Figure 6.6 samples B are made of pure TiO_2, samples C are moderately iron-doped films (*ca.* 0.6% Fe) and samples D are highly doped films (*ca.* 2.5% Fe). Absorption edges of Fe-doped TiO_2 films are clearly shifted to the visible region with increasing iron concentration, indicating the greater light absorbance of Fe-doped films. Highly doped films (sample D) show the lowest transmittances (at 550 nm) for both glass and polycarbonate substrates. Scattering loss instead is higher for iron-doped samples, probably because of their high roughness. This characteristic could also be responsible for the low transmittances when compared with pure TiO_2 films.

Rhodamine B was degraded by using all samples as the catalysts under UV irradiation. By using both substrate materials, the photocatalytic activity of

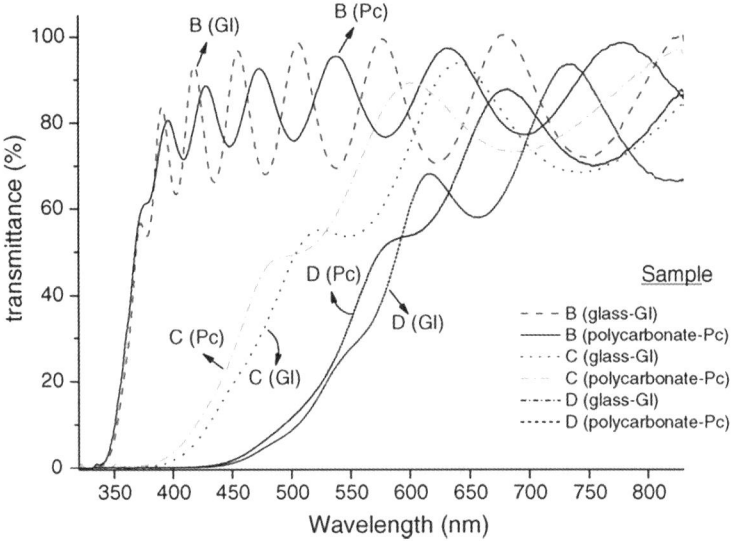

Figure 6.6 UV–Vis transmittance of DC magnetron sputtering deposited TiO_2 films on polycarbonate and glass. Sample B indicates pure TiO_2; samples C and D have increasing iron content. (Reproduced with permission from ref. 8.)

pure TiO_2 films deposited at low sputtering pressure was prominent. Such films exhibit small grain size, indicating that the lifetime of free electron–hole pairs are particle-size dependent. Hence the number of free charges on the TiO_2 surface should be reduced, resulting in improved photocatalytic activity. Moderate concentrations of iron dopant yield a slight increase in photocatalytic activity with respect to pure TiO_2 films deposited under the same experimental conditions. Conversely, a high iron concentration is detrimental for photocatalytic efficiency. Finally, comparison between the photocatalytic activity of TiO_2 films deposited on polycarbonate and glass substrates shows that the polymeric material is more active. Hence, polycarbonate appears a promising material in terms of its photocatalytic properties, with a view to industrial application as a low-weight "self-cleaning" material.

As mentioned briefly already, a key aim when preparing a photocatalytic polymer is to retain its mechanical properties. Films deposited through room temperature magnetron sputtering on polycarbonate have been tested by applying a test velocity of $dl/dt = 0.2\,\text{mm min}^{-1}$ on the sample strips. Similarly, some samples were prepared for the photocatalytic experiments after strain deformations of 0.5%, 1% and 5%.[9] Figure 6.7 shows typical stress–strain curves for polycarbonate substrates. Young's modulus for polycarbonate substrate was measured as $\sim 8.4\,\text{MPa}$ and for the composites materials 8.9–9.5 MPa. Hence, the mechanical performance of the composite materials improve, because the Young's modulus of TiO_2 is rather high if compared to

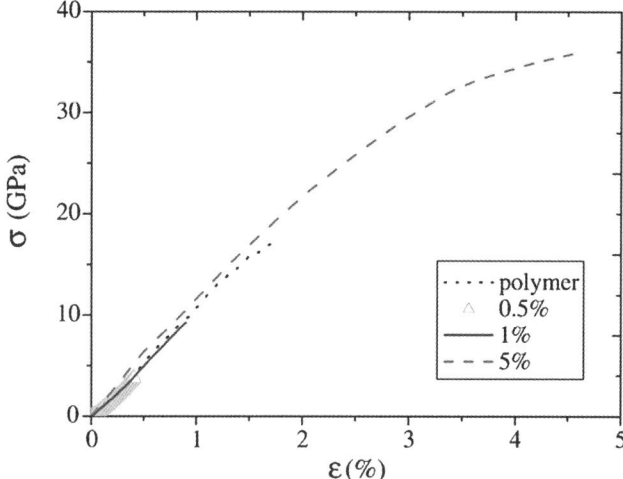

Figure 6.7 Stress–strain curves for polycarbonate samples nude and functionalized by TiO$_2$. The different curves correspond to the maximum strain deformation applied to each of the samples for the photocatalytic experiments. (Reproduced with permission from ref. 9.)

polycarbonate one. Its value ranges in fact between 0.7 and 1.8 GPa for deposited films that are *ca.* 600 nm thick and poorly crystalline.

Variations in microstructure with the applied axial strain were observed by an optical microscope connected to a CCD camera (Figure 6.8). In all cases visible cracks only appeared for samples deformed at strains larger that 0.75%. Samples prepared with oxygen flow rates before the threshold conditions (A-type) had a better mechanical performance than those prepared with a composition slightly above it (B-type). In the A-type samples the cracks appeared, indeed, at quite larger deformation ($\sim$1.5%) than for B-type samples ($\sim$0.76%).

Interestingly, in the degradation of Rhodamine B the most deformed samples showed the highest degradation rates. In fact, as the material is deformed and cracks propagate it exposes more surface area than previously and, thus, the UV-generated electron–hole pairs have more sites at which to reduce/oxidize the pollutant.

A multilayer spin-coating procedure has been applied to deposit TiO$_2$ on thermoplastics such as high-density polyethylene (HDPE) and poly(vinyl chloride) (PVC).[10] These polymers are among the most widespread worldwide: HDPE is processed, by blow molding or injection molding, for instance, into bottles, toys, food containers and plastic bags, while PVC finds use in buildings and construction (>50% of all PVC), automotive industry, and medical devices. The prominent resistance to degradation of PVC is the reason why it is often preferred in long-life applications such as outdoor tubes and floor coverings.

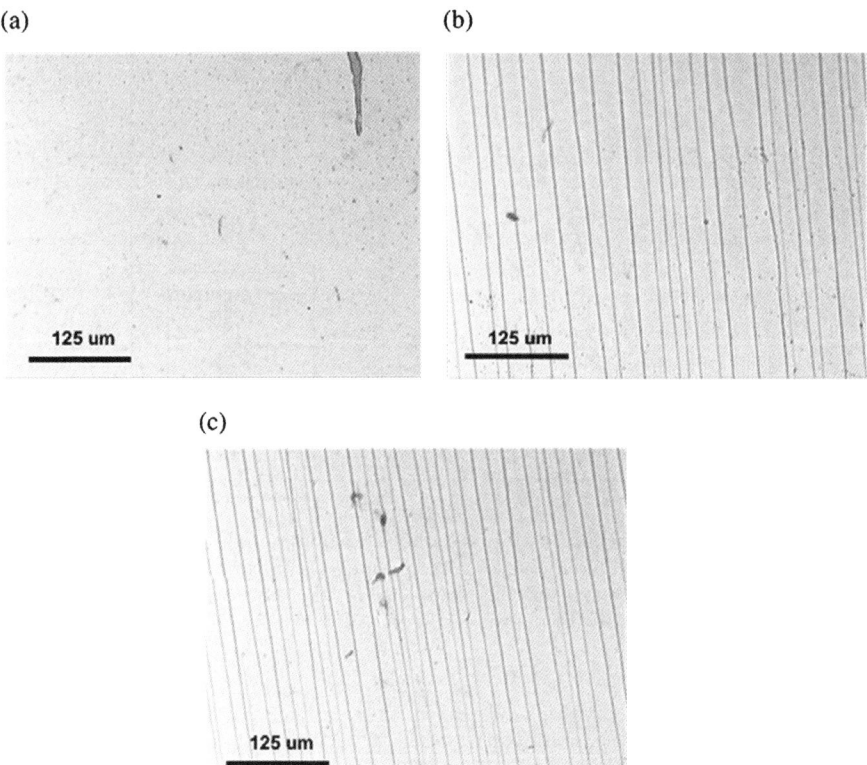

Figure 6.8 Images of surfaces of an A-type sample for 0.5% (a) and 5% (b) strain deformations, and for 5% strain deformations in a B-type sample (c). (Reproduced with permission from ref. 9.)

The multilayer structure consisted of a polyurethane protective layer over the substrate and two layers of TiO_2 on the protective layer, deposited by spin-coating of a Degussa P25 suspension. Finally, immobilized TiO_2 particles bound in a diluted polyurethane dispersion were applied to fix the underlying particles and improve the mechanical resistance of the TiO_2 layer. To obtain photocatalytically active materials reactive oxygen-plasma surface etching of the fabricated coatings was performed. In the absence of this final treatment, the reactivity of such films was either very low or absent. This treatment, kept quite short (15 s), gave rise to etching of only polyurethane binder on the outer surface in order to expose a higher amount of TiO_2 particles to the reaction ambient.

The so-obtained samples (Figure 6.9) were characterized by SEM. It can be seen that the plasma treatment promotes a high exposure of TiO_2 particles and a prominent porosity of sample surfaces. Extending the plasma treatment over a certain time does not result in a higher photocatalytic degradation rate, probably because light can only penetrate through a limited film thickness.

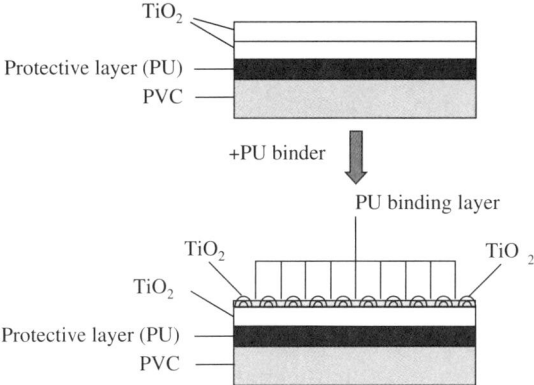

Figure 6.9 Scheme of multilayer films deposited on PVC (PU stands for polyurethane).

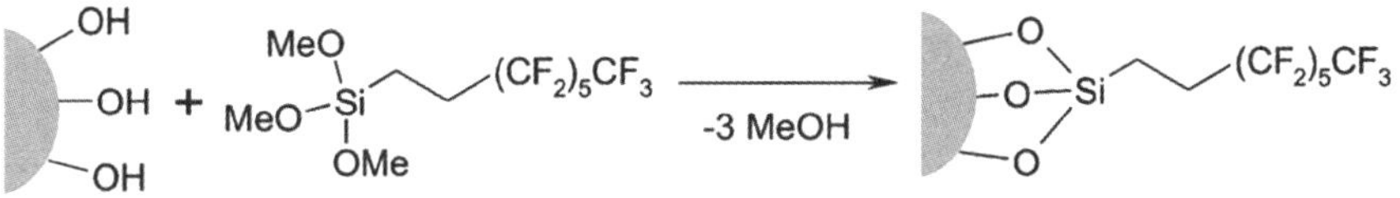

Figure 6.10 Chemical reaction leading to surface modification of nanoparticulate titanium dioxide. (Reproduced with permission from ref. 11.)

Samples were tested in the degradation of palmitic acid in the gas phase under UV irradiation. HDPE samples exhibited a higher activity than PVC ones, which can be related to photocatalytic inhibition caused by PVC plasticizers.

Nanostructured titanium dioxide, modified on its surface with silanes containing organic or fluoro-organic side chains, constitutes a smart transparent coating applicable on the surface of plastics.[11]

The coating can be obtained starting from a TiO_2 anatase sol (obtained through the hydrolysis of titanium tetraisopropoxide by a conventional method); further drying of the powder by rotovapor, dispersion in toluene, followed by mixing with a fluorosilicate and centrifugation to separate the particle agglomerates affords a fluorosilane-modified TiO_2 with moderate polarity (Figure 6.10). These particles are then added to a binding sol (made of pre-hydrolyzed methyl-silica) in methyl ethyl ketone. The coating can be applied on PVC through spray, dip, spin or flow coating. The best performing thickness of the resulting film was 1–2 μm. It should be underlined that the transparency of the coating is guaranteed only if the particle size of the anatase is significantly below 20 nm (since TiO_2 has a refractive index of *ca.* 2.5). Accordingly the so-prepared sol affords particle sizes of 5–10 nm.

Interestingly, while coating-sols containing modified TiO_2 nanoparticles look completely homogenous, after application of the coating sol and subsequent evaporation of the solvents from a wet film a gradual change of the

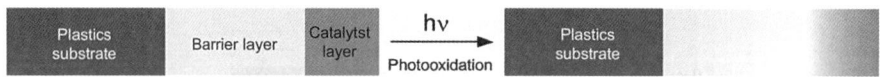

Figure 6.11 Modified TiO$_2$ nanocoating before and after activation. (Adapted with permission from ref. 11.)

balance between polar and nonpolar compounds in the film take place that, finally, leads to a thermodynamically driven diffusion of TiO$_2$ nanoparticles towards the interface between coating and air.

During the first hours of UV irradiation, film activation takes place, in which the organic side chains at the surface of the TiO$_2$ particles are oxidized, revealing the active form of the particles. During activation the contact angle is reduced from 90–100° to below 10°, *i.e.*, to superhydrophilic behavior. This leads to the formation of a SiO$_2$ protective layer between the polymeric substrate and the catalyst (Figure 6.11).

6.2 Photocatalytic Textiles

Having clothes coated with a thin transparent layer of TiO$_2$ can give major added value, by decomposing the organic matter deposited on the textiles. The smell of smoke on smokers' clothes can also be eliminated, along with pathogens, such as bacteria, that under normal conditions can survive on textile surfaces for up to three months. Self-cleaning textiles could be generally applied also for sportswear, military uniforms and carpets (indoors and outdoors).[12]

Coating cotton is quite easy, whereas fibers like wool, silk and hemp have proved less amenable. These fibers, indeed, are made of a protein called keratin, which does not have any reactive chemical groups on its surface to bind with TiO$_2$; consequently, their surface should be modified to establish a chemical bond with TiO$_2$.

Thus, a dispersion of Degussa P25 particles and polyglycol in water can be padded on cotton woven fabrics, by using a reactive amino-silicone additive for padding TiO$_2$ particles on the fabrics.[13] The cotton has to be previously treated in a boiling bath containing sodium carbonate and soap. Afterwards the fabric samples were padded twice with a treatment solution on a laboratory padding mangle.

Another way to support TiO$_2$ on cotton is a simple coating process. In this way mixing TiO$_2$ with self-crosslinking acrylic binders yields a coating formulation that can be applied after treating the textile with a water repellent agent.

Figure 6.12 shows SEM images of untreated (a), TiO$_2$-padded (b) and TiO$_2$-coated cotton (c) fabrics. The structure of cotton fiber is a typical round and smooth surface. The morphological change in the appearance of cotton fiber after TiO$_2$-padding can be clearly seen in Figure 6.12(b). The cotton fiber surface is covered with TiO$_2$ particles, becoming rough and uneven. The loading of TiO$_2$ on cotton fibers was, thereby, proved, although their

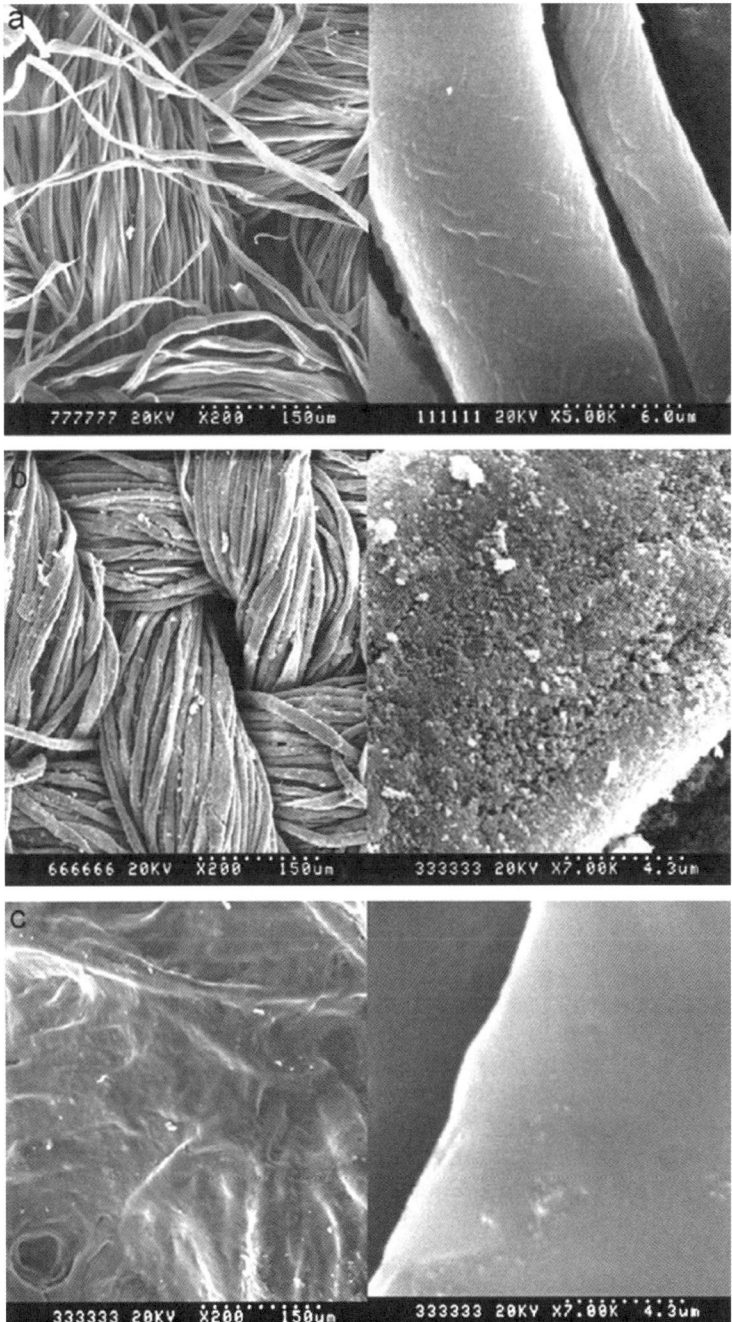

Figure 6.12 SEM images of untreated (a), TiO_2-padded (b) and TiO_2-coated (c) cotton fabrics. (Reproduced with permission from ref. 13.)

distribution on the fiber surface was not quite even, probably owing to aggregation of some fine TiO_2 particles. The morphological features of TiO_2-coated cotton fabric surface differ greatly from those of the TiO_2-padded cotton fabric, with few TiO_2 particles since they probably aggregate with the binder layer.

Gaseous ammonia decomposition was proven to take place by using these materials. Both padded and coated fibers were active, but in the former case the TiO_2 loading had a primary influence on reaction rate, probably because of the much greater exposed surface than in the coated material. Hence, the padded textile was *ca.* 2.5-fold more effective than the coated one, in terms of kinetic constants. On the other hand, the TiO_2-coated cotton fabric showed relatively strong adsorption of ammonia molecules due to the presence of the acrylic binder layer partially covering TiO_2 particles on the surface.

Sol-gel methods at low temperature ($<100\,°C$) can be surely applied in the preparation of TiO_2-modified cotton.[14] Cellulose fibers, previously cleaned with acetone, can be covered by using a titanium isopropoxide (TIP) sol, prepared through mixing with isopropanol and triethylamine as stabilizer of the solution. The addition of an acidic aqueous solution then yields a TiO_2 transparent sol, with which the cellulose fibers can be impregnated. After drying, a boiling treatment in water can remove all the unattached TiO_2 particles from the surface of the fibers.

SEM images of such materials (Figure 6.13) show how the folds running parallel to the elongation direction, present in the pure fibers, are covered by a continuous and homogeneous TiO_2 thin film, which obscures the surface folds below. The formation of some aggregates is also clearly visible. The presence of a badly crystallized anatase phase was shown by X-ray diffraction.

Accordingly these TiO_2-covered cellulose fibers show high photocatalytic efficiency in decomposing adsorbed Methylene Blue and heptane-extracted bitumen fraction under solar-like light. Furthermore, the TiO_2 film, although highly efficient in pollutant degradation, does not promote simultaneous fiber degradation.

A metal doping agent, such as gold, can be introduced in the above preparation by soaking in a $HAuCl_4$ aqueous solution with subsequent irradiation to promote reduction of Au cations, thus yielding Au nanoparticles on a TiO_2 film.[14b] The resulting cotton fibers are, hence, covered by a thin Au/TiO_2 film consisting of anatase nanocrystallites that strongly adhere to the support, with a purple color and a significantly improved photocatalytic activity under solar light with respect to undoped samples.

Fibrous proteins, such as keratins, which are also the main structural constituents of animal tissues, are present for instance in wool, silk, hemp and spider silk. Self-cleaning keratin fibers can be realized following a bottom-up nanotechnology approach in which anatase nanocrystals of TiO_2 are prepared and carefully applied to the fibers *via* a low temperature sol-gel process to maintain their intrinsic properties, which would be drastically affected by warming at high temperature. The sol-gel process consists of a procedure similar to the previously described ones, based on titanium tetraisopropoxide

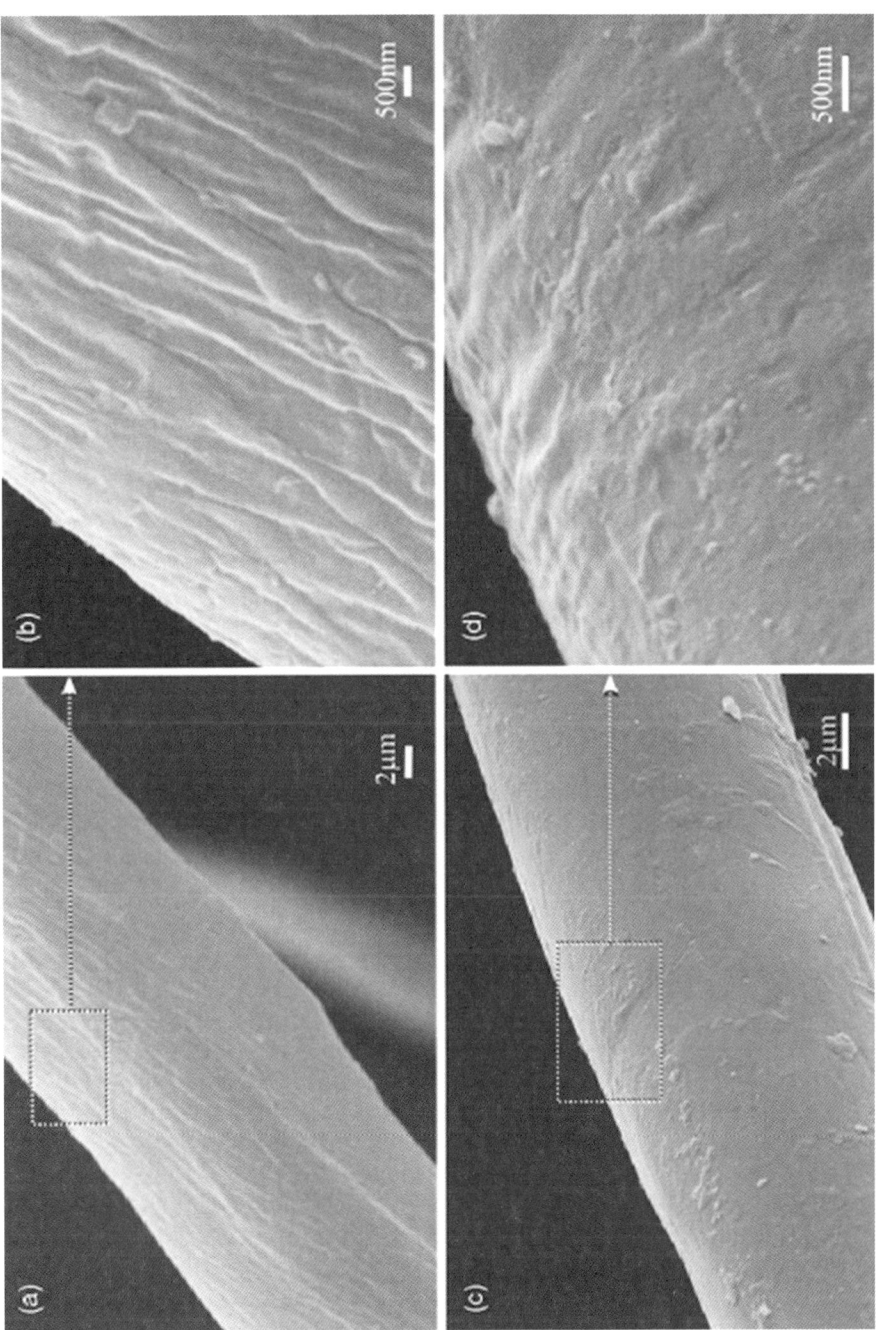

Figure 6.13 SEM images of pure cellulose fiber (a, b), and TiO$_2$-impregnated fiber (c, d). (Reproduced with permission from ref. 14a.)

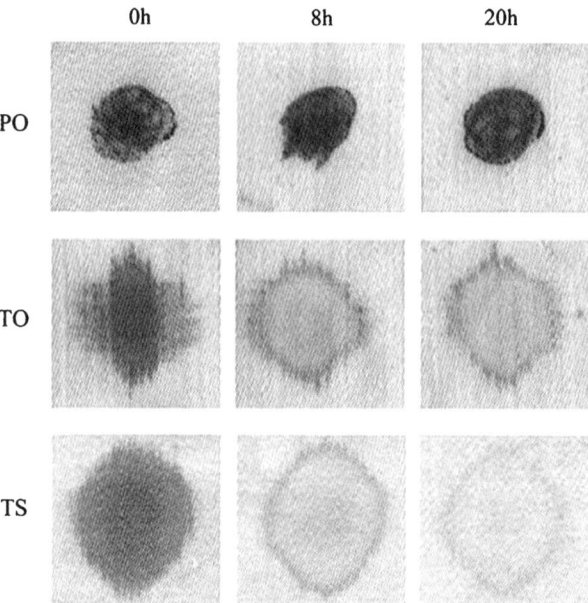

Figure 6.14 Red wine stain degradation on pristine keratin fibers (PO) and on two TiO$_2$-modified fibers (TO and TS) under UV irradiation. (Reproduced with permission from ref. 15.)

hydrolysis. The functionalization confers both self-cleaning properties and self-protection against UV degradation.[15] Exposure of such proteins to UV-light indeed would normally result in a major change in their physicochemical and hence mechanical properties. The photocatalytic self-cleaning properties of the prepared samples were investigated by the decomposition of Methylene Blue and degradation of food stains. Figure 6.14 shows the degradation of red wine stains under UV irradiation in the case of pristine and TiO$_2$-modified fibers. TS in Figure 6.14 indicates a sample treated by succinylation before deposition of TiO$_2$. This treatment allows for an increased presence of TiO$_2$ on the surface and hence a higher photocatalytic activity, since succinic anhydride can introduce additional carboxylic groups by acylation, resulting in enhanced bonding between TiO$_2$ and the fibers (Figure 6.15). Interestingly, the described method allows us to obtain many self-cleaning textiles, such as wool.

Electrospun polymeric nanofibers with high photocatalytic activity can be prepared by absorption of negatively charged colloidal TiO$_2$ nanoparticles applying a layer-by-layer (LbL) deposition with positively charged polyhedral oligosilsesquioxane (POSS) molecules. Application fields of this method are protective clothing systems, photocatalysis, sensors and electrodes.[16] Electrospinning is a popular method for obtaining ultrafine fibers with micrometer diameters from various polymer solutions or melts, whereas the LbL assembly process involves sequential adsorption of oppositely charged species, resulting

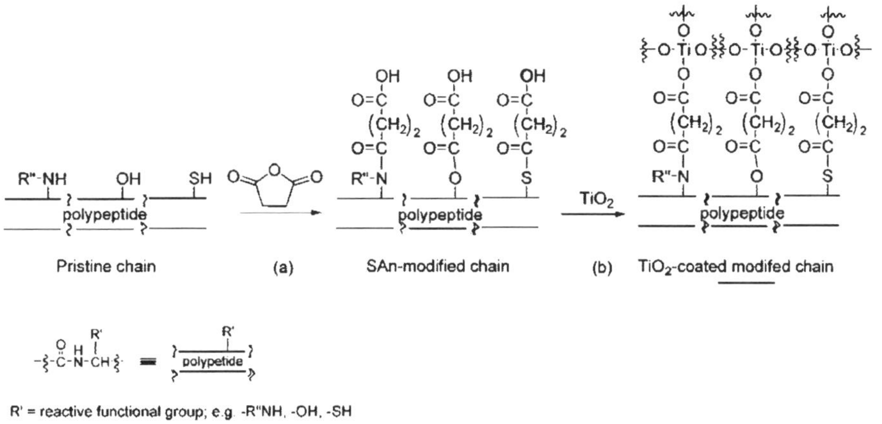

Figure 6.15 Succinylation of keratins and the binding action of TiO$_2$ to the modified agent. (Reproduced with permission from ref. 15.)

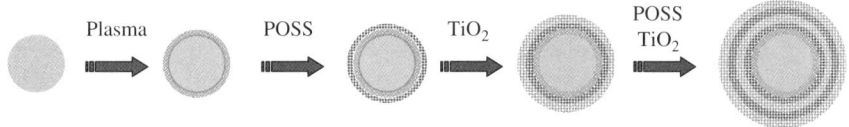

Figure 6.16 Scheme of electrospinning process followed by LbL TiO$_2$ deposition.

in the deposition of a thin, uniform film. Electrospinning of nanofibers from polystyrene, polyacrylonitrile, blends of poly(methyl methacrylate) (PMMA) and poly(ethylene oxide) (PEO), and poly(dimethylsiloxane-*b*-ether-imide) (PSEI) can be carried out, followed by a LbL assembly, showing that the method is of general interest in depositing TiO$_2$.

Figure 6.16 shows the scheme of the electrospinning preparation method. In practice a distribution of fiber diameters is obtained (400–1300 nm), along with a random orientation of the fibers. This is typical of electrospun polymers. The layer thickness of particles on the fiber is approximately 25 nm. Photocatalytic abatement of the toxic allyl alcohol (2-propen-1-ol) in vapor phase can be carried out with satisfactory results.

Lyocell fibers, produced from wood pulp cellulose, can be functionalized by TiO$_2$ through two different methods.[17] TiO$_2$ nanoparticles can be (i) used and bound by a SiO$_2$ nanocoating on the fiber surface or (ii) produced directly on the fiber surface by a sol-gel procedure similar to that previously analyzed. Red beet sap and red wine were used to test the self-cleaning properties of the resulting materials. Samples prepared in both ways demonstrated photocatalytic activity, even though a higher degradation of organics was observed when using TiO$_2$–SiO$_2$ nanocoating compared to TiO$_2$ coating. Clearly, the longer the treated fabric was exposed to daylight the more intensive was the change in the stain. Table 6.1 shows that the most relevant changes are those corresponding to bending, compression and surface properties, whilst the changes in tensile and shear properties are smaller. The coatings on the fibers moreover reduce the shear rigidity (indicated by G) slightly. In addition, a greater decrease in hysteresis of the shear force at 0.5° (2HG) and 5° (2HG5) is present. The low shear hysteresis (2HG) has an advantageous influence on handle and indicates a greater softness of the fabric, whereas the value of (2HG5) influences the fabric fitting and appearance of the garment or similar 3D textile products. Generally speaking it is possible to state that nanocoating influences not only surface morphology but also the mechanical properties of the presented textiles. The greatest influence of nanocoatings results in changes in the fabrics' bending, compression and surface properties. By using a TiO_2–SiO_2 composite nanocoating an increase in fabric flexibility and soft feeling, as well as fullness and softness, is perceived, while TiO_2 nanocoating slightly reduces these hand characteristics.

6.3 Photocatalytic Paper

Among the physical properties of a paper, its visible appearance is very important, especially to the printer and to the final reader. For this reason, there have been many efforts to produce paper at high whiteness levels. Incident light can be transmitted, scattered, reflected or absorbed and all these phenomena may occur separately or in combination. A paper sheet appears completely white if it totally reflects light, and scatters diffusively at all wavelengths of the visible spectrum.

TiO_2 is usually used in paper products to improve opacity and whiteness. However, its higher cost compared to clay and calcium carbonate limits conventional its usage to high-added-value printing papers.

The field of photocatalytic paper, including the study of its related mechanisms and technologic development, is in its early stages. Few patents and research have been reported. The main applications described are the destruction of organic molecules and sterilization.[18]

A key point is the investigation of methods to fix TiO_2 to cellulose substrates to minimize photochemical damage to the paper. Another approach is based on the use of (i) zeolites, to give higher surface areas and hence enhanced mineralization, and (ii) metal-doping for enhanced photocatalytic disinfection.

The first article reporting a paper with photocatalytic properties was by Matsubara *et al.* in 1995, wherein the authors described the catalytic destruction of acetaldehyde in the vapor phase with photocatalytic paper and weak fluorescent light illumination.[19]

There are two main approaches for adding a photocatalyst to paper in conventional papermaking: *wet-end addition*, where the TiO_2 is deposited onto individual fibers before sheet formation (*i.e.*, the filtration step), and *size press treatment*, where the dry paper sheet is impregnated with a photocatalyst.[20]

Table 6.1 Low-stress mechanical properties of pristine lyocell (sample 1), modified lyocell through TiO_2 particle/SiO_2 method (sample 2), modified lyocell through TiO_2 using a sol-gel method. (Reproduced with permission from ref. 17.)

Properties	Symbols and units	Typical values	Sample 1	Sample 2		Sample 3	
Tensile properties							
Tensile energy	WT [N/m]	5–20	5.95	5.25	(−11.8%)	5.69	(−4.4%)
Tensile resilience	RT	55–70	51.98	57.70	(+11.0%)	54.60	(+5.0%)
Linearity	LT [-]	0.55–0.70	0.543	0.522	(−3.9%)	0.554	(+2.0%)
Extensibility (warp)	EM1 [%]	3–5	3.03	3.00	(−1.0%)	2.88	(−5.1%)
(weft)	EM2 [%]	More than 4	5.71	5.20	(−8.9%)	5.49	(−3.9%)
Shear properties							
Shear rigidity (warp)	G1 [N/(m-1°)]	0.6–0.9	0.37	0.34	(−8.1%)	0.37	(0.0%)
(weft)	G2[N/(m-1°)]	0.6–0.9	0.34	0.33	(−2.9%)	0.33	(−2.9%)
Shear stress at 0.5°	2HG [N/m]		0.42	0.27	(−35.7%)	0.32	(−23.8%)
Shear stress at 5°	2HG5 [N/m]	1–3	1.41	1.03	(−27.0%)	1.07	(−24.1%)
Bending properties							
Bending rigidity (warp)	B1 [x10^{-4} N-m]	0.04–0.10	0.090	0.059	(−34.4%)	0.075	(−16.6%)
(weft)	B2 [x10^{-4} N-m]	0.04–0.10	0.052	0.042	(−19.2%)	0.059	(+13.5%)
Bending moment	2HB[x10^{-2}N]	0.015–0.50	0.043	0.014	(−67.4%)	0.024	(−44.2%)
Compression properties							
Compression energy	WC [N/m]	0.1–0.5	0.135	0.135	(0.0%)	0.101	(−25.2%)
Compression resilience	RC [%]	35–60	40.85	47.29	(+15.8%)	51.37	(+25.8%)
Linearity	LC[-]		0.265	0.301	(+13.6%)	0.337	(+27.2%)
Fabric thickness at 49 N/m²	T_o [mm]		0.332	0.396	(+19.3%)	0.270	(−18.7%)
Fabric thickness at 4900 N/m²	T_m [mm]		0.130	0.216	(+66.2%)	0.150	(+15.4%)
Surface properties							
Coefficient of friction	MIU [-]	0.15–0.30	0.173	0.351	(+202.9%)	0.280	(+61.9%)
Mean deviation of MIU	MMD [-]	0.010–0.05	0.0121	0.0148	(+22.3%)	0.0157	(+29.8%)
Geometrical roughness	SMD [mm]	2–15	2.855	3.421	(+19.8%)	3.298	(+15.5%)

Though ensuring a complete distribution of TiO_2 throughout the sheet, the major challenge of wet-end addition of TiO_2 is the retention of nanometer-scale, anionic and stable colloidal TiO_2 particles in the paper sheet during the rapid filtration process on a paper machine. The implementation of high molecular weight cationic water-soluble polymers make, on the one hand, the TiO_2 particles deposit on the wood fibers before paper sheet formation, while, on the other hand, they can also cause the TiO_2 particles to flocculate with themselves and with other fillers, giving rise to the formation of large aggregates (5 µm). Owing to the covering effect of the cationic polymers the paper surfaces are likely to have a lower TiO_2 content than the inner part of the sheet. Wet-end addition can be performed though by synthesizing TiO_2 particles in the presence of fibers, heating to only 97 °C, *via* titanium tetraisopropoxide hydrolysis.[20]

In size press treatment, dry paper is impregnated with TiO_2 by passing it through a size press where the paper surface is exposed to an aqueous suspension of TiO_2 and a binder, typically starch.

The papermaker can control the extent of penetration of the size press dispersion into the paper, by varying the base paper hydrophobicity and porosity. This process allows for TiO_2 concentration near the paper surfaces; because TiO_2 is coated with a binder, the local environment around the TiO_2, along with porosity and hydrophobicity extents, can be strictly controlled and differentiated from the untreated regions of the paper sheet.[21]

The presence of TiO_2 can be responsible for major photodegradation of paper compared with TiO_2-free papers. Methods to protect cellulose have been reported and are based on the co-precipitation of TiO_2 with colloidal silica,[21a] aluminium silicates or polymeric binders.[22]

Although information on paper composition and the influence of structure on photocatalytic activity has not been completely clarified, it is known that lignin works as photosensitizer and thus could perhaps be used to enhance TiO_2 efficiency.[23]

Photoresistant non-woven paper, produced by the company Ahlstrom[21] by using TiO_2 Degussa P25 and Millennium PC500, has used to degrade diuron in aqueous solution.[24] This flexible photocatalytic paper is prepared by using colloidal silica as binder. Figure 6.17 reports the experimental set-ups used to analyze TiO_2 supported both on paper and in slurries; for comparison purposes the same amount of TiO_2 was employed in each case. The reaction mixtures were irradiated by means of a 125W UV-lamp in both cases.

The nature of the TiO_2 particles (P25 or PC500) deposited on Ahlstrom paper NW10 does not seem to influence the kinetics of the photocatalytic degradation of diuron, in contrast to results obtained with suspended powders (Figure 6.18). The efficiency of C500 deposited on paper is, however, very close to that obtained with unsupported PC500. The difference in shape between the two disappearance curves in the first tens of minutes of degradation could be caused by additional diuron adsorption in the dark on the paper and on the silica binder. Conversely, P25/NW10 presents an efficiency that is two times lower than that of P25 powder. These differences in behavior for P25 and

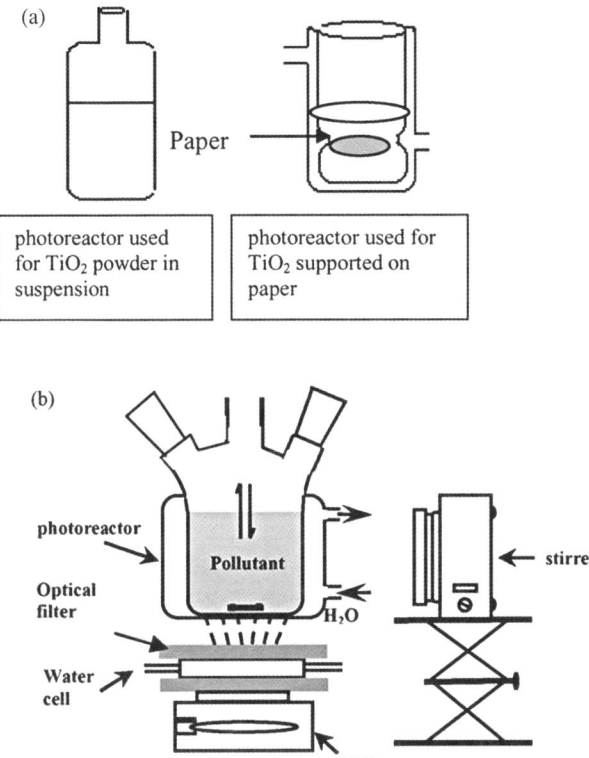

Figure 6.17 Photoreactor scheme used to test the activity of TiO$_2$ supported on paper (a) and of the corresponding slurries for comparison (b). (Reproduced with permission from ref. 24.)

PC500 supported on NW10 could be explained by the better dispersion of PC500 on NW10 due to its smaller particle size.

6.4 TiO$_2$ on Metals

Metals such as stainless steel, aluminium alloy and copper alloy, widely employed in household, construction and industrial contexts, can be functionalized by means of TiO$_2$ to gain specific properties, such as photocatalytic activity, surface superhydrophilicity and antibacterial activity. A protective effect against corrosion can be achieved at the same time by uniformly covering metals with TiO$_2$ films.[25]

Thick TiO$_2$ films can be prepared by integrating the preparation of a sol from titanium tetraisopropoxide hydrolysis in the presence of diethanolamine along with its mixing with Degussa P25 TiO$_2$ particles. Films obtained from the modified sol-gel are about ten times thicker for a single dip coating/heat

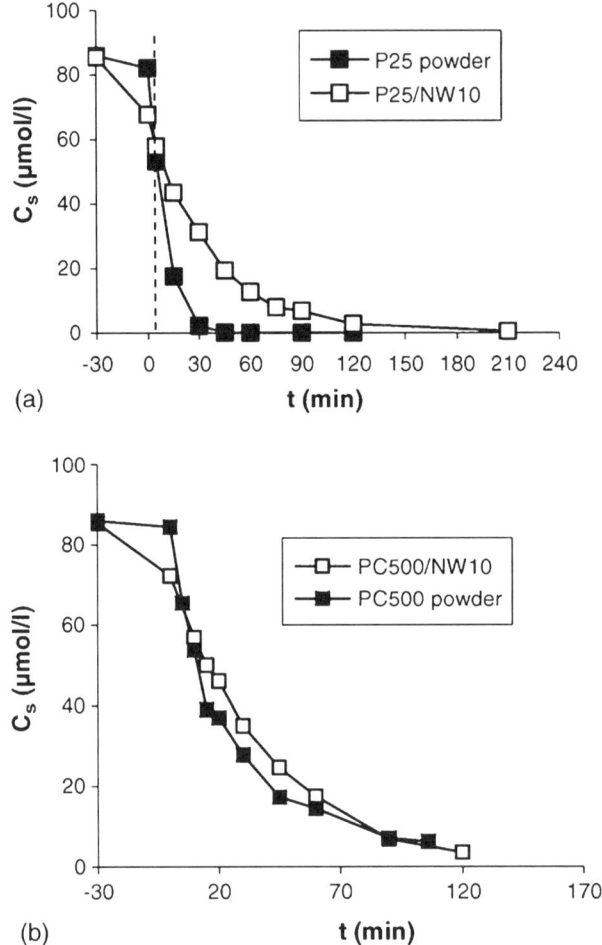

Figure 6.18 Photodegradation of diuron. Comparison of the efficiencies of P25 (a) and PC500 (b) in powder and supported on NW10 paper. (Reproduced with permission from ref. 24.)

treatment cycle than films obtained from the sol without powder addition. The prepared thick films were smooth and free of macrocracking and fracture. The aggregate size of these films was in the range 100–150 nm and films containing both anatase and rutile phases can be obtained. Moreover, the so-prepared films are much harder than films obtained from unmodified sols, and displayed excellent adhesion to the substrate.[26] These catalysts were found to be photocatalytically active in the degradation of 4-chlorobenzoic acid.

An efficient nanoparticle deposition system (NPDS) applicable to metal surfaces consists in spraying nano- and micro-sized TiO_2 powders through a supersonic nozzle at room temperature under low vacuum conditions.[27] The accelerated particles can be thus deposited on many substrates without thermal treatments. Figure 6.19 shows the components of an NPDS, including the air compressor, powder supplier, vacuum chamber, vacuum pump and controllers. The nozzle accelerates particles to a supersonic flow, which then impacts against the substrate. Commercial rutile TiO_2 powder was used to apply TiO_2 coatings on stainless steel (Figure 6.20) and on polymeric substrates with a thickness of *ca.* 300 nm. The TiO_2-covered stainless steel showed hardness and modulus values *ca.* 10% higher than the uncovered ones. This method was found to be applicable also to Al and Cu alloys.

A uniform and active TiO_2 film can be coated on stainless steel substrates by using a microemulsion of an organic surfactant (Triton X-100) in cyclohexane–water solvents mixed with titanium isopropoxide. Finally, the TiO_2 sol (contained in water droplets) can be deposited by dip-coating.

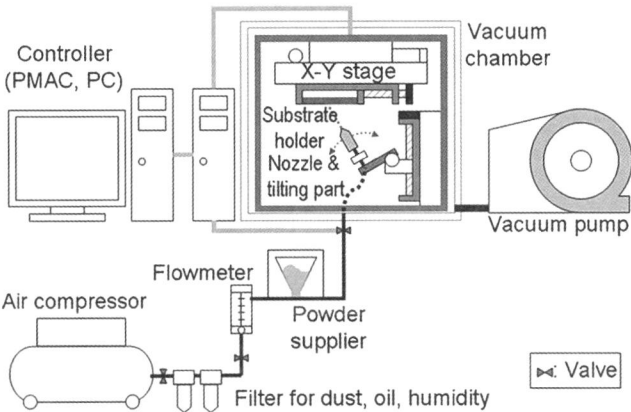

Figure 6.19 Scheme of a nanoparticle deposition system. (Reproduced with permission from ref. 27.)

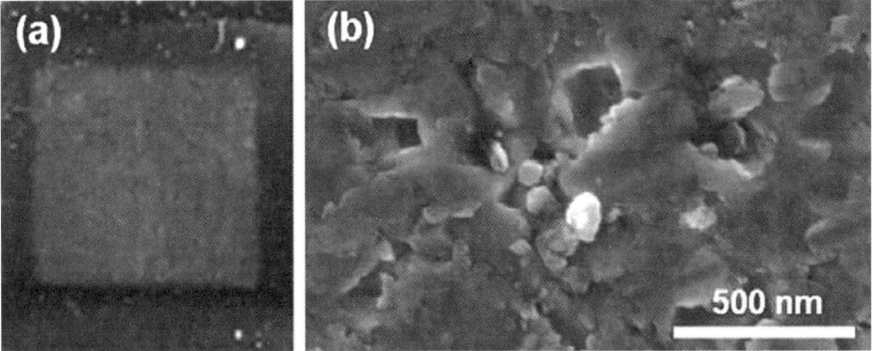

Figure 6.20 Images of a 5×5 mm TiO_2 coating on stainless steel: (a) optical image and (b) SEM image. (Reproduced with permission from ref. 27.)

Calcination at 500 °C to both remove organics and crystallize TiO_2 completes the procedure.[28] Increasing amounts of Fe(II) and Fe(III) ions diffuse from the stainless-steel substrate to TiO_2 films with increasing calcination time.

The films were superhydrophilic. Water contact angles for freshly prepared anatase TiO_2 films on stainless steel were indeed about 15–20° due to the

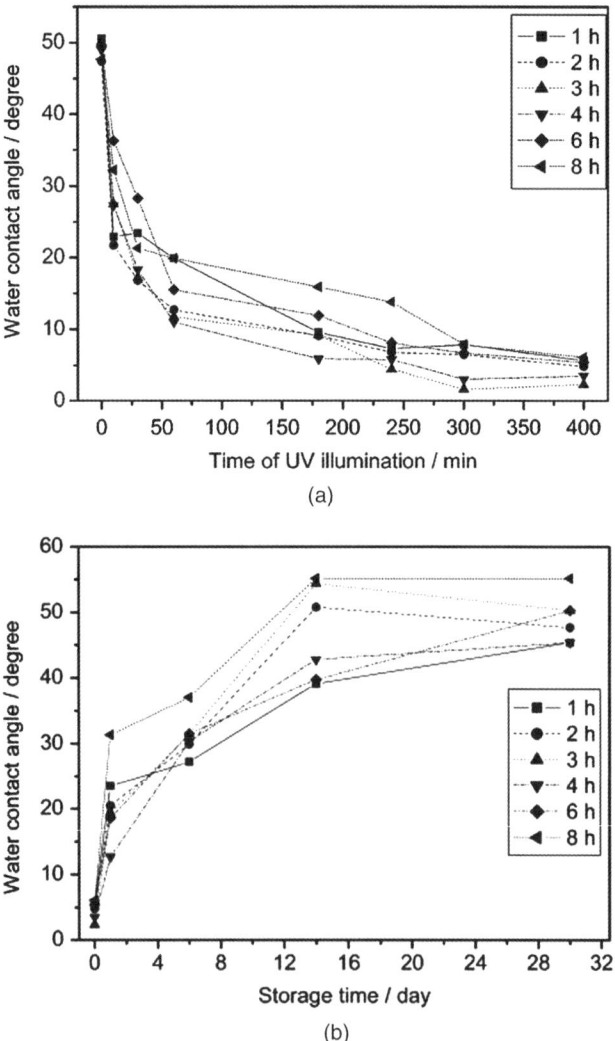

Figure 6.21 (a) Contact angle of TiO_2 films coated on stainless steel with different calcination times *versus* UV illumination time (540 mW cm^{-2}, 22 °C, RH 80%, in air); (b) change in water contact angles *versus* storage time in the dark under ambient conditions. (Reproduced with permission from ref. 28.)

TiO$_2$ on Plastic, Textile, Metal and Paper   191

presence of some surface defects. However, when these samples were kept in the dark, the water contact angles increased to 50–60°. To turn these slightly hydrophilic films into photo-induced hydrophilic ones, they were illuminated with a UV lamp under normal conditions. Figure 6.21(a) shows that water contact angles sharply decrease from 45–50° to 10–18° within 60 min and then gradually decrease to 2–8° after 300 min irradiation. Keeping the same films in the dark for 20–30 days results in the re-establishment of the previous

Figure 6.22 TiO$_2$-treated plastic films are stain-proof and can be cleaned easily with plain water. (Photograph courtesy of Nippon Soda Corporation.)

Figure 6.23 A photocatalytic tarpaulin storage tent is easily cleaned by rainwater. (Photograph courtesy of Nippon Soda Corporation.)

conditions (see Figure 6.21b). This behavior could be ascribed to the fact that the surface defective sites can be healed or replaced gradually by oxygen atoms, which causes the highly hydrophilic surface to be converted into a slightly hydrophilic one.

In addition, these films were tested as antibacterial devices for the sterilization of *Bacillus pumilus*. The efficiency found was remarkably higher than that without TiO_2 films; UV light alone did not cause sterilization. Moreover, in the absence of UV illumination the survival ratio was not reduced, so that one can conclude that the films itself was not toxic to *Bacillus pumilus*. Thus the bactericidal effect was due to the photocatalytic reaction of the TiO_2 films.

6.5 Practical Applications

Commercial examples of polymeric materials are still few and not as widespread as for glass, cement or ceramics. However, the field of photocatalytic plastics has been deeply explored, yielding interesting results, such those of the Nippon Soda Corporation (Figure 6.22). The transparent stain-proof plastic film is ideal for objects exposed to oily pollutants such as those present in the kitchens.[29]

The same company produced a storage tent (Figure 6.23) made of TiO_2-modified tarpaulin material, which has remained clean for many years, being washed only by rainwater.

Other polymeric photocatalytic materials have been produced by Sekisui Jushi Corporation for application as road signals and soundproof barriers (Figure 6.24). These products remain clean and are able to degrade the deposited smog.[30]

Commercial TiO_2-coated PVC has been developed by Taiyo Kogyo Corporation in collaboration with Nippon Soda Co. (Figure 6.25). PTFE (polytetrafluoroethylene) was also made photocatalytic by mixing the photocatalyst material with resin in the top surface layer of the PTFE, since direct coating of the photocatalyst to PTFE was found to be very difficult due to the differences

Figure 6.24 TiO_2-treated polycarbonate for application in road signals and soundproof road barriers. (Photograph courtesy of Sekisui Jushi Corporation.)

TiO_2 on Plastic, Textile, Metal and Paper

Figure 6.25 Two examples of TiO_2-coated PVC: (left-hand side) an ecology museum in Taiwan; (right-hand side) an external chapel in Osaka. (Photograph courtesy of Taiyo Kogyo Corporation.)

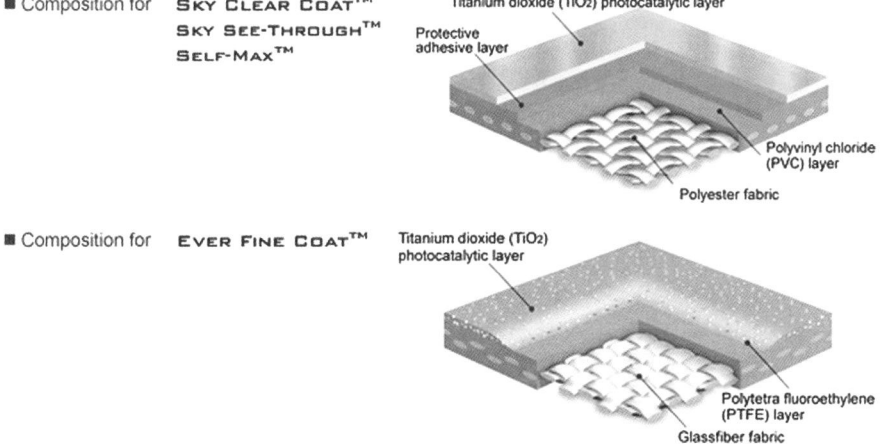

Figure 6.26 Scheme of the four photocatalytic products proposed by Taiyo Kogyo. (Photograph courtesy of Taiyo Kogyo Corporation.)

in physicochemical properties. An experiment carried out by the two companies has shown that $10\,000\,m^2$ of photocatalyst-coated fabric decomposes the major air pollutant nitrogen dioxide (NO_2) expelled by 1.7 four-ton trucks.

Recently (2008) the same company presented a new product line, which consists of a photocatalytic membrane, in four types: "Ever Fine Coat," "Sky Clear Coat," "Sky See-Through" and "Self-Max".[31] The first product is based on PTFE, where as the other three products are based on PVC. "Ever Fine Coat," in particular, is available in different colors (white, gray, ivory, blue,

Figure 6.27 Photographs and SEM image of fibers for textiles by Kuraray Trading Co. (Photograph courtesy of Kuraray Trading Co.)

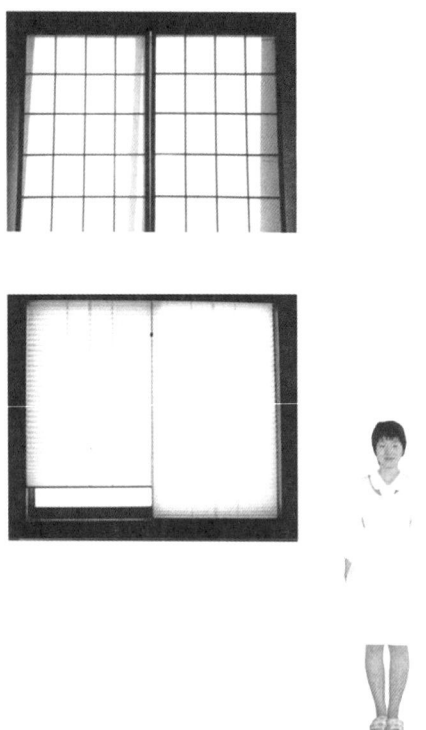

Figure 6.28 Self-cleaning and self-sterilizing paper applied in window blinds and in dresses. (Photograph courtesy of MOLZA Corporation.)

Figure 6.29 An aluminium-siding coated with TiO_2. (Photograph courtesy of YKK Corporation.)

etc.) and has superb self-cleaning performances. The other products are white, except for "Sky See-Through" that shows a high transmittance (*ca.* 65%). Figure 6.26 shows the structure of these membranes.

Kuraray Trading Co., experts in polyester filament manufacture, has developed unique polymer products such as an antibacterial deodorizing fiber and the ethylene vinyl alcohol-based "Sophista." These products were the results of extensive and original development of composite materials incorporating Kuraray polyester and other fibers, along with secondary products.[32] Figure 6.27 shows some pictures of the self-cleaning textiles developed and a SEM image of the fiber used to manufacture them.

Even in the case of photocatalytic paper materials, the examples are very few and mainly by Japanese companies. Ahlstrom started working on photocatalytic nonwoven products in 1996, describing them in conferences and to the press and now they propose these products as efficient devices with which to purify air and water.[33] Nippon Paper (Tokyo, Japan), on one other hand, developed a few years ago the "Light Catalytic Newsprint" (PDASH™) with Yomiuri, one of Japan's largest newspaper publishers.[34] These newsprints (containing TiO_2) deodorize the surrounding air when irradiated by UV and ambient light. The Japan Synthetic Textile Inspection Institute Foundation confirmed that PDASH™ successfully led to a reduction of more than 99% in the concentration of acetaldehyde, *i.e.*, the main component of cigarette odors, in *ca.* 20 h. Moreover, the developed technology allows for the production of coated paper that is compatible with any type of printing machine in high-speed printing. The virgin pulp source to produce this paper is made from PEFCC (Pan-European Forest Certification Council) certified woodchips.

Figure 6.28 shows other commercial products produced from photocatalytic paper by the MOLZA Corporation.[35] Finally, Figure 6.29 shows an aluminium-siding coated with TiO_2 by the YKK Corporation, in a building constructed in 1999 in Sendai.[36]

References

1. A. Fujishima, X. Zhang and D. A. Tryk, TiO_2 photocatalysis and related surface phenomena, *Surf. Sci. Rep.*, 2008, **63**, 515–582.
2. (a) M. P. Paschoalino, J. Kiwi and W. F. Jardim, Gas-phase photocatalytic decontamination using polymer supported TiO_2, *Appl. Catal. B: Environ.*, 2006, **68**, 68–73; (b) K. O. Awitor, A. Rivaton, J.-L- Gardette, A. J. Down and M. B. Johnson, Photo-protection and photo-catalytic activity of crystalline anatase titanium dioxide sputter-coated on polymer films, *Thin Solid Films*, 2008, **516**, 2286–2291; (c) A. Kubacka, M. Ferrer, M. L. Cerrada, C. Serrano, M. Sánchez-Chaves, M. Fernández-García, A. de Andrés, R. J. Jiménez Riobóo, F. Fernández-Martín and M. Fernández-García, Boosting TiO_2-anatase antimicrobial activity: Polymer-oxide thin films, *Appl. Catal. B: Environ.*, 2009, **89**, 441–447.
3. M. Birnie, M. Gillott and S. Riffat, The immobilization of titanium dioxide on organic polymers, for a cost effective and energy efficient means of improving indoor air quality, *Int. J. Green Energy*, 2006, **3**, 101–114.
4. J.-H. Yang, Y.-S. Han and J.-H. Choy, TiO_2 thin-films on polymer substrates and their photocatalytic activity, *Thin Solid Films*, 2006, **495**, 266–271.
5. M. Langlet, A. Kim, M. Audier and J. M. Hermann, Sol-gel preparation of photocatalytic TiO_2 films on polymer substrates, *J. Sol-Gel Sci. Technol.*, 2002, **25**, 223–234.
6. M. S. Vohra and K. Tanaka, Enhanced photocatalytic activity of Nafion-coated TiO_2, *Environ. Sci. Technol.*, 2001, **35**, 411–415.
7. Y. Zhiyong, D. Laub, M. Bensimon and J. Kiwi, Flexible polymer TiO_2 modified film photocatalysts active in the photodegradation of azo-dyes in solution, *Inorg. Chem. Acta*, 2008, **361**, 589–594.
8. J. O. Carneiro, V. Teixeira, A. Portinha, A. Magalhães, P. Coutinho, C. J. Tavares and R. Newton, Iron-doped photocatalytic TiO_2 sputtered coatings on plastics for self-cleaning applications, *Mater. Sci. Eng.*, 2007, **138**, 144–150.
9. C. J. Tavares, S. M. Marques, S. Lanceros-Méndez, V. Sencadas, V. Teixeira, J. O. Carneiro, A. J. Martins and A. J. Fernandes, Strain analysis of photocatalytic TiO_2 thin films on polymer substrates, *Thin Solid Films*, 2008, **516**, 1434–1438.

10. J. Kasanen, M. Suvanto and T. T. Pakkanen, Self-cleaning, titanium dioxide based, multilayer coating fabricated on polymer and glass surfaces, *J. Appl. Polym. Sci.*, 2009, **111**, 2597–2606.
11. H. Schmidt, M. Naumann, T. S. Müller and M. Akarsu, Application of spray techniques for new photocatalytic gradient coatings on plastics, *Thin Solid Films*, 2006, **502**, 132–137.
12. (a) K. Sunada, Y. Kikuchi, K. Hashimoto and A. Fujishima, Bactericidal and detoxification effects of TiO_2 thin film photocatalysts, *Environ. Sci. Technol.*, 1998, **32**, 726–728; (b) K. Qi, W. A. Daoud, J. H. Xin, C. L. Mak, W. Tang and W. P. Cheung, Self-cleaning cotton, *J. Mater. Chem.*, **16**, 4567–4574.
13. Y. Dong, Z. Bai, R. Liu and T. Zhu, Decomposition of indoor ammonia with TiO_2-loaded cotton woven fabrics prepared by different textile finishing methods, *Atmos. Environ.*, 2007, **41**, 3182–3192.
14. (a) M. J. Uddin, F. Cesano, F. Bonino, S. Bordiga, G. Spoto, D. Scarano and A. Zecchina, Photoactive TiO_2 films on cellulose fibres: synthesis and characterization, *J. Photochem. Photobiol. A: Chem.*, 2007, **189**, 286–294; (b) M. J. Uddin, F. Cesano, D. Scarano, F. Bonino, G. Agostini, G. Spoto, S. Bordiga and A. Zecchina, Cotton textile fibres coated by Au/TiO_2 films: synthesis, characterization and self cleaning properties, *J. Photochem. Photobiol. A: Chem.*, 2008, **199**, 64.
15. W. A. Daoud, S. K. Leung, W. S. Tung, J. H. Xin, K. Cheuk and K. Qi, Self-cleaning keratins, *Chem. Mater.*, 2008, **20**, 1242–1244.
16. J. A. Lee, K. C. Krogman, M. Ma, R. M. Hill, P. T. Hammond and G. C. Rutledge, Highly reactive multilayer-assembled TiO_2 coating on electrospun polymer nanofibers, *Adv. Mater.*, 2009, **21**, 1252–1256.
17. N. Veronovski, A. Rudolf, M. S. Smole, T. Kreže and J. Geršak, Self-cleaning and handle properties of TiO_2-modified textiles, *Fibers Polym.*, 2009, **10**, 551–556.
18. (a) R. Pelton, X. Geng and M. Brook, Photocatalytic paper from colloidal TiO_2-fact or fantasy, *Adv. Colloid Interface Sci.*, 2006, **127**, 43–53; (b) L. Ye, C. D. M. Filipe, M. Kavoosi, C. A. Haynes, R. Pelton and M. A. Brook, Immobilization of TiO_2 nanoparticles onto paper modification through bioconjugation, *J. Mater. Chem.*, 2009, **19**, 2189–2198.
19. H. Matsubara, M. Takada, S. Koyama, K. Hashimoto and A. Fujishima, Photoactive TiO_2 containing paper: preparation and its photocatalytic activity under weak UV light illumination, *Chem. Lett.*, 1995, 767–768.
20. W. A. Daoud, J. H. Xin and Y.-H. Zhang, Surface functionalization of cellulose fibers with titanium dioxide nanoparticles and their combined bactericidal activities, *Surf. Sci.*, 2005, **599**, 69–75.
21. (a) P. Escaffre, P. Girard, J. Dussaud and L. Bouvier, Photocatalytic composition having binding agent containing colloidal silica, 1999, *Pat. Appl. WO99-FR748 9951345*; (b) A. Aguedach, S. Brosillon, J. Morvan and E. K. Lhadi, Photocatalytic degradation of azo-dyes reactive black 5 and reactive yellow 145 in water over a newly deposited titanium dioxide, *Appl. Catal. B: Environ.*, 2005, **57**, 55–62.

22. N. Kimura, S. Abe, T. Yoshimoto and S. Fukayama, Photocatalyst-carrying structure and photocatalyst coating material, 2001, *US Pat. 6228480 B1*.
23. L. R. C. Barclay, M.-C. Basque and M. R. Vinqvist, Singlet-oxygen reactions sensitized on solid surfaces of lignin or titanium dioxide: product studies from hindered secondary amines and from lipid peroxidation, *Can. J. Chem.*, 2003, **81**, 457–467.
24. M. El Madani, C. Guillard, N. Pérol, J. M. Chovelon, M. El Azzouzi, A. Zrineh and J. M. Herrmann, Photocatalytic degradation of diuron in aqueous solution in presence of two industrial titania catalysts, either as suspended powders or deposited on flexible industrial photoresistant papers, *Appl. Catal. B: Environ.*, 2006, **65**, 70–76.
25. (a) Y. Zhu, L. Zhang, L. Wang, Y. Fu and L. Cao, The preparation and chemical structure of TiO_2 film photocatalysts supported on stainless steel substrates via the sol-gel method, *J. Mater. Chem.*, 2001, **11**, 1864–1868;23.(b) P. Evans and D. W. Sheel, Photoactive and antibacterial TiO_2 thin films on stainless steel, *Surf. Coat. Technol.*, 2007, **201**, 9319–9324.
26. (a) G. Balasubramanian, D. D. Dionysiou, M. T. Suidan, V. Subramanian, I. Baudin and J.-M. Laîné, Titania powder modified sol-gel process for photocatalytic applications, *J. Mater. Sci*, 2003, **38**, 823–831; (b) G. Balasubramanian, D. D. Dionysiou, M. T. Suidan, I. Baudin and J.-M. Laîné, Evaluating the activities of immobilized TiO_2 powder films for the photocatalytic degradation of organic contaminants in water, *Appl. Catal. B: Environ.*, 2004, **47**, 73–84.
27. D. M. Chun, M. H. Kim, J. C. Lee and S. H. Ahn, TiO_2 coating on metal and polymer substrates by nano-particle deposition system (NPDS), *CIRP Ann.-Manuf. Technol.*, 2008, **57**, 551–554.
28. J. C. Yu, W. Ho, J. Lin, H. Yip and P. K. Wong, Photocatalytic activity, antibacterial effect, and photoinduced hydrophilicity of TiO_2 films coated on a stainless steel substrate, *Environ. Sci. Technol.*, 2003, **37**, 2296–2301.
29. Nippon Soda Corporation: http://www.nippon-soda.co.jp
30. Sekisui Jushi Corporation : http://www.jislon.com/contents/top/top.html.
31. Details of products from Taiyo Kogyo Corporation are available online at http://www.makmax.com/business/tio2_fabric.html
32. Details of photocatalytic products from Kuraray Trading Co. are available online at http://www.kuraray.co.jp/en/release/2003/031219.html
33. Details of photocatalytic paper produced by Ahlstrom can be found online at http://www.ahlstrom.com
34. Details about photocatalytic paper produced by Nippon Paper can be found online at http://www-japan.org/trends/07_sci-tech/sci080131.html
35. Details about photocatalytic paper produced by Molza Corporation can be found online at http://www.molza.co.jp
36. Details about photocatalytic paper produced by YKK Corporation can be found online at http://www.ykk.com

CHAPTER 7
Devices for Water and Air Purification

7.1 Devices for Water Purification

Photocatalytic oxidation[1–6] is a simple and effective method for the conversion and/or destruction of many organic and inorganic compounds with a measurable rate up to negligible concentration levels. On these grounds, it is very suitable for solving various problems of environmental interest in water purification.

The process is not complex: irradiation of the semiconductor surface with light of band gap (or higher) energy generates very reactive species able to break the molecular bonds of chemicals until only carbon dioxide and water are left. Titanium dioxide, a very safe, common and inexpensive material, has proven to be the most effective photocatalyst for breaking down a wide range of chemicals, without producing harmful by-products. Owing to these favorable features, this method is starting to be used to solve real problems raised by the presence of pollutants in water. In the following some field applications are described.

7.1.1 Pesticide Degradation in a Solar Photoreactor

Solar photocatalysis is today one of the most successful applications of this advanced oxidation process. In Europe the development and scale-up of solar photocatalytic applications is successfully carried out at the Plataforma Solar de Almería (PSA, Spain), which, due to such strong activity carried out, has become a benchmark worldwide. One of the relevant projects has been the SOLARDETOX (EC-DGXII BRITEEURAM-III program) on solar photocatalytic detoxification of water.[7] The project objective was to develop a simple, efficient and commercially competitive water-treatment technology, based on compound parabolic solar collectors (CPCs) and TiO_2 photocatalysis to make possible easy design and installation. The result of that project was the design,

---
Clean by Light Irradiation: Practical Applications of Supported TiO_2
By Vincenzo Augugliaro, Vittorio Loddo, Mario Pagliaro, Giovanni Palmisano and Leonardo Palmisano
© V. Augugliaro, V. Loddo, M. Pagliaro, G. Palmisano and L. Palmisano 2010
Published by the Royal Society of Chemistry, www.rsc.org

Figure 7.1 View of the solar CPC collector field (Arganda del Rey, Madrid, Spain). (Reproduced with permission from ref. 7.)

set-up and operation of the first European industrial solar detoxification treatment plant at Arganda del Rey (Madrid, Spain). The irradiated surface of the CPCs is of $100\,m^2$; Figure 7.1 shows a view of the demonstration plant.

The SOLARDETOX research has also allowed the construction and operation of one of the first commercial plants, implemented by the Spanish company Albaida. With the aim of recycling the huge number of pesticide bottles needed for greenhouse agriculture activity in Almería province (Spain), the Albaida plant treats the washing water of the plastic bottles.

Figure 7.2 shows the conceptual design of the solar photocatalytic treatment plant. The water containing pesticides coming from the bottle-washing process is decontaminated in a batch solar photoreactor until 80% mineralization of TOC is achieved. Solar decontamination is carried out by coupling photocatalysis and photo-Fenton methods.[8] The treated water is then transferred to the post-process treatment (*i.e.*, sedimentation and recuperation of photocatalyst, iron precipitation) and then reused for the washing process or discharged for irrigation. Figure 7.3 shows the system used for recuperation of TiO_2 and Fe^{2+}, while Figure 7.4 shows the solar collector field of the Albaida plant.

7.1.2 Cyanide Degradation in a Pilot Plant Photoreactor

To degrade free cyanide ions contained in power plant wastewater, a continuously stirred tank photoreactor at pilot plant level has been designed, constructed and tested. The working power plant, located at Puertollano (Spain), utilizes an integrated gasification combined cycle based on a coal gasification process.[9] The photoreactor for cyanide degradation was arranged in parallel with the chemical treatment performed on the gross contaminated

Devices for Water and Air Purification

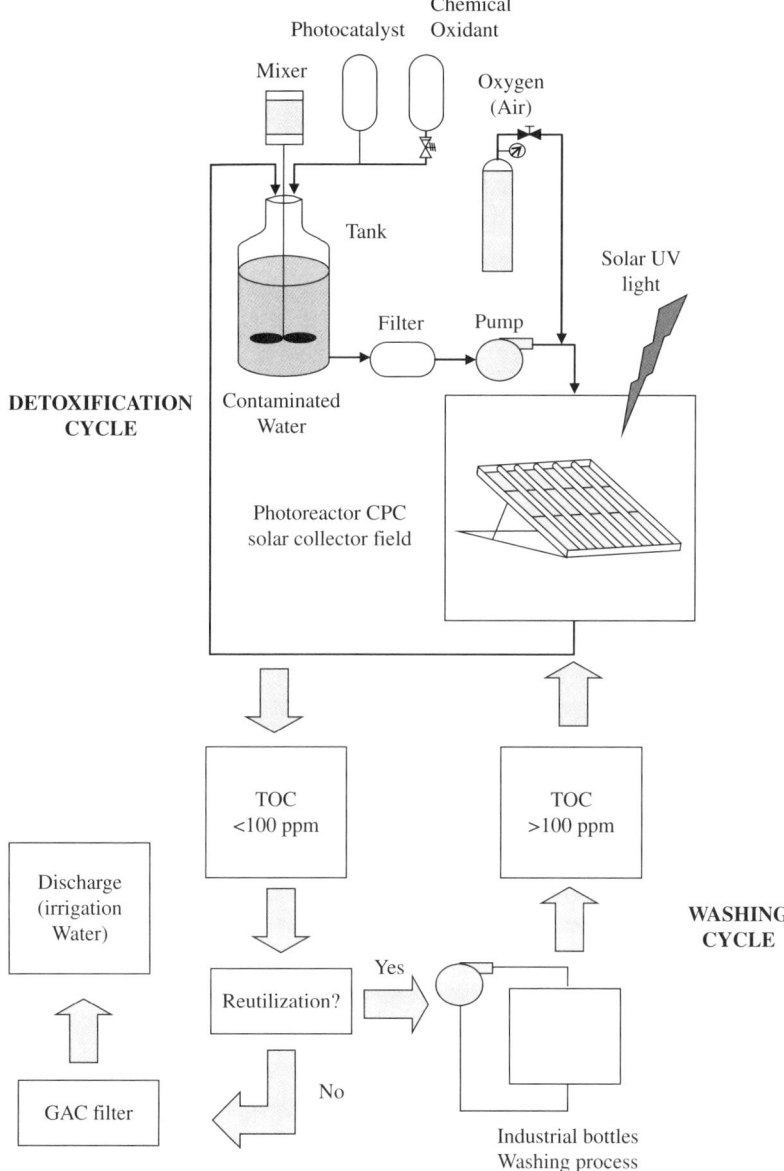

Figure 7.2 Conceptual design of the Albaida solar photocatalytic process.

effluent of the industrial plant. Figure 7.5 gives a block diagram of the decontamination process; blocks other than the photoreactor represent existing units of the water treatment system of the plant. The photoreactor was designed to treat a flow rate of $1\,m^3\,h^{-1}$ of the liquid effluent feeding the ozonization reactor and to perform a 98% conversion of cyanides.

Figure 7.3 TiO_2/Fe^{2+} recuperation system. (Reproduced with permission from ref. 8.)

The reactor configuration chosen to perform cyanide oxidation at the pilot plant scale was that of a continuous slurry stirred tank reactor with immersed lamps. The choice of this photoreactor is based on the following considerations: (i) by increasing the catalyst concentration, a maximum of activity, *i.e.*, of local volumetric reaction rate, can be reached; (ii) mass transport resistance between the bulk of the liquid phase and the irradiated surface of the catalyst can be minimized depending on the agitation degree of the system; and (iii) the irradiation efficiency, *i.e.*, the ratio between the total photon flow emitted by the lamp and that absorbed by the whole suspension, can approximate unity as the catalyst concentration of the suspension can be chosen in order that no radiation is transmitted by the suspension.

The main operative drawback of a continuous slurry reactor is that the exit flow consists of the internal reacting suspension so that wash out of the catalyst eventually occurs. It is therefore necessary to employ solid–liquid separation equipment coupled with the reactor to recover the catalyst and a solid conveyor for recycling the catalyst to the photoreactor. To avoid the need of the

Figure 7.4 Industrial solar photocatalytic water treatment plant for pesticide-bottle recycling (La Mojonera, Almería, Spain) (150 m^2 of total collector surface).

separation–transportation unit, the photoreactor has been designed with an inside settling zone so that the exit flow does not contain solid particles. This choice allows one to simplify the photoprocess operation but means that the catalyst particles must possess certain features; in particular the particles should have a high settling velocity.

Literature information on the kinetics of cyanide photodegradation and on the quantum yield of the photoreaction[10–12] was used to calculate the reacting volume of the photoreactor and the irradiation power. The following values were used for the photoreactor set-up: reactor "design" volume: 0.84 m^3; security factor of 2.5; with previous values the reacting volume of the photoreactor is 2 m^3. Figure 7.6 shows the scheme of the photoreactor. The reacting volume is cylindrical (1500 mm in diameter and 1700 mm high). In the scheme of the photoreactor an outer annular zone is allowed to separate the catalyst particles from the liquid solution leaving the reactor. In this way the outer

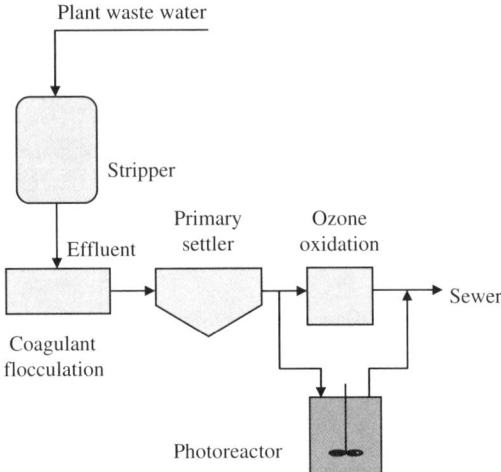

Figure 7.5 Block diagram of the decontamination process; blocks other than that for the photoreactor represent existing units of the water treatment system of the IGCC plant.

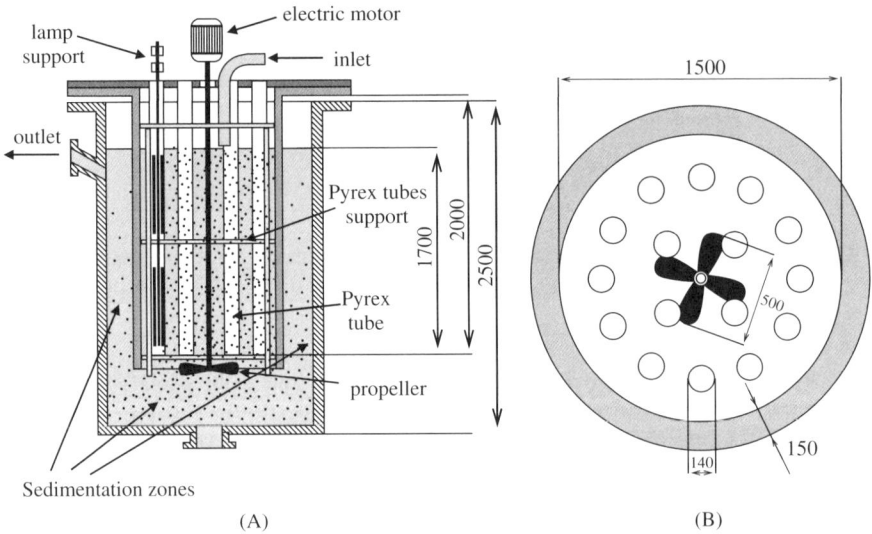

Figure 7.6 (a) Scheme of the pilot plant photoreactor; (b) photoreactor top view, showing the distribution of Pyrex tubes containing the lamps. The figures indicate millimetres.

diameter of the photoreactor is 1800 mm. A rotating pump feeds the liquid flow to the photoreactor while the exit stream leaves the photoreactor by gravity. The irradiating system consists of 16 Pyrex cylindrical tubes (*ca.* 2 m long), each of which contains eight fluorescent actinic lamps (40 W, 60 cm high). The

Devices for Water and Air Purification

reacting mixture is mixed by a turbine impeller (500 mm diameter) with inclined blades in order to obtain a high axial flow. The impeller is positioned at about 1/6 of the liquid height from the bottom in order to better suspend the catalyst particles. Figures 7.7 and 7.8 are photographs of the photoreactor and of some details.

Figure 7.7 Picture of (a) the photoreactor; (b) view of the external vessel and jacket of the photoreactor; (c) the photoreactor cover; (d) view of the outlet of the photoreactor.

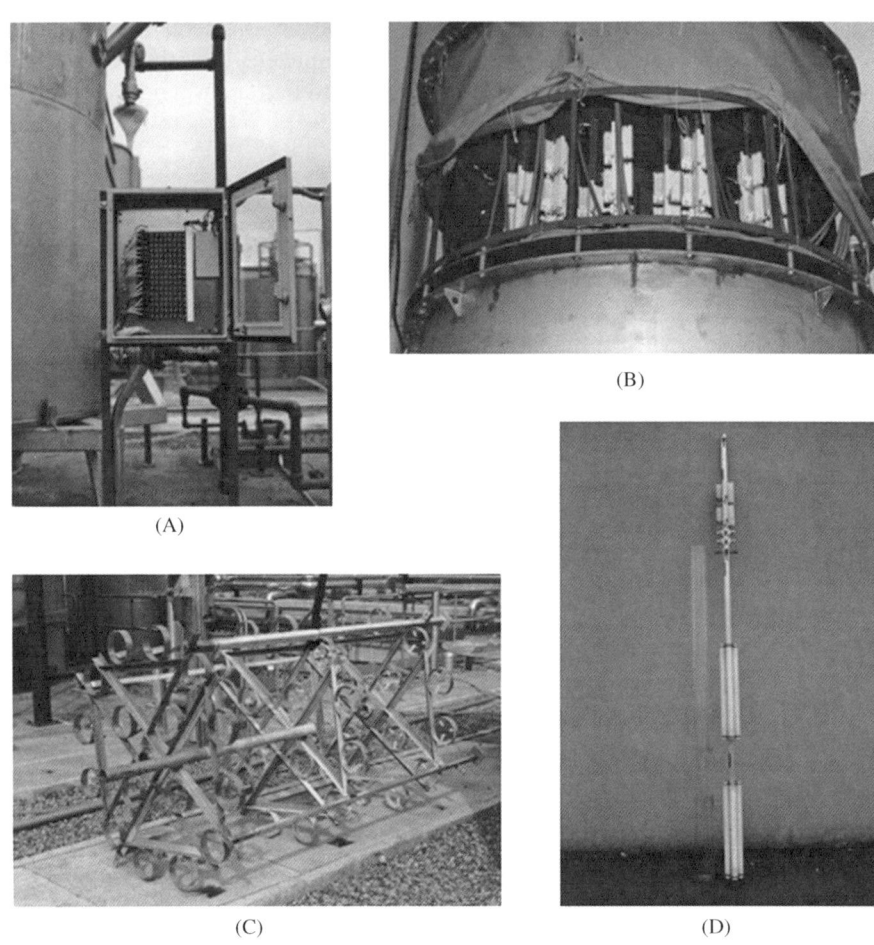

Figure 7.8 Picture of (a) the control system panel used to check the correct working order of the lamps; (b) view of the opening that allows the reacting volume to be in contact with the atmosphere; (c) stainless steel support of the Pyrex tubes; (d) Pyrex tube with the support containing eight lamps.

The catalyst was TiO_2 supported on Al_2O_3 particles. The supported catalyst had a TiO_2 content of 53 wt%; it exhibited a high reactivity while maintaining long-term mechanical stability. The reactivity runs carried out on the power plant wastewater showed a cyanide conversion higher than 98%.

The main merit of this pilot plant investigation has been the demonstration of the feasibility of a process for cyanide degradation by mean of heterogeneous photocatalytic method at the industrial level. The good performance of the pilot plant photoreactor allows one to obtain a liquid effluent with a concentration of free cyanide far below the 0.5 ppm that is the maximum allowable value according to EC environment protection law.

7.1.3 Photo-CREC-Water Reactors

Researchers at the Chemical Reactor Engineering Centre (CREC)[13] of the University of Western Ontario (Canada) have designed and constructed photocatalytic reactors (either for gas–solid or liquid–solid regime) under the optimality criteria of improved efficiency and improved irradiation of the photocatalyst.

Three photocatalytic reactors for water treatment have been produced:

1. Photo-CREC-Water I with immobilized TiO_2;
2. Photo-CREC-Water II, with suspended TiO_2 and artificial irradiation;
3. Photo-CREC-Water III, with suspended TiO_2 and simulated solar irradiation.

The original Photo-CREC-Water I reactor[14,15] has been modified with respect to the original design by Serrano and de Lasa.[16] Figure 7.9 is a schematic representation of the photocatalytic reactor.

The Photo-CREC-Water I reactor is characterized by an annular channel with 16 baskets positioned at 45° angles. Stainless steel spacers placed between the baskets secure the baskets in position.

The near-UV lamp is located in the central channel, providing monochromatic light at 365 nm. Water is circulated in a downward flow; water at the photoreactor outlet is discharged into an air bubbler that guarantees saturation with oxygen.

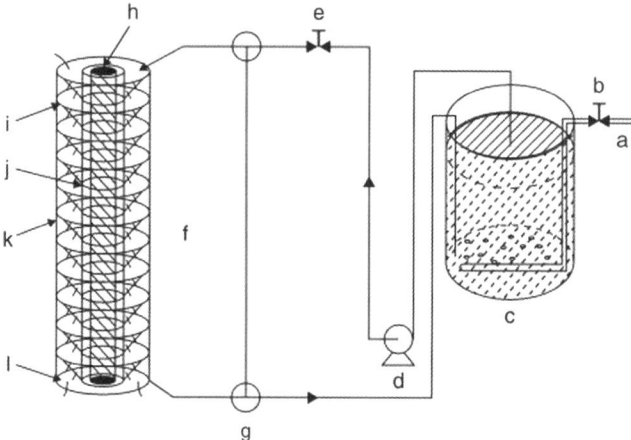

Figure 7.9 Schematic representation of the Photo-CREC-Water I reactor: (a) compressed air, (b) air gas regulator, (c) oxygenator with air pipe distributor, (d) water pump, (e) valve, (f) concentric Photo-CREC unit with lamp placed in the center and 16 conical baskets spaced throughout the unit, (g) three-way valve, (h) lamp, (i) basket, (j) Pyrex glass tube, (k) external Plexiglas tube and (l) annular channel. (Reproduced with permission from ref. 16.)

The core of the Photo-CREC Water I is the photocatalytic basket containing the TiO_2-mesh. Before TiO_2 impregnation, the fiber-glass mesh is treated with a nitric acid solution and afterwards it is calcined at 550 °C for a few hours. The mesh is then mounted on the inner face of each of the conical baskets and "*in situ*" impregnated by circulating a TiO_2 slurry through it. Strong attachment of TiO_2 particle on the fiberglass mesh was established[16] when translucent water was obtained after the first washing of meshes and in all subsequent experiments, confirming that the Photo-CREC-Water I reactors do not require particle recovery.

The Photo-CREC Water-II reactor consists of two concentric Pyrex tubes with a TiO_2 suspension flowing in the concentric channel; a UV lamp irradiates the catalyst from the inner tube. Figure 7.10 illustrates the components of this reactor.

The inner tube may be changed so varying the dimension of the annular cross-section and the fluid dynamics of the circulating suspension. A stirred tank is connected in series with the photoreactor. The tank allows sampling, pollutant and catalyst addition, pH and temperature monitoring and oxygen saturation of suspension. The unit is equipped with windows along the outer tube wall in order to perform radiometric and spectro-radiometric measurements. A four-point flow distributor injector at the reactor entrance guarantees uniform injection and intense mixing.

With respect to the Photo-CREC Water II, the Photo-CREC Water-III reactor (Figure 7.11) shows the following main features: (i) external illumination simulating solar irradiation; (ii) larger irradiation area; (iii) presence of

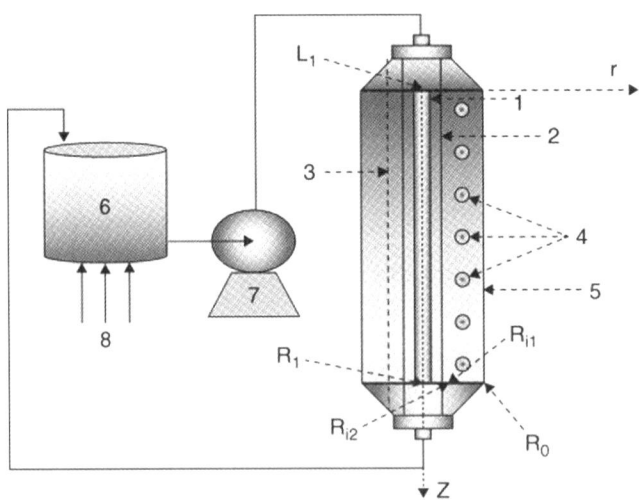

Figure 7.10 Schematic representation of the Photo-CREC Water-II Reactor: (1) lamp, (2) and (3) replaceable inner tube, (4) fused-silica windows, (5) UV-opaque outer tube, (6) stirred tank, (7) pump and (8) air injector. (Reproduced with permission from ref. 16.)

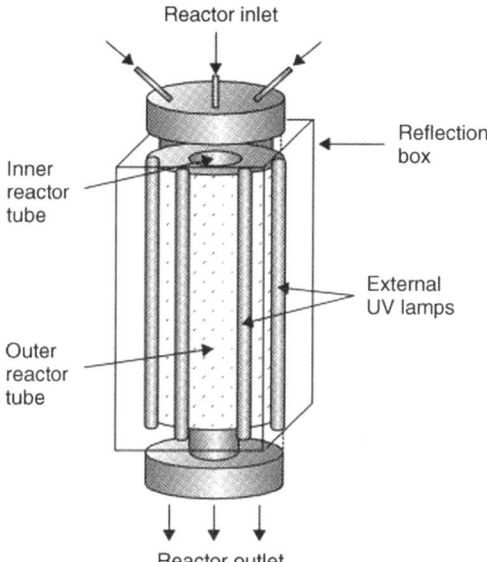

Figure 7.11 Schematic representation of the Photo-CREC Water-III reactor. (Reproduced with permission from ref. 16.)

internal sensor to measure the radiation field; and (iv) internal irradiation to make uniform the radiation field in radial directions. The reactor is irradiated externally by UV lamps. The external illumination permits the simulation of solar irradiation whereas the combination of internal and external illumination allows the increase of irradiation efficiency.

The bench-scale Photo-CREC reactors maximize the catalyst–liquid contact and the efficiency of light–TiO_2 interaction;[17] on this basis they facilitate the study of important reaction engineering parameters such as the adsorption and the reaction rates of model compounds, the photodegradation mechanisms and the photocatalyst performance.

Photo-CREC Water reactors have been successfully used in the photoconversion of many organic and inorganic pollutants, as well as the inactivation of microorganisms contained in water. A semi-commercialized version of the Photo-CREC unit (Figure 7.12) has a water treatment capacity of 300 l, addressing the basic daily needs of a community of 500 people (around 1200 l day^{-1}). Research is underway to increase the Photo-CREC reactor's capacity, with the goal of supplying drinking water to communities of 3000 people.

7.1.4 UBE Photocatalytic Fiber Reactor

UBE Group,[18] using technologies created in the development of an ultra-high temperature resistant ceramic, has successfully developed a photocatalytic fiber with a large surface area and good fiber strength. The fiber contains titania photocatalyst so that it is able to oxidize and decompose organic substances.[19]

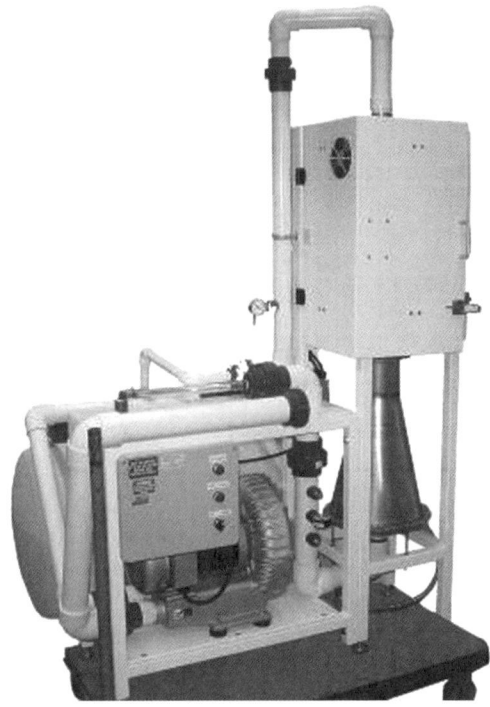

Figure 7.12 Semi-commercialized version of the photo-CREC water reactor. (Reproduced with permission from ref. 16.)

The photocatalytic fiber is arranged in a very simple purifier with a module composed of the cone-shaped felt material (made of the photocatalytic fiber) and a UV lamp. Figure 7.13 shows the set-up of the total recirculation photo-reacting system. The contaminated water, contained in a reservoir, is continuously fed to the photoreactor, where the contaminants are destroyed. The photocatalytic fiber can purify not only the many types of wastewater but can also destroy bacteria present in wastewaters. Many bacteria (common bacterium, *Legionella pneumophila* and coliform)[20] that exist in bathwater before purification are rapidly decomposed into CO_2 and H_2O. Figure 7.14 shows the behavior of cell concentration *versus* irradiation time for a deactivation run carried out with the UBE photoreactor on wastewater contaminated by *Legionella* and aerobic bacteria. Figure 7.15 shows a UBE photocatalytic device with a fiber module.

7.2 Devices for Air Purification

People inhale 15–18 kg of air every day. If living in contaminated areas, a person also inhales on average a 15–100 mg daily dose of poisonous substances such as carbonic oxide, formaldehyde, benzopyrene and many other

Devices for Water and Air Purification

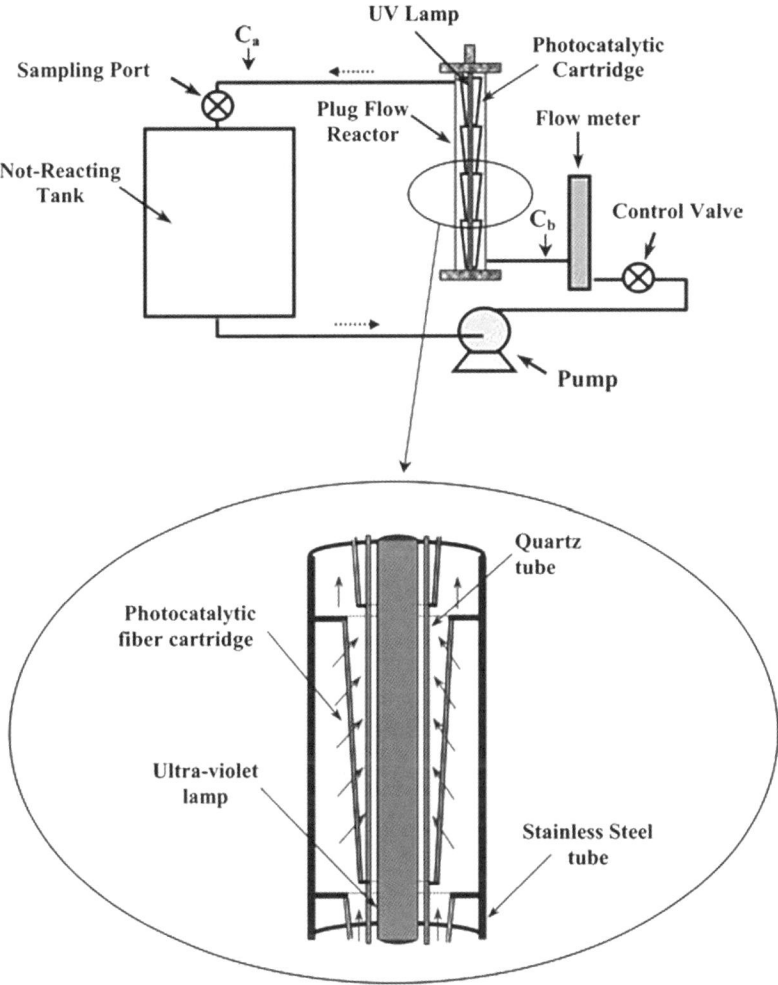

Figure 7.13 Set-up of the UBE photoreacting system. (Reproduced with permission from ref. 18.)

detrimental compounds that in big cities are in large excess over permissible rates. It is, therefore, of great concern to purify, as much as possible, the air of enclosed sites.

Heterogeneous photocatalysis is one of the most efficient and cost-effective air purification technologies, and scientists predict that it would be the main method of air purification in the near future. Photocatalytic methods for indoor air purification have considerable advantages over other methods. Unlike air purification by adsorption, photocatalytic oxidation leads to complete neutralization of air pollutants, forming CO_2 and H_2O as final products, and no regular regeneration of the photocatalyst is usually required. Almost any

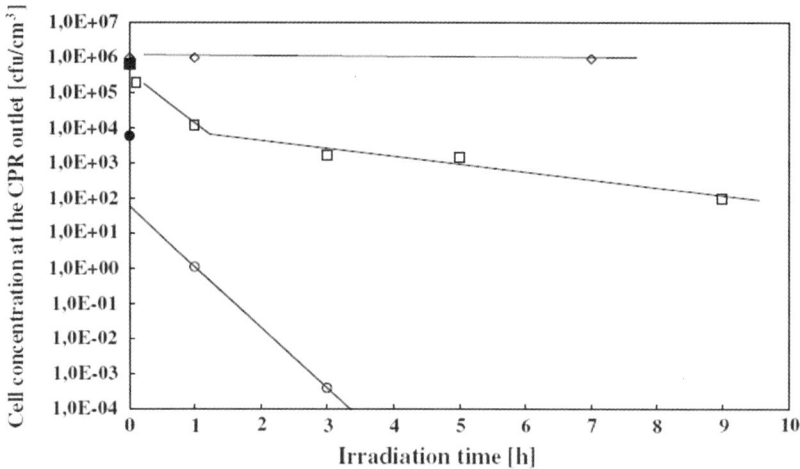

Figure 7.14 Cell concentrations *versus* irradiation time for a run with *Legionella* without catalyst (◇) and for photocatalytic runs with *Legionella* (○) and with aerobic bacteria (□); ● and ■ indicate the initial concentration of *Legionella* and aerobic bacteria, respectively. (Reproduced with permission from ref. 20.)

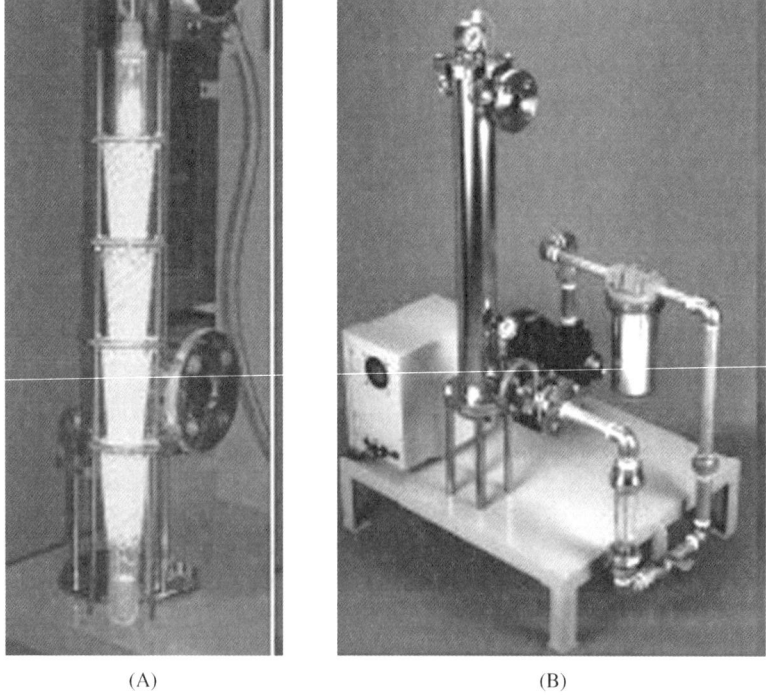

(A) (B)

Figure 7.15 UBE photocatalytic device: (a) fiber module; (b) fiber cartridge. (Reproduced with permission from ref. 18.)

organic substance, even at low concentration, can be oxidized under ambient conditions using air oxygen. Owing to these favorable features, the photocatalytic method has attracted the attention of industry devoted to producing apparatuses for air conditioning; here the objective is purify the indoor air of homes, hospitals, farms, *etc*.

In some purifying apparatuses the photocatalytic method is coupled with the generation of negative or positive air ions.[21] These charged particles, such as O^{2-}, O^-, OH^-, $(H_2O)^{x-}$, N_2^+, O^+, O_2^+, NO^+, CH_4^+, NH_4^+, and so on, are widely distributed in natural air and are generated by radioactive elements in the soil, cosmic rays, radioactive aerosols, UV radiations, and many other phenomena. Temperature and humidity strongly affect the mean lifetimes of air ions; in most cases lifetimes vary from several seconds to thousands of seconds.[22] This feature is beneficial as a high concentration of negative air ions is conducive to microbial sterilization. Moreover, it is known that air ions may affect the moods of people; a high proportion of negative ions in the air makes people feel lively, uplifted and enthusiastic, while too many positive ions will leave people feeling depressed, lethargic and full of aches, pains and complaints. Negative air ions are considered to be a living resource for human beings, leading to the widespread use of air ionizers. The following presents some commercial apparatuses, both photocatalytic and photocatalytic/ionizer.

7.2.1 Photo-CREC-Air Reactor

For air treatment, researchers at the Chemical Reactor Engineering Centre (CREC)[13] of the University of Western Ontario (Canada) have designed and constructed the Photo-CREC-Air unit with a Venturi design and TiO_2 anchored on an irradiated mesh.[23-26] Figure 7.16 shows a schematic diagram of the unit set-up.

The main body of the closed-loop system is made of zinc-plated pipes connected with aluminized-steel 90° elbows and a stainless steel Venturi section. External lamps irradiate the reaction section. Radiation penetrates through windows placed in the divergent section of the Venturi. Pen-Ray lamps are symmetrically placed around the reaction section and housed in reflectors for uniform irradiation distribution. The reaction section is formed by a basket supporting a fiber-glass mesh impregnated with TiO_2.

The Photo-CREC-Air unit operates in batch mode with a given amount of model pollutant injected in a set volume of air. Model pollutants are vaporized almost instantaneously and mixed intimately with the air stream. A favorable feature of this reactor is that it provides the highest gas velocity in the near-window region, ensuring a high degree of window sweeping and preventing TiO_2 particle deposition on the windows.

The bench-scale Photo-CREC-Air unit shows the same useful features as the Photo-CREC-Water reactors (high mixing and catalyst–gas contact, efficient light–TiO_2 interaction) and it is highly suited for investigation purposes.

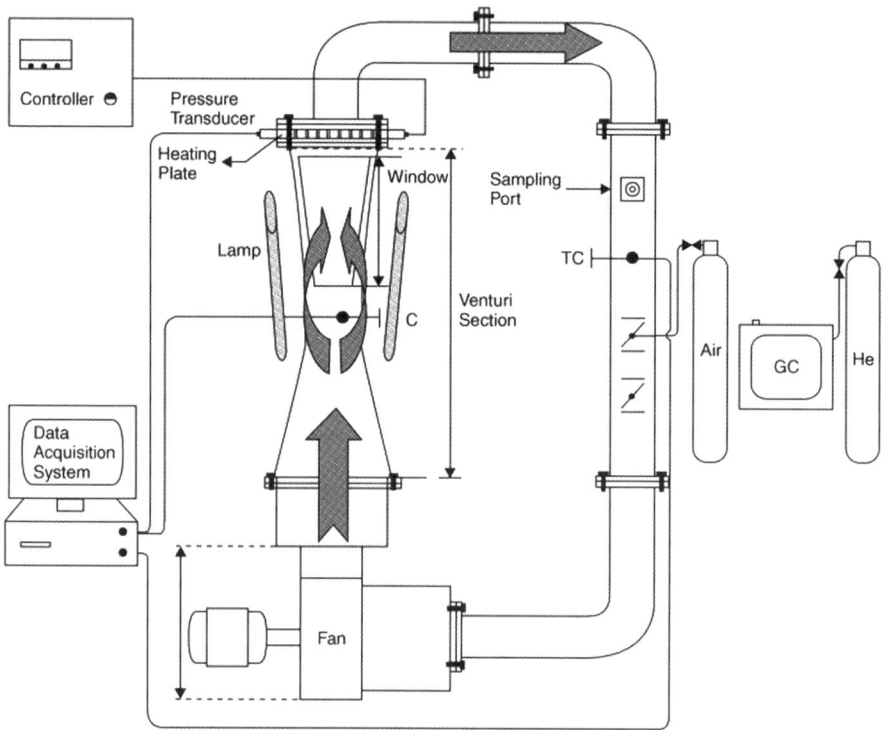

Figure 7.16 Schematic diagram of the Photo-CREC-Air unit set-up. (Reproduced with permission from ref. 25.)

7.2.2 AirSteril Purifier

The AirSteril air purifier[27] (Figure 7.17) uses ultraviolet germicidal light (UVC, 253.7 nm) together with a photocatalyst (nanostructured TiO_2 supported on plates) to kill airborne pathogens (viruses, bacteria, fungi, *etc.*) and at the same time to decompose odors as well as harmful gases. Indoor air pollutants are converted into harmless substances such as CO_2 and water vapor by the multistage nano-titanium dioxide coated base plates. Negative ions generated by the strong UVC irradiation of air can also remove airborne positively charged dust particles and pollen through attraction rendering them too heavy to remain airborne.

Table 7.1 describes some characteristics of AirSteril device.

7.2.3 Airlife Purifier

Figure 7.18 shows the working scheme and a photograph of the Airlife air cleaner.[28] These photocatalytic air purifiers are equipped with a special porous carrier with a TiO_2 photocatalyst, which adsorbs the light quanta and through which the air flows. The main function of the TiO_2 photocatalyst is to adsorb

Devices for Water and Air Purification 215

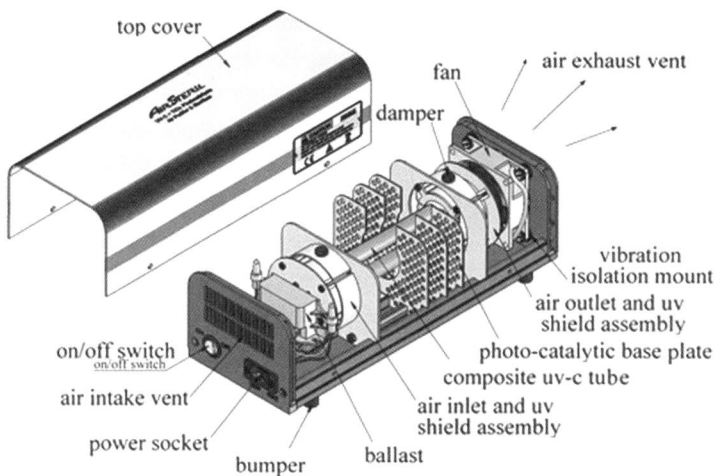

Figure 7.17 Details of the AirSteril air purifier. (Reproduced with permission from ref. 27.)

Table 7.1 Some features of the AirSteril device.

Feature	Specification
UVC lamp tube	13 W composite tube
UVC intensity inside chamber	$10\,\text{mW}\,\text{cm}^{-2}$
UVC lamp tube life	8000 h
Number of photocatalyst base plates	3
Dimensions (length × width × height)	$350 \times 130 \times 1000\,\text{mm}$
Air flow rate	$20\,\text{m}^3\,\text{h}^{-1}$
Power consumption	25 W
Covering area	$25\text{--}50\,\text{m}^2$ (hospital/clinic $10\text{--}25\,\text{m}^2$)
Installation	Wall mount/table top
Quantity of ozone produced	$8\text{--}80\,\text{mg}\,\text{h}^{-1}$

organic and inorganic contaminants that are then decomposed into CO_2 and H_2O under the influence of the UV lamp, which irradiates with mild ultraviolet light at a wavelength less than 300 nm (A-range). The reaction runs at room temperature.

Airlife Purifiers have been tested in degradation experiments of organic and inorganic substances at the Novosibirsk Institute of Catalysis. In addition, experiments on the pathogenic viruses and bacteria elimination have been carried out at the Novosibirsk Scientific Research Institute of Tuberculosis.

7.2.4 Daikin Purifier

Daikin air purifiers, designed for use in residences and small commercial premises, utilize a two-stage electrostatic plasma ionizer combined with a

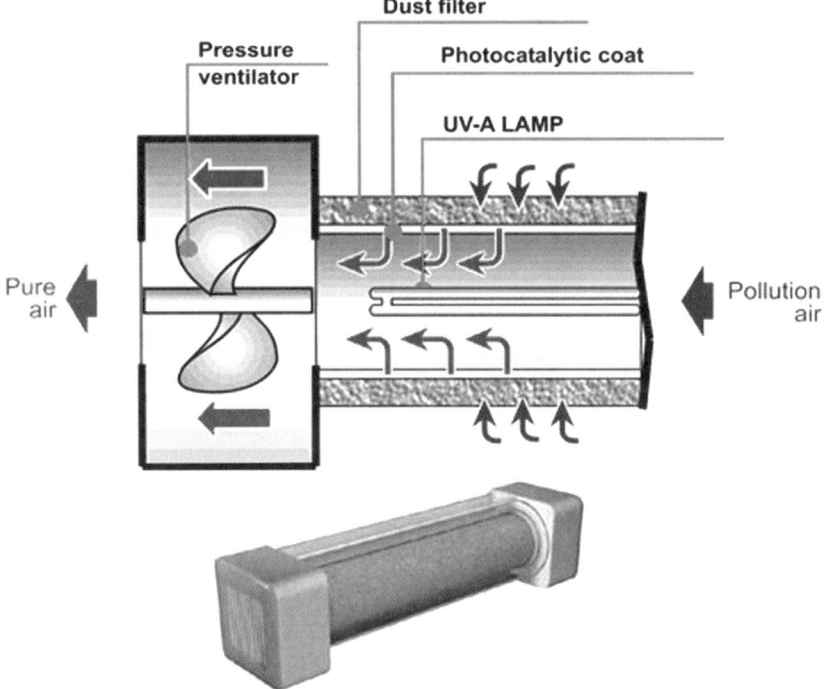

Figure 7.18 Working scheme and photograph of the Airlife Purifier device. (Reproduced with permission from ref. 28.)

titanium dioxide photocatalytic filter and UV light.[29] The indoor climate is enhanced still further by means of negative ion generation, allowing purified air to be refreshed and circulated throughout the conditioned space, adding a natural and spring-like quality to the atmosphere.

Daikin air purifiers consist of (i) a prefilter in which large dust particles and pet hairs are captured and the bacteria and mould spores attached to them are removed by a polyphenolic antioxidant (catechin); (ii) a bio-antibody filter in which airborne viruses are quickly adsorbed by the bio-antibodies and then rapidly removed; (iii) a flash streamer consisting of a plasma ionizer in which fine dust and pollen are positively charged; (iv) an electrostatic dust collection filter in which the positively charged substances are adsorbed by a negatively charged filter; (v) a titanium apatite photocatalytic filter in which organic substances, mould spores and viruses are adsorbed and removed; and (vi) a streamer deodorizing catalyst in which odors are removed.

Figure 7.19 shows photographs of Daikin air cleaners and Figure 7.20 gives the scheme of their working principle. Figure 7.21 illustrates how the flash streamer unit works.

Devices for Water and Air Purification 217

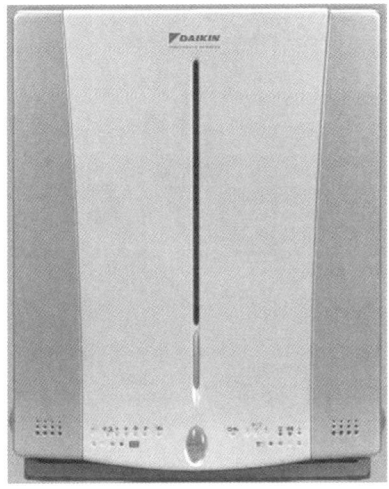
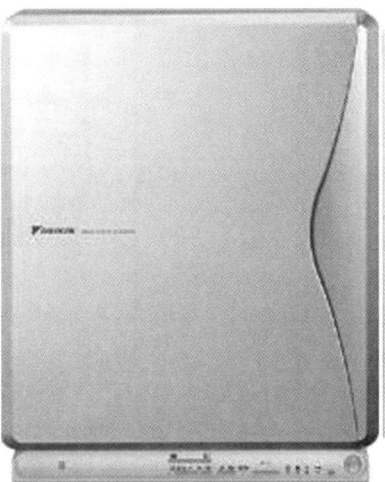

Daikin MC704VM   Daikin MC707B

Figure 7.19 Daikin air purifier models. (Reproduced with permission from ref. 29.)

7.2.5 Genesis Air Purifier

Genesis Air[30] incorporates the Genesis Air Photocatalyst Gap™ technology to oxidize airborne biologics and volatile organic compounds.

To achieve air quality a synergistic approach is realized in a three-step process: (i) capture by filtration of particles such as pollen, mold, mildew, house dust, bacteria, *etc.*, (ii) cleaning by ultraviolet lamps and (iii) photocatalytic conversion that produces hydroxyl radicals (oxidant primary species) able to oxidize biologics in the air stream and volatile organic compounds. Figures 7.22–25 show some Genesis purifiers and their applications.

7.2.6 Airwise® Purifiers

Airwise[31] is based on a technology that does not need mechanical filtration methods. Airwise® technology oxidizes odors, fungi, mold and parasites, and toxic chemical gases. At the same time, it settles dust and other large particles out of the air and destroys microorganisms like bacteria and viruses.

Figure 7.26 shows the scheme and a photograph of this purifier. Airwise® creates inductive warmth, drawing humidity into the purifier. When humidified air enters the target area, a photocatalytic reaction occurs and superoxide ions and hydroxyl radicals are created. As an added benefit, positively charged indoor pollutants are treated by active negative ions generated at an accelerated rate, thereby enhancing the speed at which dust and other airborne particles drop out of the breathing space. Airwise® technology works "outside of the box" so it is not necessary for polluted air to actually enter the purifier. This process significantly reduces allergens such as pollen and it destroys viruses,

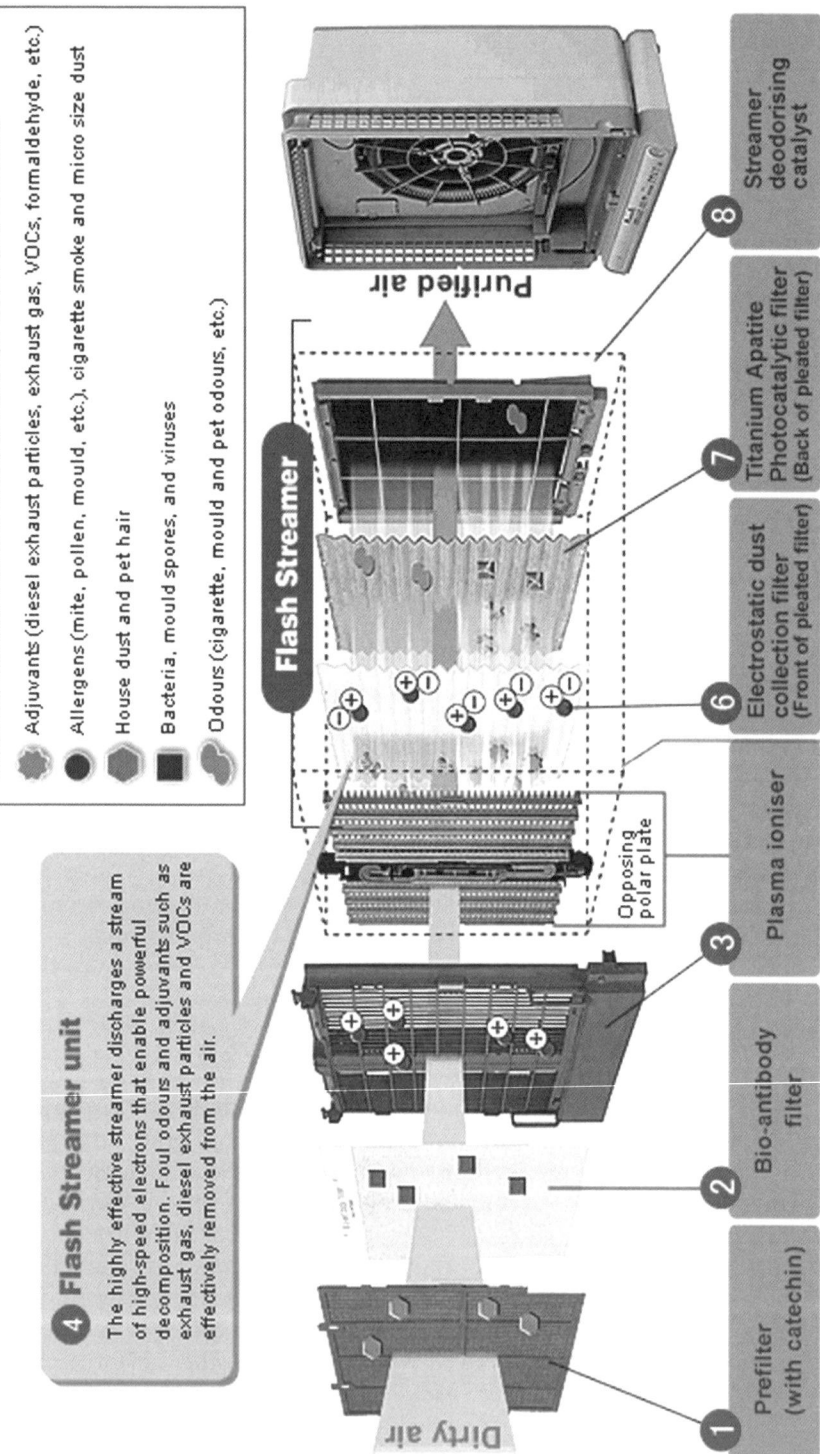

Figure 7.20 Scheme of the working principle of Daikin air purifiers. (Reproduced with permission from ref. 29.)

Devices for Water and Air Purification 219

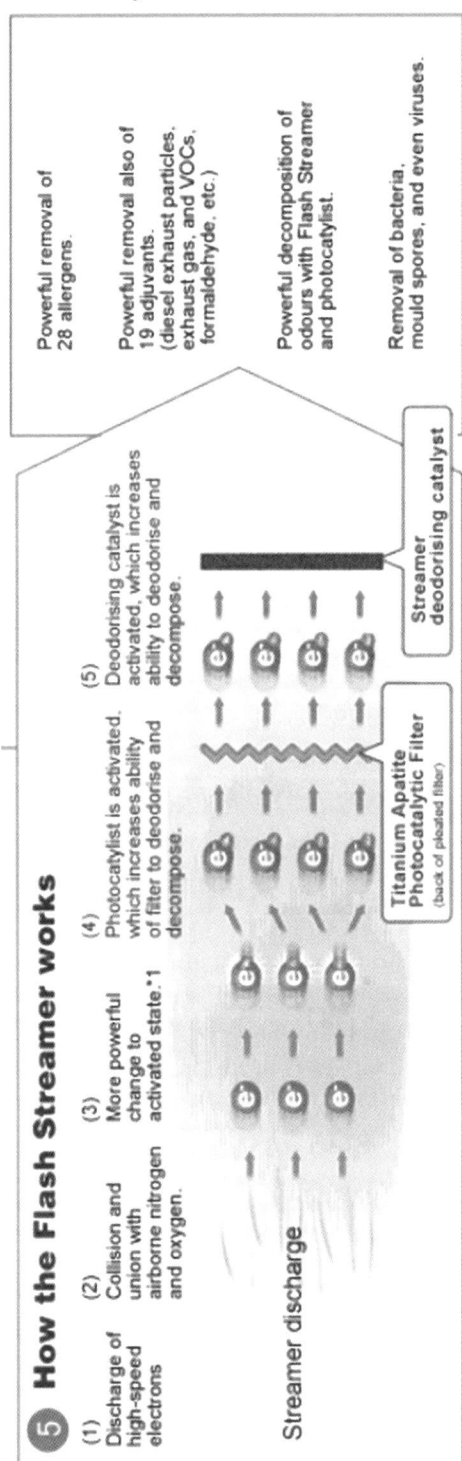

Figure 7.21 Scheme of the flash streamer unit of Daikin air purifiers. (Reproduced with permission from ref. 29.)

Figure 7.22 Stand-alone system designed to fit in a grid ceiling. (Photograph courtesy of ref. 30.)

bacteria, mold and mildew. Airwise® needs no fan and it has no grid or filter, so it requires no cleaning or filter replacement.

7.2.7 "Luch" Series Cleaners

Air cleaners of the "Luch" series[32] are intended for the purification and disinfecting of air in living quarters and administrative rooms, schools, medical establishments, libraries, *etc*. Table 7.2 shows the characteristics of the devices. Figures 7.27 and 7.28 show the working scheme of Luch-22 and Luch-60 Air cleaners, respectively (photographs given in Figure 7.29).

The unit operation is based on the photocatalytic oxidation of organic admixtures on the photocatalyst surface under mild UV-radiation.[33] In this way it can neutralize various toxic compounds and microorganisms that are not

Devices for Water and Air Purification

Figure 7.23 Standard photocatalytic purifiers that may be installed in supply or return ducts. (Photograph courtesy of ref. 30.)

amenable to traditional (adsorptive, electrostatic, catalytic) methods. The range of pollutant concentrations for effective destruction is 1 to 10 admissible concentration limits.

7.2.8 Aero Super Element Cleaners

The photocatalytic air cleaner Aero Super Element[34] is an ecologically safe device for clearing and decontaminating air. Figure 7.30 shows the device.

The Aero Super cleaner is particularly recommended for use in antitubercular and other medical institutions, including premises with a normalized level content (surgeries, dressing examinations, intensive care units, labor

Figure 7.24 Wall mounted or portable reduced size gaming special unit. (Photograph courtesy of ref. 30.)

wards, laboratories, drugstores, isolation wards), and also in all other premises that require a high degree of decontamination and cleaning of air.

Aero Super devices are designed for continuous working and they guarantee a high efficiency of decontamination and cleaning of air. The decontaminating agent provides a high degree of air decontamination, two and more times larger than existing methods of air decontamination (bactericidal lamps). The ultraviolet radiation band (320–400 nm) used and low noise level (up to 32 dB) permit comfortable use of device for 24-hour action.

7.2.9 Zand-Air Cleaners

The ZAND-AIR™ 200 system[35] incorporates leading-edge photocatalysis and high-efficiency particle filtration, and converts volatile organic compounds and chemically active compounds into water and carbon dioxide, while capturing sub-microscopic-particles.

This system is designed for in-line, ducted multiphase filtration systems and it may be incorporated in new or existing construction and renovations to provide protection for people and laboratory environments against indoor air pollutants. The uniqueness of this product lies in its ability to clean and eliminate all pollutants within the capacity of existing blowers and filtration equipment. Figure 7.31 shows the set-up of the apparatus.

The first step of the purification process is the adsorption of toxic chemicals and gases: the activated carbon filter in the front position adsorbs organic

Devices for Water and Air Purification 223

Figure 7.25 Genesis unit designed to be ceiling mounted or portable on casters. (Photograph courtesy of ref. 30.)

hydrocarbons, volatile organic compounds, chemically active compounds and other harmful agents. Microparticles are then removed by a high efficiency particulate air (HEPA) filter, capable of removing at least 99.97% of airborne particles (pollen, mold, bacteria, *etc.*) 0.3 μm in diameter. The next step is the destruction of toxic chemicals and odors by photocatalytic oxidation: the oxidation converts malign toxic compounds (even carbon monoxide and nitrous oxide) into benign constituents such as H_2O and CO_2. The catalyst is such that it does not wear out or lose its effectiveness as a result of its actions. The last step consists of purification by ultraviolet light: the ultraviolet light attacks the molecular structure of viruses and bacteria, which are too small to be filtered out by the HEPA filter, thus rendering them harmless.

Figure 7.32 shows, viewed from below, three ZAND-AIR™ 200 units installed in frames as a battery below the roof, above a drop ceiling, mounted

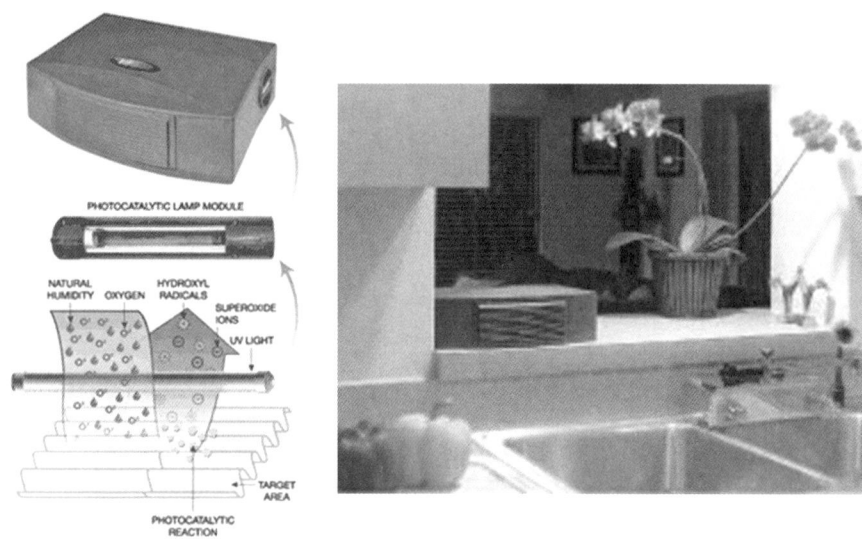

Figure 7.26 Airwise® air purifier. (Photograph courtesy of ref. 31.)

Table 7.2 Specifications of air cleaners of the Luch series.

Parameter	LUCH-22	LUCH-60	LUCH-400
Power consumption (W)	25	100	1100
Air flow ($m^3 h^{-1}$)	23	40/85/130	600–1500
Noise level (dB)	Up to 34	40–55	30–70
Photocatalyst lifetime (years)	5	10	10
Weight (kg)	1	7	62
Dimensions (mm)	Diameter 260; height 74	345 × 175; height 500	1625 × 555; height 855

into the air conditioning enclosure. The units may be unplugged for removal and servicing outside of their frame structures.

Figure 7.33 shows a system suitable for clean air inside laboratories.

7.2.10 Comefresh Electronic Industry Cleaner

The air cleaner of Comefresh Electronic Industry[36] uses the technology of a four-stage filtration process: (i) HEPA filter; (ii) an ultra-low particulate air (ULPA) filter theoretically able to remove from the air at least 99.999% of dust, pollen, mold, bacteria and any airborne particles 120 nm or larger in size; (iii) UV light, photocatalyst filter; and (iv) activated carbon filter. In this way the cleaner ensures the removal of particles, harmful vapors, odors and volatile organic compounds; moreover, it traps pollen and kills bacteria and viruses. Figure 7.34 shows a photograph of the Comefresh air cleaner.

Devices for Water and Air Purification 225

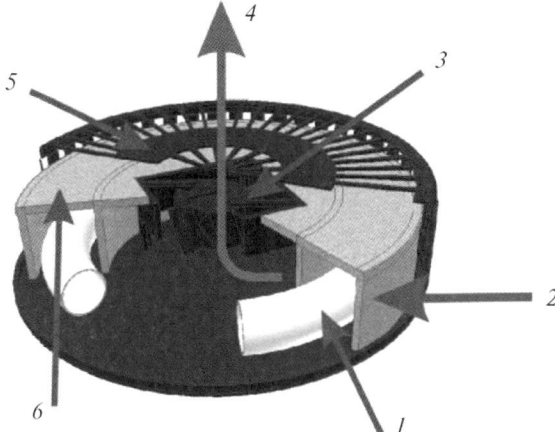

Figure 7.27 Luch-22 photocatalytic air cleaner produced at the Luch NPO (Novosibirsk, Russia): 1, UV lamp; 2, contaminated air; 3, ventilator; 4, purified air; 5, case; and 6, filter with photocatalyst. (Reproduced with permission from ref. 32.)

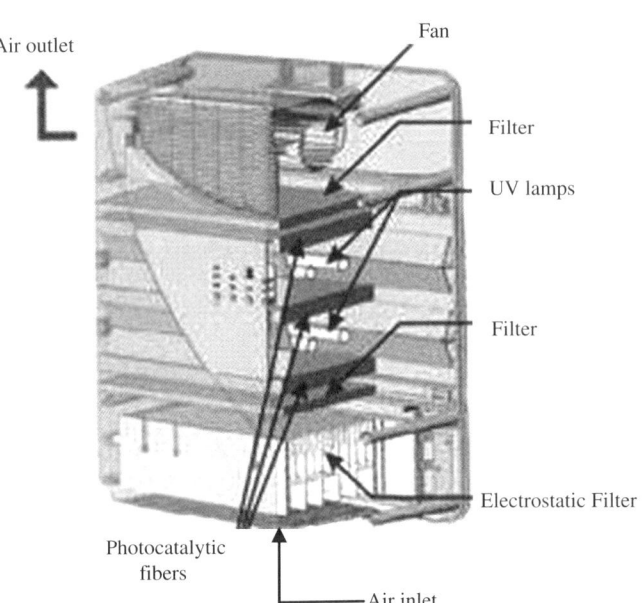

Figure 7.28 Working scheme of Luch-60 Air cleaner. (Reproduced with permission from ref. 32.)

Figure 7.29 Photograph of Luch-22 and Luch-60 Air cleaners. (Photograph courtesy of ref. 32.)

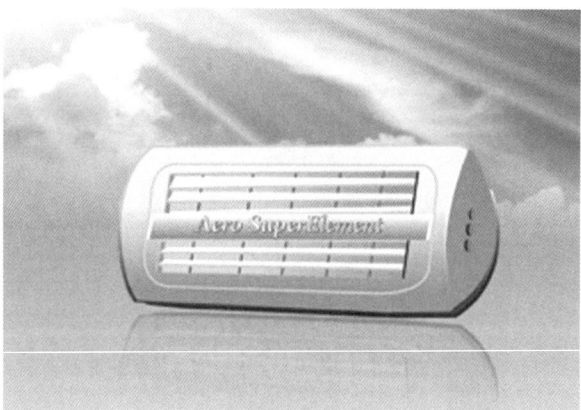

Figure 7.30 Photograph of an Aero Super Element air cleaner. (Photograph courtesy of ref. 34.)

7.2.11 Airpura Purifiers

Airpura air purifiers[37] give particular attention to air flow. While in other systems air moves around different chambers and changes direction, Airpura simplifies air flow for quieter more efficient performance, so it reduces noise. All of the air is filtered and no particles or gases escape the filters so that the

Devices for Water and Air Purification

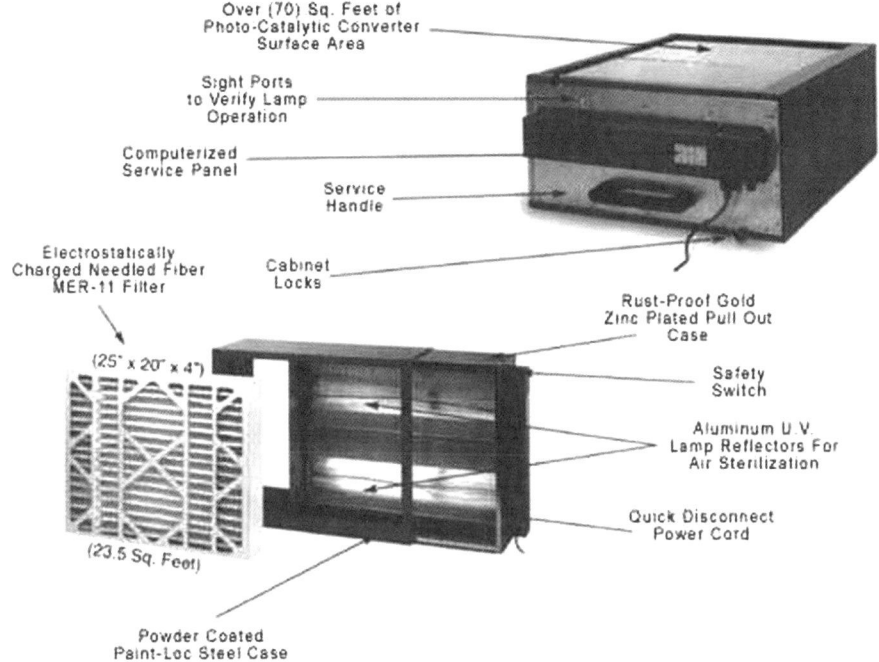

Figure 7.31 Set-up of ZAND-AIR™ 200. (Photograph courtesy of ref. 35.)

Figure 7.32 Typical ZAND-AIR™ 200 installation. (Photograph courtesy of ref. 35.)

Figure 7.33 A zIVF-AIRe 100C CLEAN AIR™ unit, a system that provides air filtration and purification for IVF laboratories. (Photograph courtesy of ref. 35.)

claimed 99.97% HEPA efficiency may be achieved. Figure 7.35 gives a picture and working scheme of an Airpura 600 Series purifier.

The core of the apparatus is the TitanClean reflector, the surface area of which is coated with an optimized amount of TiO_2. The angled reflector design maximizes the range of photocatalytic oxidization within the filter chamber. The location of the TitanClean reflector and the UV light in the centre of the filter chamber allows them to work in concert with the HEPA filter. Particulate pollution is stopped by the HEPA filter before reaching the reflector. This keeps the coated surface cleaner and more effective.

The UV germicidal lamp does not generate ozone and it maintains 98% of its antigen and pathogen destruction effect due to the interior position of the reflector. Contact time of airborne chemicals is increased as they slow down on passing through the HEPA. Diffusion of the germicidal dosage from the UV lamp is enhanced in the confined chamber.

Figure 7.34 Photograph of a Comefresh air cleaner. (Photograph courtesy of ref. 36.)

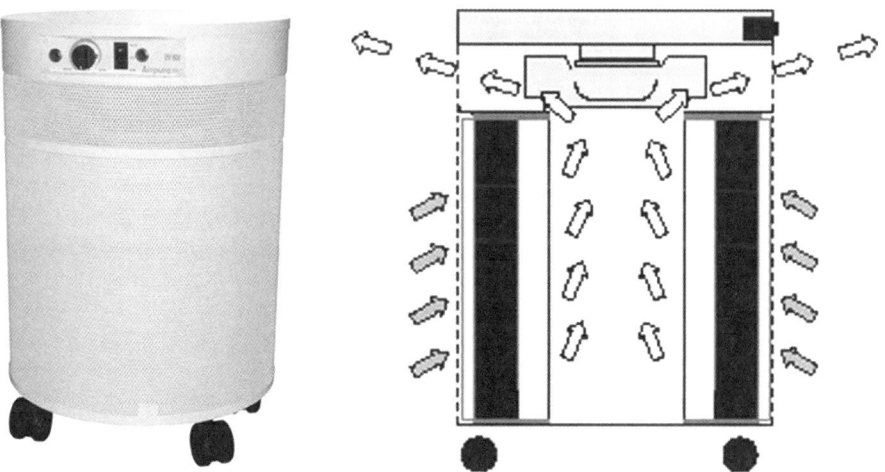

Figure 7.35 Picture and working scheme of an Airpura 600 Series purifier. (Photograph courtesy of ref. 37.)

7.2.12 Air Oasis™ Purifier

In Air Oasis™ purifier[38] apparatuses the photocatalytic method is coupled with the generation of negative air ions.[21]

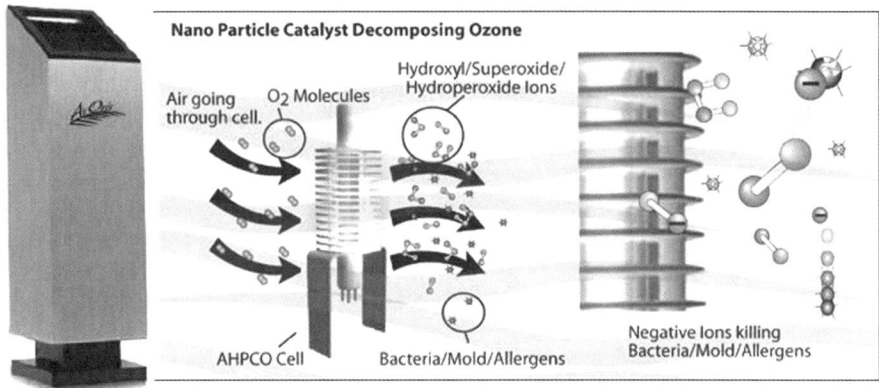

Figure 7.36 Picture and working scheme of an Air Oasis™ purifier. (Photograph courtesy of ref. 38.)

Figure 7.37 Picture of an Air Sterilizer "Medicare." (Photograph courtesy of ref. 39.)

The Air Oasis™ purifiers utilize a high-performance TiO_2 catalyst placed on an advanced hydrated photocatalytic oxidation (AHPCO™) cell. The AHPCO cell consists of a high-intensity broad spectrum UV tube in a hydrated catalytic matrix cell that generates hydroperoxides, superoxide ions, hydroxyl ions, negative ions and ozonide ions that are released to air circulating through the cell, providing protection against airborne fungi, bacteria, viruses, odors and volatile organic compounds. Low-level ozone is produced in the cell, most of which is converted into airborne hydroperoxides and hydroxyl ions via the metallic catalyst. The photogenerated ions treat indoor air by suppressing bacteria and fungi growth, volatile organic compounds and reducing odors. Figure 7.36 shows a picture and working scheme of an Air Oasis™ purifier.

7.2.13 Air Sterilizer "Medicare"

This photocatalytic purifier[39] uses a honeycomb photocatalyst for sterilizing air. The fiber structure of the photocatalyst maximizes the surface in contact with air, thus giving a high efficiency to the system. In this way it can remove ordinary bacteria, dust, toxic substances and offensive odors and prevent respiratory diseases. Figure 7.37 shows an Air Sterilizer "Medicare."

7.2.14 Photocatalytic Cold Fluorescent Lamp

A titanium dioxide layer coats the bulb of these lamps,[40,41] which have a standard base socket (Figure 7.38). The fluorescent light bulb uses fluorescent light and TiO_2 to transform tiny air particles and toxic gases into safer substances. In this way fresh, clean air is created through a product

Figure 7.38 A photocatalytic cold fluorescent lamp. (Photograph courtesy of ref. 40.)

that is maintenance-free as there are no air filters to replace or collecting grids to clean.

References

1. M. Schiavello (ed.), *Photocatalysis and Environment. Trends and Applications*, Kluwer Academic Publishers, Dordrecht, 1988.
2. D. F. Ollis and H. El-Ekabi (eds), *Photocatalytic Purification and Treatment of Water and Air*, Elsevier Science Publishers, New York, 1993.
3. M. Schiavello (ed.), *Heterogeneous Photocatalysis*, Wiley Series in Photoscience and Photoengineering John Wiley & Sons, **vol. 3**, Chichester, 1997.
4. M. Schiavello (ed.), *Photoelectrochemistry, Photocatalysis, and Photoreactors, Fundamentals and Developments, Reidel*, Dordrecht, 1985.
5. E. Pelizzetti and N. Serpone (eds), *Homogeneous and Heterogeneous Photocatalysis*, Reidel, Dordrecht, 1986.
6. N. Serpone and E. Pelizzetti (eds), *Photocatalysis, Fundamentals and Applications*, John Wiley & Sons, Chichester, 1989.
7. S. Malato, J. Blanco, A. Vidal, P. Fernández, J. Cáceres, P. Trincado, J. C. Oliveira and M. Vincent, New large solar photocatalytic plant: set-up and preliminary results, *Chemosphere*, 2002, **47**, 235–240.
8. J. Blanco, S. Malato, P. Fernández, D. Alarcón, W. Gernjak and M. I. Maldonado, Solar energy and feasible applications to water processes, in *Proceedings 5th European Thermal-Sciences Conference*, ed. G. G. M. Stoffels, T. H. van der Meer and A. A. van Steenhoven, ISBN 978-90-386-1274-4, Eindhoven, 2008. Available online at http://www.eurotherm2008.tue.nl/
9. V. Augugliaro, J. C. Conesa, E. García López, V. Loddo, M. J. López Muñoz, G. Marcí, L. Palmisano, M. Schiavello and J. Soria, Pilot photoreactor for cyanides degradation, *Chim. Ind.*, 2001, **83**, 34.
10. V. Augugliaro, V. Loddo, G. Marcì, L. Palmisano and M. J. López-Muñoz, Photocatalytic oxidation of cyanides in aqueous titanium dioxide suspensions, *J. Catal.*, 1997, **166**, 272–283.
11. V. Augugliaro, J. Blanco Gálvez, J. Cáceres Vázquez, E. García López, V. Loddo, M. J. López Muñoz, S. Malato Rodríguez, G. Marcì, L. Palmisano, M. Schiavello and J. Soria Ruiz, Photocatalytic oxidation of cyanide in aqueous TiO_2 suspensions irradiated by sunlight in mild and strong oxidant conditions, *Catal. Today*, 1999, **54**, 245–253.
12. V. Augugliaro, E. García López, V. Loddo, G. Marcì and L. Palmisano, Degradation kinetics of iron(III) cyanocomplexes in irradiated systems, *Adv. Environ. Res.*, 1999, **3**, 179–188.
13. H. de Lasa, B. Serrano and M. Salaices, *Photocatalytic Reaction Engineering*, Springer, 2005.

14. J. Valladares, A new photocatalytic reactor for the photodegradation of organic pollutants in water. Ph.D. Dissertation, University of Western Ontario, London, Canada, 1995.
15. H. I. de Lasa and J. Valladares, Photocatalytic reactor, *US Pat. No. 5683589*, 1997.
16. B. Serrano and H. I. de Lasa, Photocatalytic degradation of water organic pollutants. Kinetic modeling and energy efficiency, *Ind. Eng. Chem. Res.*, 1997, **36**, 4705–4711.
17. B. Serrano, A. Ortíz, J. Moreira and H. I. de Lasa, Energy efficiency in photocatalytic reactors for the full span of reaction times, *Ind. Eng. Chem. Res.*, 2009, **48**, 9864–9876.
18. http://www.ube-ind.co.jp/english/rd/business.htm
19. T. Ishikawa, Photocatalytic fiber with gradient surface structure produced from a polycarbosilane and its applications, *Int. J. Appl. Ceram. Technol.*, 2004, **1**, 49–55.
20. J. M. Coronado, J. Soria, J. C. Conesa, R. Bellod, C. Adán, H. Yamaoka, V. Loddo and V. Augugliaro, Photocatalytic inactivation of *Legionella pneumophila* and an aerobic bacteria consortium in water over TiO_2/SiO_2 fibres in a continuous reactor, *Top. Catal.*, 2005, **35**, 279–286.
21. J. Zhang and Z. Yu, Experimental and simulative analysis of relationship between ultraviolet irradiations and concentration of negative air ions in small chambers, *Aerosol Sci.*, 2006, **37**, 1347–1355.
22. L. Aare and P. Tiia, Evolution of negative small air ions at two different temperatures, *J. Atm. Solar-Terrestrial Phys.*, 2002, **64**, 763–774.
23. H. Ibrahim, Photocatalytic reactor for the degradation of airborne pollutants, photoconversion efficiency and kinetic modeling, Ph.D. Dissertation, University of Western Ontario, London, Canada, 2001.
24. H. Ibrahim and H. I. de Lasa, Photo-catalytic conversion of air borne pollutants. Effect of catalyst type and catalyst loading in a novel photo-CREC-air unit, *Appl. Catal. B: Environ.*, 2002, **38**, 201–213.
25. H. Ibrahim and H. I. de Lasa, Photo-catalytic degradation of air borne pollutants. Apparent quantum efficiencies in a novel photo-CREC-air reactor, *Chem. Eng. Sci.*, 2003, **58**, 943–949.
26. H. Ibrahim and H. I. de Lasa, Kinetic modeling of the photocatalytic degradation of air-borne pollutants, *AIChE J.*, 2004, **50**, 1017–1027.
27. http://www.airpurifier.com.hk/product-en.htm
28. http://www.vozdyx.ru/eng/foto.shtml
29. http://www.daikin-air-purifiers.co.uk
30. http://www.genesisair.com/products.html
31. http://www.simply-natural.biz/airwise.php
32. http://www.en.catalysis.ru/block/index.php?ID=27&SECTION_ID=294.
33. I. A. Baturov, A. V. Vorontsov and D. V. Kozlov, Regularities of decomposition of organic vapors using a photocatalytic air cleaner, *Russ. Chem. Bull., Int. Ed.*, 2005, **54**, 1866–1873.
34. http://www.eleline.com/en_air
35. http://www.zander-air-purification.com/zand-air200-air-filter.html

36. http://comefresh.manufacturer.globalsources.com/si/6008821537308/Homepage.htm
37. http://www.nontoxic.com/airpura/airpura.html
38. http://www.peakpureair.com/airoasis.htm
39. http://www.tradekorea.com/product-detail/P00047177/Air_Sterilizer_Medicare_.html
40. http://www.purion.ie/component/virtuemart/category/6/bulbs.html
41. http://www.germfreebulb.com/howitworks.html

CHAPTER 8
Standardization

8.1 Introduction

Standardization is defined as "the activity consisting of the processes of formulating and implementing documents (the standards) that provide rules or characteristics for activities or their results, aimed at the achievement of the optimum degree of order in a given context." Science, technology and experience contribute to issue standards, whose main benefits are improvement of the suitability of products, processes and services, prevention of technical barriers to trade and facilitation of technological cooperation.

Three bodies are responsible for the planning, development and adoption of International Standards: the International Electrotechnical Committee (IEC) is responsible for electrotechnical sector, the International Telecommunication Union (ITU) for most telecommunications technologies and the International Organization for Standardization (ISO) for all the other sectors, including heterogeneous photocatalysis.

ISO[1] is a legal association, the members of which are the National Standards Bodies representing social and economic interests at international level. The principal deliverable of the ISO is the International Standard, the main features of which are openness and transparency, consensus and technical coherence. These principles are followed by the ISO Technical Committee (ISO/TC), which is the technical body responsible for the planning and execution of the work.

In the field of heterogeneous photocatalysis, even though the photocatalysis industry has grown very much in the last 10 years, the lack of standard evaluation methods for photocatalytic products hinders consumer understanding of the performance of photocatalytic products and the ability to compare correctly the products from several producers. The establishment of standard evaluation methods for photocatalytic products is therefore a challenge for increasing their diffusion in industrial and private fields.

Japan has been a pioneer in this field, working since 2002 through the Photocatalysis Standardization Committee dedicated to proposing standard evaluation methods for (i) self-cleaning products; (ii) air purification products;

Clean by Light Irradiation: Practical Applications of Supported TiO_2
By Vincenzo Augugliaro, Vittorio Loddo, Mario Pagliaro, Giovanni Palmisano and Leonardo Palmisano
© V. Augugliaro, V. Loddo, M. Pagliaro, G. Palmisano and L. Palmisano 2010
Published by the Royal Society of Chemistry, www.rsc.org

(iii) water purification products; and (iv) antibacterial products. At present, however, the prominent national standards bodies – European Committee for Standardization (CEN), British Standards Institute (BSI), American Society for Testing Materials (ASTM), Japanese Industrial Standards Committee (JISC), *etc.* – are involved in proposing procedures for photocatalytic systems evaluation.

One of the main initiatives, set to address the issue of standardization in heterogeneous photocatalysis, is the "Photocatalytic Technologies and Novel Nanosurfaces Materials" (PHONASUM) Committee founded by the European Community. The ISO committee working on fine ceramics (TC 206) has recently published several photocatalyst test methods, originating from the JISC, including ones for assessing their air-purification performance via the removal of NO_x (ISO/DIS22197-1) and the measurement of water droplet contact angle (ISO/WD 27448-1).

8.2 Efficiency Parameters of Photocatalytic Systems

8.2.1 Quantum Yield

Heterogeneous photocatalytic processes, carried out under whatever conditions (suspension, film, batch or continuous), can be considered as a case of radiative transfer in participating media and in their modeling the radiant energy balance equation must be added to the usual equations of mass, heat and momentum balance. The obtained set of conservation equations is very difficult to solve because (i) the radiation field is a function of wavelength and concentration of all the radiation absorbing species present in the participating medium; (ii) the fluidodynamics of a fluid–solid system is highly non-ideal so that the specific system needs to be a priori modeled; and (iii) the kinetics of a heterogeneous photocatalytic reaction are affected in a complex way by the concentration of the reacting species and by the radiation field.

Provided these difficulties are overcome, the solution of conservation equations allows one to know the values of radiation intensity, temperature and concentration of all species at each point of a reacting mixture. This information is very valuable as it permits us to determine the parameters needed for correctly comparing the performances of different photocatalytic systems,[2] *i.e.*, the rate of photon absorption (rpa) defined as:

$$\text{rpa} = \frac{\text{absorbed photons}}{(\text{time}) \cdot (\text{surface area})} \quad \left[\frac{\text{einstein}}{\text{s} \cdot \text{m}^2}\right] \quad (8.1)$$

and the intrinsic reaction rate (irr) defined as:

$$\text{irr} = \frac{\text{reacted molecules}}{(\text{time}) \cdot (\text{surface area})} \quad \left[\frac{\text{mol}}{\text{s} \cdot \text{m}^2}\right] \quad (8.2)$$

The ratio between irr and rpa allows the determination of the "quantum yield" (φ_λ) for monochromatic radiation:

$$\varphi_\lambda = \frac{\text{reacted molecules}}{\text{absorbed photons}} = \frac{\text{irr}}{\text{rpa}} \qquad (8.3)$$

There is a general agreement in the literature that the quantum yield, a parameter in the form of a "number ratio," is the most convenient estimator of photocatalytic efficiency. The other parameters used for quantifying photocatalytic activity (such as the time needed to achieve a certain percentage of degradation, the degradation rate, the half-life of a model compound, *etc.*) are strongly dependent on the experimental conditions used so that it is difficult to employ them for comparing different photocatalytic systems.

Importantly, even the quantum yield is not an absolute parameter. In fact in the quantum yield definition [Equation (8.3)] both the terms of the ratio, the rate of reaction and the rate of photon absorption, depend on intrinsic and operative features of the photoreacting system. The reaction rate depends heavily on parameters such as the intensity of the irradiation, the amount of the catalyst, the nature and concentration of the reacting molecules, the concentration of dissolved oxygen, *etc.* Only for zero-order kinetics should the reaction rate be uniquely defined at a given wavelength; often, to simplify the evaluation, the quantum yield is based on the initial rate of molecule photoconversion. The absorbed photon flow depends on photoreactor geometry, the relative position of photoreactor with respect to radiation source, the physical and chemical characteristics of the absorbing particles, the particle agglomeration determined by the properties of the surrounding fluid, *etc.*

Proper definition of quantum yield requires careful assessment of photons absorbed by the photocatalyst in the heterogeneous reactor. Even if the photoreactor set-up is standardized, the photon absorbed flow is a difficult parameter to evaluate due to the reflection, transmission and scattering phenomena occurring inside the photoreactor due to the presence of the semiconductor particles.[3] The extent of light scattered by the particulate matter in the dispersion may be very significant; according to some accounts, it may range from 20 to 80% of the total incident photon flow. Methods based on chemical actinometry are limited as they provide only the total rate of photons entering the reactor and do not account for various light scattering losses. In fact semiconductor surfaces are highly reflective and therefore the light incident on them can easily be back-scattered or forward-scattered. To overcome the difficulties encountered in the determination of the number of photons absorbed by the photocatalyst, and given that the rate of photons entering the reactor is a parameter much simpler to measure, researchers frequently consider an "apparent quantum yield" or a "photonic efficiency" as the parameter to be evaluated in standardization procedures.

Photonic efficiency is defined as the number of reacting molecules divided by the number of photons, at a given wavelength, incident inside the front window of the cell (flat parallel windows). Given that the rate of photons entering the reactor is always smaller than the rate of photons absorbed, the apparent quantum yield provides an underestimated value of the quantum yield. Even if the photonic efficiency overcomes the difficulty of determining the absorbed photon flow, it is a parameter that other investigators can use for comparing their results with those of others provided that reactor geometry and light source, together with the properties of the photocatalytic material used, are stipulated.

To standardize the efficiencies of mineralization process of various organic substrates for a given set of conditions, Serpone et al.[4] propose to refer all results to an equivalent experiment carried out under identical conditions for a standard process and standard photocatalyst. The efficiency parameter used to cross-reference the photocatalytic experiments is the "relative photonic efficiency," which is defined as the rate of disappearance of the substrate divided by the rate of disappearance of a standard molecule in a standardized process. Phenol photodegradation was suggested as the standard process and Degussa P25 TiO_2 as the standard photocatalyst. The method consists of determining photon efficiencies by measuring the initial degradation rate of a test substrate and the rate of photons incident on the front window of the reactor. When the photonic efficiencies for the test substrate, and for the standard one (i.e., phenol), are obtained under identical experimental conditions there will be no need to measure the absorbed photon flow. Moreover, if the quantum yield of a standard process is determined correctly, knowledge of relative photonic efficiency would allow the determination of the quantum yield of the test process.

This method has the advantage of simplicity and it overcomes all the difficulties encountered in the definition of heterogeneous photocatalytic process (determination of the absorbed photon flow, utilization of different light sources, different reactor geometries, etc.). The main drawback of this methodology is that the quantum yield of a standard process would vary enormously with different reaction conditions, and the value would have to be determined for each set of specific reaction conditions.

To compare different photocatalytic systems the use of quantum efficiencies requires calculation of the incident photons in reactor media, while the use of quantum yields requires assessment of the absorbed photons by TiO_2. Quantum yields and quantum efficiencies entail the implicit hypothesis that all the energy of the used photons contributes to the formation of •OH groups. Careful consideration of this matter is required for a proper evaluation of the energy efficiencies.

8.2.2 Experimental Method for the Determination of Absorbed Photons

In the absence of a solution of the integro-differential radiation transfer equation, the photon flow absorbed by the photocatalyst particles can be estimated by using a method reported in the literature.[5-11] By considering the scheme of Figure 8.1, the following macroscopic energy balance can be

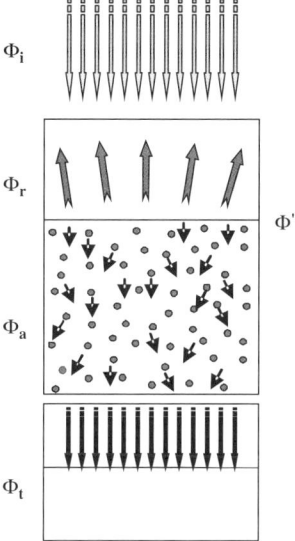

Figure 8.1 Scheme of photon balance for a simple photoreactor.

performed on the whole suspension under the assumption that it does not lose radiation through the lateral wall:

$$\Phi_i = \Phi_r + \Phi_a + \Phi_t \tag{8.4}$$

where Φ_i is the incident photon flow, Φ_r the backward scattered photon flow, Φ_a the absorbed photon flow and Φ_t the transmitted photon flow.

It was found[7,8] that the dependence of transmitted photon flow on mass of catalyst is best expressed by the following relationship when the mass of catalyst is higher than the critical one (the amount of photocatalyst sufficient to cover the cross section of the photoreactor):

$$\Phi_t = \Phi' exp(-Em_{cat}) \tag{8.5}$$

where Φ' is a constant whose value is always less than that of Φ_i, E the napierian extinction coefficient and m_{cat} the mass of catalyst.

The energy balance under the limiting condition of $m_{cat} = 0$ allows the determination of Φ_r:

$$\Phi_r = \Phi_i - \Phi' \tag{8.6}$$

where Φ_r represents the photon flow backward reflected by a medium so optically thick that the reflectance may be hypothesized to be independent of the medium depth, *i.e.*, of the catalyst mass.

8.2.3 Photochemical Thermodynamic Efficiency Factor

Serrano and De Lasa[12] first introduced the "photochemical thermodynamic efficiency factor" (PTEF), which evaluates the performance of a photocatalytic system on a thermodynamic basis. A macroscopic point of view is adopted for defining PTEF. In a photocatalytic reactor containing the irradiated catalyst a fraction of energy emitted by the lamp is absorbed by the photocatalyst, Q_{abs}. A fraction of this absorbed energy is used for ·OH radical formation, Q_{used}, while the remainder is dissipated as thermal energy and/or through parasite reactions. The relevance of Q_{used} and Q_{abs} for the overall performance of photoprocess is quantified through the definition of a dimensionless parameter, the PTEF:

$$\text{PTEF} = \frac{Q_{used}}{Q_{abs}} \tag{8.7}$$

By considering that the catalyst contained in the photoreactor consumes the Q_{used} energy to produce ·OH radicals, the PTEF definition may be transformed into the following:

$$\text{PTEF} = \frac{r_{rad} \Delta H_{rad} W}{Q_{abs}} \tag{8.8}$$

where r_{rad} represent the formation rate of ·OH radicals per unit mass of catalyst, ΔH_{rad} the formation enthalpy of an ·OH radical and W the total mass of catalyst.

The PTEF is a function of the formation rate and of the formation enthalpy of ·OH radical. The latter was estimated by Serrano and de Lasa[12] in aqueous media, starting from H_2O and dissolved oxygen, as 94.6 kJ mol^{-1}. Concerning r_{rad}, by considering that the ·OH radical species is very reactive, it may be assumed that its rate of formation is equal to its rate of consumption as determined by the reaction with both the adsorbed model pollutant and the adsorbed intermediates. At the very beginning of the photoconversion the surface concentration of all chemical species equals the surface concentration of the model compound, which acts as the only ·OH scavenger; under these conditions it may be assumed that the rate of ·OH consumption is equal to the consumption rate of model pollutant, so the following relationship holds:

$$\frac{r_{rad}}{v_{rad}} = \frac{r^0_{MP}}{v_{MP}} \tag{8.9}$$

where r^0_{MP} represents the rate of disappearance of the model pollutant at time $t=0$ (a directly measurable parameter) and v_{rad} and v_{MP} are the stoichiometric coefficients for the consumption of radical and model pollutant, respectively.

The PTEF increases by increasing the initial rate of disappearance of the model pollutant, i.e., by increasing the pollutant concentration. The driving force of the disappearance rate of model pollutant on the catalyst surface is its fractional site coverage, whose highest value is one. Therefore, an upper limit

Standardization

exists for the PTEF values; this upper value is an intrinsic characteristic of the photodegraded pollutant as well as of the photoreactor. The PTEF_{max} may be considered to be the product of two yields:

$$\text{PTEF}_{max} = \varphi^0_{max}\eta_{rad} \tag{8.10}$$

where φ^0_{max} represents the maximum fraction of photons (absorbed by the photocatalyst at initial conditions) that give rises to the formation of ·OH radicals, and η_{rad} the fraction of the photon energy used in the formation of ·OH radicals. The product of these two parameters provides an assessment of the energy efficiency of a photoreacting system.

The PTEF is a dimensionless quantity, as required by thermodynamic consistency; it can be broadly applied, covering various kinetic models and being appropriate for various photochemical reactors, either homogenous or heterogeneous.

8.2.4 Technological Parameters

A prerequisite of any technology for water cleaning is that it be able to remove the relevant pollutant together with the conversion of total organic carbon into inorganic carbon. The oxidation processes, including the photocatalytic ones, must guarantee the complete mineralization of all organic carbon to insure that the substrate(s) and any intermediate product(s) formed during the process have also been degraded.

That prerequisite being satisfied, in selecting a specific waste-treatment technology several important factors must be considered, such as economics, regulations, eluent quality standards, operation *etc*. Among these factors economics is generally the most important. Since photocatalytic degradation of aqueous organic pollutants is an electric-energy intensive process (processes under solar irradiation are not considered here), electric energy can represent a major fraction of the operating costs.

If the efficiency of the photocatalytic process determines its economic viability, the key design variables, *i.e.*, exposure to lamp radiation and order of magnitude of contaminant concentration removal, are combined into parameters that for a fixed volume of contaminated water give a measure of the treatment rate as a function of applied energy.

Two figures-of-merit are proposed:[13–15] one suitable for high organic concentrations and the other for low concentrations. The "electric energy per mass" (EEM) is defined as the electric energy required to obtain the degradation of a unit mass of a contaminant in aqueous bodies or atmosphere. The EEM parameter is most useful when the concentration is high because under this condition the rate of removal of the contaminant generally follows zero-order kinetics with respect to contaminant concentration and therefore the removal rate is directly proportional to the rate of electric energy use.

The EEM value for a batch system can be calculated from the following relationship:

$$\text{EEM} = \frac{Pt}{VM(C_i - C_f)} \tag{8.11}$$

where P is the electric power supplied to the photocatalytic system, t the irradiation time needed to reach the desired concentration variation, V the volume of treated water or air, M the molar mass of contaminant species and C_i and C_f are the initial and final mass concentrations of contaminant, respectively. The removal efficiencies are higher for lower EEM values.

The other figure-of-merit is the "electric energy per order" (EEO),[14] which is defined as the number of kWh of electrical energy required to reduce the concentration of a pollutant species in a unit volume of contaminated water or air by one order of magnitude (90%). The EEO parameter is best used for low concentration values of contaminant because under this condition the amount of electric energy required to reduce the contaminant concentration by one order of magnitude is independent of contaminant concentration. In fact for low contaminant concentration the degradation rate generally follows first-order kinetics with respect to contaminant concentration and so the time needed for any order of reduction is independent of the initial concentration.

The EEO value for a batch system can be calculated from the following relationship:

$$\text{EEO} = \frac{Pt}{V \log_{10}\left(\frac{C_i}{C_f}\right)} = \frac{Pt}{V \log_{10}\left(\frac{C_i}{0.1 C_i}\right)} = \frac{Pt}{V} \tag{8.12}$$

where P, t, V, C_i and C_f have the same definitions as in Equation (8.11). For first-order kinetics the following relationship holds:

$$\ln\left(\frac{C_i}{C_f}\right) = 2.3026 \log_{10}\left(\frac{C_i}{C_f}\right) = k_1 t \tag{8.13}$$

where k_1 is the first (or pseudo-first) order rate constant for the decay of the pollutant concentration. For a 90% variation of final concentration of pollutant the required time of irradiation is:

$$t = \frac{2.306}{k_1} \tag{8.14}$$

Substituting Equation (8.14) into Equation (8.12) gives:

$$\text{EEO} = \frac{2.3026 P}{V k_1} \tag{8.15}$$

which is a parameter that may be easily calculated provided that the reaction kinetics and the kWh of electricity required to reduce the pollutant concentration by one order of magnitude are measured.

Efficient photocatalytic processes exhibit low values of EEO, indicating that less energy is required to achieve a fixed drop in concentration of model pollutants.

The EEO is an overall parameter based on electrical lamp power requirements and, as such, it does not take into account the transformation efficiency of electrical into useful radiative energy, radiation absorption by the photoreactor walls, *etc.* Furthermore, de Lasa *et al.*[16] outline that (i) the EEO efficiency parameter is not dimensionless, as required by thermodynamics, (ii) the EEO definition holds only for first-order reaction kinetics and (iii) the EEO is focused on the decrease of the model pollutant concentration and it neglects the photoconversion of stable intermediate species, thereby underestimating photocatalytic efficiency.

To carry out an economic analysis of various water pollutant degradation processes, the figure of merit "energetic efficiency of degradation" (EED) has been proposed.[17] The EED is defined as:

$$\text{EED} = \frac{\text{parts per million of degraded organic carbon in irradiated solution}}{\text{kWh of used electrical energy}}$$

(8.16)

The EED describes the efficiency of a given process and it is a useful parameter to determine the economic viability of different processes involving different reactors, different light sources, *etc.*

In an industrial environment, process efficiencies described by EEO, EEM and EED are useful in comparing the economics of different industrial strategies. They are macroscopic parameters, which, however, do not give information on the efficiency of an absorbed photon in inducing the photocatalytic process, as the quantum yield does in homogeneous photochemistry and heterogeneous photocatalysis. Trends in EEO (or the equivalent EEM) and EED inversely reflect the trends in quantum yields; however, they fail to provide a relatively simple method to establish photon efficiencies.

8.3 Experimental Comparison of Photocatalytic Systems

The photocatalytic method is capable of degrading a broad range of pollutants, both of organic and inorganic. Therefore, several substances are suitable for assessing the photocatalytic efficiency of a system. The literature shows that the broad field of model pollutants and respective test procedures can be sub-classified into three categories according to the materials used for analysis, namely:

1. dyestuffs,[18,19]
2. organic compounds like volatile organic compounds (VOCs),[20]
3. inorganic gases.[21,22]

Dyes are degraded by TiO_2 under the influence of UV or solar light. The decomposition is assessed by decolorization measurements as well as chromatographic investigations. The widespread application of dyes originates from the mainly nontoxic and convenient use of these dyes. However, the decomposition process of dyes is still a subject of discussion as these substances only show a limited resistance to UV light, independently of the existence of a catalyst. Therefore, the decolorization of a dye is not an unambiguous criterion for the assessment of photocatalytic efficiency.

VOCs are prominent representatives for modeling the degradation of polluted air for indoor situations. Commonly used organic compounds are:

1. trichloroethylene (C_2HCl_3);[23]
2. acetone (C_3H_6O), 1-butanol ($C_4H_{10}O$), butyraldehyde (C_4H_8O) and m-xylene (C_8H_{10});[24]
3. 1,3-butadiene (C_4H_6) and toluene ($C_6H_5CH_3$);[25]
4. formaldehyde (CH_2O).[24,25]

The photocatalytic method can decompose VOCs to harmless substances like CO_2 and H_2O. This also includes the removal of odors, which makes this method especially attractive for treating air in house ventilation systems.

8.3.1 Photocatalytic Films

The activity of photocatalytic TiO_2 films can vary considerably and depends on many factors, such as film thickness, substrate, roughness, crystallite size, deposition temperature, *etc*. Owing to the difficulty of defining activity parameters that explicitly take into account all the previous factors, it has been proposed to standardize the photoactivity through experiments carried out under well-defined conditions.

The experimental evaluation of photocatalytic activities of TiO_2 films is generally carried out with different methods.[26] Testing methods based on the decolorization of different dyes, such as Methylene Blue, are probably the most commonly used as they provide an excellent visual representation of the process evolution. In this test, a change in the color of a dye is used as an indicator of the photocatalytic activity of the surface. In almost all cases these tests are based upon the slow photooxidation of the dye so that they can be very long lasting processes and therefore with limited suitability as test in the field.

Other techniques are based on the photooxidation of organic solid films such as stearic acid deposited on the surfaces. The photocatalytic destruction of such solid compounds is of practical interest since it represents a reasonable model system for the type of solid organic films that are often deposited on exterior glass surfaces. Therefore, such treatment is proposed as a standard method for determination of activities of self-cleaning surfaces.

The drawback of all these methods is that they are not absolute methods and a standard photocatalytic surface must be evaluated first. The relative comparison of the photo-efficiency of the probe to that of the standard specimen requires constant conditions. This requirement generates an important problem as the probe conditions may differ very much between laboratories. Moreover, many of these methods require the use of complex apparatus and as such are suitable only for use in the laboratory, and therefore cannot be adapted to a practical end use application.

A method recently proposed to determine the activity of photocatalytic surfaces is based on the measurement of contact angle changes. This evaluation method, based on ISO/TC206, consists in measuring the changes of contact angle of a drop of distilled water put onto the TiO_2 film with time under irradiation.

8.3.1.1 Self-cleaning Glasses

An important field of successful application of photocatalytic TiO_2 films is represented by the so-called "*self-cleaning*" glasses. Thin films of anatase TiO_2 deposited on glass can mineralize most of the organic pollutants present in the atmosphere and responsible for the dirt and grime that deposit on window glass. The TiO_2 films are also hydrophilic under near-UV irradiation so that they hinder the adhesion of hydrophobic organic pollutants to the surface (Figure 8.2).

Comparison of the effectiveness of different self-cleaning glasses must be made by means of standard methods capable of assessing the photocatalytic activity. These methods should possess specific features before being adopted as standard ones. The method should be very fast; this requirement is not easy to accomplish because most commercial self-cleaning glasses utilize only a very thin layer of TiO_2 and therefore the photomineralization kinetics are very slow. It can take hours or days to destroy completely a test compound under typical solar UVA conditions.

Moreover, the method should be apt for making measurements in the field without the need for sophisticated analytical equipment and a technician trained to run it. The test appropriate for use in the field should involve a simple, fast physical change, *e.g.*, a color change within a few minutes of UV exposure. Determination of such a set of standard methods is currently in

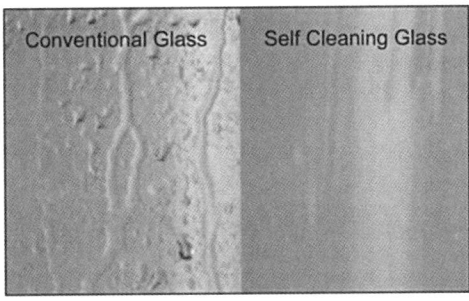

Figure 8.2 Self-cleaning, water-repelling glass.

progress at the ISO (TC206). The different tests proposed and used to date are discussed in the following.

Stearic Acid Test. The stearic acid (SA) test[27–30] consists in depositing a thin layer of SA onto the photocatalytic film under test and in monitoring its degradation rate under strictly defined operative conditions. The disappearance rate of SA is the parameter used for comparing the performances of different samples. The stoichiometry of SA mineralization is:

$$CH_3(CH_2)_{16}COOH + 26\,O_2 \xrightarrow[h\nu]{TiO_2} 18\,CO_2 + 18\,H_2O \qquad (8.17)$$

The photocatalytic mineralization of the SA film does not generate stable intermediates, volatile or otherwise; the only major species present during the photomineralization process being SA and CO_2. Therefore, the mineralization process can be easily monitored by following either the CO_2 generation by gas chromatography or the SA disappearance by infrared absorption spectroscopy.

This test presents the following advantages: (i) SA may be considered as a model compound representative of environmental contaminants forming solid deposits on glass surfaces; (ii) SA is a very stable compound and its degradation occurs only by photocatalysis; (iii) the deposition of a monolayer SA film from an organic solution is very easy; (iv) because SA is a solid its concentration does not change during degradation so that the observed kinetics are independent of the SA film thickness; and (v) the SA degradation process is easily monitored.

Despite these very positive features, the SA test suffers from two major drawbacks. First, with the usual thicknesses of TiO_2 films deposited on glasses (nanometer range), the test is very slow. For rapid assessment, the TiO_2 films should have a thickness in the micron range. Second, it requires a skilled operator in both the analytical method and the sample preparation to conduct the test, and so it is relatively labor intensive and expensive to run.

Methylene Blue Test. The Methylene Blue (MB) test[31–37] is a widely used method for assessing the photocatalytic activity of TiO_2 films or powders. MB, a cationic dye, is mineralized through a photobleaching process described by the stoichiometry:

$$2\,C_{16}H_{18}N_3S^+ + 51\,O_2 \xrightarrow[h\nu]{TiO_2} 32\,CO_2 + 6\,HNO_3 \\ + 2\,H_2SO_4 + 2\,H^+ + 12\,H_2O. \qquad (8.18)$$

When the MB test is used for assessing photocatalytic activity, generally only the dye decoloration, which is assumed to be due to dye oxidation, is monitored.

Standardization

This method is simple, easy and able to provide an impressive demonstration of the efficacy of semiconductor photocatalysis through dye decoloration (Figures 8.3 and 8.4).

The main drawback of this method is that a measurable degradation of Methylene Blue may occur[38] within a few hours of UV exposure in the absence of any catalyst. Moreover, the MB test method has been questioned[39] because MB,

Figure 8.3 Methylene Blue test on un-coated and TiO_2-coated solid surfaces. (Photograph courtesy of http://www.hkapc.org/Products/TiO_2)

Figure 8.4 Methylene Blue test on an aqueous TiO_2 suspension. (Photograph courtesy of http://www.reinbiotech.com)

which is an indicator for redox reactions, may be easily decolorized by the reduction reaction to leuco-Methylene Blue (LMB) which is colorless. As this reaction is reversible, the reduced Methylene Blue can also be oxidized again and obtain its original color. This reduction is thermodynamically possible because the redox potentials of the MB/LMB couple and of photogenerated electrons are *ca.* 0.53 and 0.32 V, respectively, *versus* NHE at pH 0. The relative importance of MB photooxidation with respect to photoreduction to LMB is affected by kinetic factors. In neutral or slightly alkaline solution, the MB photobleaching is dominantly determined by photooxidation. Under acidic conditions the photooxidation reaction becomes slow and, as a result, the photobleaching of MB by TiO_2 is more likely to show some evidence of LMB production. MB does not work in air since its reduced form, leuco-Methylene Blue, is highly oxygen sensitive.

Since MB shows no absolute resistance to UV-induced degradation and decolorization by reduction reaction, it appears not to be an ideal model pollutant. The ambiguity existing for the MB test must be solved before this method of assessment can be accepted as a standard one.

Resazurin Ink Test. The major problem with the SA and MB tests is that the photooxidation reaction is usually a slow process. Such tests are adequate for assessing the performances of very active, usually thick, titania films, where the kinetics of photomineralization are rapid but they are inappropriate for photocatalytic products that exhibit slow degradation rates.

An innovative solution to this problem has been proposed by Mills' research group.[40-43] The core of the test is an indicator ink that can be easily coated onto the semiconductor film and which changes color within a few minutes upon irradiation of the photocatalytic film.

The formulation consists of a redox dye, resazurin, a sacrificial electron donor, glycerol, and a polymer, hydroxy ethyl cellulose, all dissolved in water. Whether wet or dry, upon irradiation of the ink on a photocatalytic surface with ultrabandgap radiation the photogenerated electrons reduce the blue resazurin to the pink resorufin (an irreversible reaction, see Figure 8.5), while the photogenerated holes oxidize glycerol, which acts as a hole-trap preventing electron–hole recombination, to glyceraldehyde. The reduction process is accompanied by a color change from blue to pink; the color change indicates completion of the photoreduction reaction and it occurs in a few minutes even for very thin TiO_2

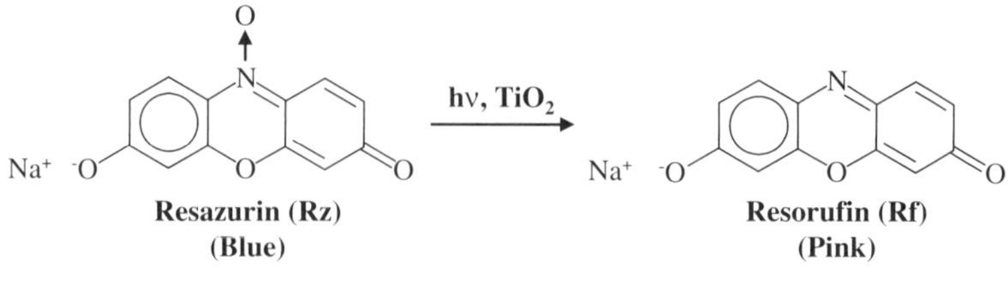

Figure 8.5 Structure of the redox dye resazurin and its reduced form, resorufin.

Standardization

films. The dye is stable also upon prolonged UV irradiation in the absence of a photocatalyst; conversely, the ink changes rapidly from blue to pink upon UV irradiation if TiO_2, film or powder, is present. The process rapidity makes this test particularly suited to "on the spot" tests. Moreover, this test does not need a skilled operator or expensive analytical equipment as the eye is sufficient for a qualitative indicator, or a semi-quantitative assessment, of photocatalytic activity.

Dichloroindophenol Test. Market research indicates that customers prefer a decoloration test as the bleaching process is often related with a cleaning action. The color is associated with the presence of a (contaminant) substance while the color disappearance is associated with the decontaminating action of photocatalyst joined with light. Therefore, a photoactivity test should involve a substance that changes from a highly colored initial form to a colorless final form upon UV irradiation. On this basis a redox dye, *i.e.*, 2,6-dichloroindophenol (DCIP), has been used[44] in an ink formulation in the same manner as the resazurin ink. The resulting product arising from DCIP reduction upon UV irradiation in the presence of a photocatalyst underlayer is colorless, *i.e.*, this dye is photobleached.

In a typical ink formulation, DCIP gives a blue ink and a subsequent blue film when spun onto the surface of the photocatalyst substrate (Figure 8.6). Although the reduction kinetics under ambient conditions for the DCIP ink are slower than those of an resazurin ink, the latter is shown to be both oxygen and humidity sensitive, whereas the DCIP ink is not.

Contact Angle. When a water drop lies on a solid surface, as shown in Figure 8.7(A), the contact angle is close to $0°$ if the solid surface is strongly hydrophilic. Less strongly hydrophilic solids have a contact angle up to $90°$. If the solid surface is hydrophobic, the contact angle is larger than $90°$. A method proposed recently to determine the activity of photocatalytic surfaces is based on the measurement of contact angle changes by non-invasive image sensing with a digital camera. A layer of defined thickness of oleic acid is applied on the photocatalytic surface to be tested and then a water droplet is placed on the surface. The water contact angle is therefore continuously measured as a function of irradiation time by means of a computer-aided instrument. The photocatalytic effect is determined from the time dependence of the contact angle change and the final contact angle. The instrument is simple, as it consists of instrument base, movable sample holder and movable digital camera with necessary electronics, and so the measurement is quick, low cost, and easy to carry out.

8.3.2 Cementitious Building Materials

In the general field of self-cleaning surfaces a scale application of TiO_2 (favored by the relative low cost of this photocatalyst) is used in the building sector as an

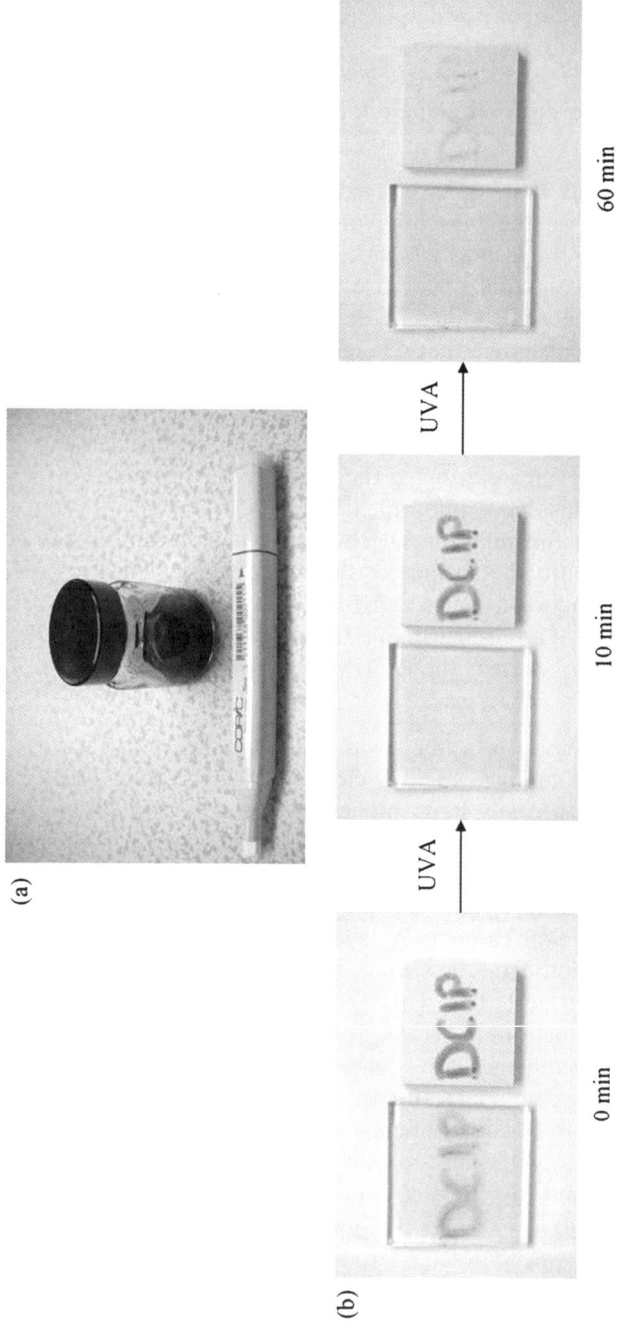

Figure 8.6 (a) 2,6-Dichloroindophenol (DCIP) ink and (b) DCIP ink test on TiO$_2$-coated (left-hand side) and un-coated (right hand side) solid surfaces. (Reproduced with permission from ref. 44.)

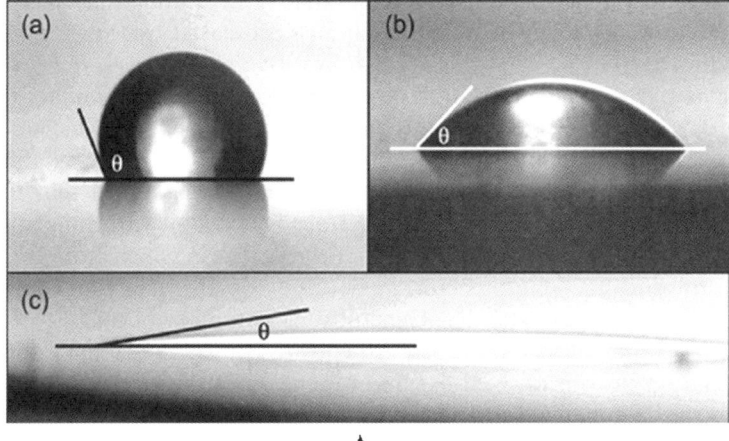

Figure 8.7 (A) Contact angles, θ, for (a) a hydrophobic surface, (b) a hydrophilic surface and (c) a strongly hydrophilic surface. (B) Contact angle goniometer with camera and software to capture and analyze the drop evolution with time. [Part (A): photograph courtesy of http://harrickplasma.com. Part (B): photograph courtesy of http://www.answers.com]

additive of cementitious materials. A European Community project (Photocatalytic Innovative Coverings Applications for Depollution Assessment, PICADA) has been started with the aim of developing and testing new photocatalytically active building materials based on TiO_2. The main result of that project has been the development of a laboratory method able to determine the specific photocatalytic activity of TiO_2 modified cementitious materials towards the degradation of typical aromatic pollutants present at trace levels in indoor and outdoor air.

In real situations the photocatalytic cementitious body is in contact through its external surface with a gaseous body, the atmosphere, which contains contaminant compounds at certain concentrations. Owing to the magnitude of atmosphere with respect to the solid body it may be assumed that the contaminant concentration near the solid surface is constant with time and independent of particular position on that surface. These considerations greatly help in choosing the photoreactor to be used for characterizing the photoactivity of cementitious materials. Generally, a continuous photoreactor (where the catalyst is present as a fixed bed or deposited onto the reactor walls) is fed with a gaseous stream of fixed organic vapor concentration. Measurement of the decrease of organic vapor concentration from the inlet to the outlet of photoreactor allows determination of the photoactivity. In these experiments, if the conversion is high ($>10\%$), the concentration of reactants varies along the reactor axis and then the reaction rate also varies along the axis. If the reactor is operated at low conversion ($<10\%$), it may be assumed that the photoreactor behaves as a differential one, *i.e.*, that the reactant concentrations are quite uniform so that the photocatalytic reaction rate is determined by that concentration (generally the arithmetic mean between inlet and outlet). The main drawback of differential reactors lies in the experimental errors related to concentration measurement due to the small difference between inlet and outlet.

A continuous stirred tank photoreactor guarantees a uniform concentration at the sample surface. This avoids the spatial gradients of flow-through reactors and allows high (but not very high) conversion levels (50% and over).

To measure the photocatalytic activity of TiO_2 dispersed in cementitious building materials, Strini *et al.*[45] have developed a method based on a specially designed stirred flow reactor. The main advantage of a mixed flow reactor is that it allows a uniform concentration of reactants at the catalytic material surface at high conversion factors, which also allows the photocatalytic activity to be measured, bypassing the limitations imposed by concentration gradients of unmixed flow reactors. During operation, following an initial equilibration period, mainly needed for reactant adsorption on sample and on reactor surfaces, the system reaches steady-state conditions. The catalytic activity is calculated from data measured after equilibrium has been reached. This ensures that the resulting activity values are obtained under defined, uniform and time-constant conditions. Under steady state conditions, all transient effects on the sample surface and on the reactor walls are equilibrated, avoiding in this way any interference with the determination of the catalytic activity.

A mixture of aromatic volatile organic compounds (benzene, toluene, ethylbenzene and *o*-xylene, BTEX) has been chosen as a typical micropollutant standard representative of air contamination. Because the BTEX concentration in typical urban outdoor situations is in the $10\text{--}100 \, \text{mg m}^{-3}$ range, BTEX concentrations in that range must be used for the photocatalytic runs;

moreover, as the humidity strongly affects the photoreactivity, the experiments must be carried out with a water vapor content typical of outdoor situations, e.g. 50% relative humidity (RH) at ambient temperature.

Special care must be devoted to preparation of the cementitious samples. The specimen is prepared by mixing white cement powder with the required TiO_2 amount and then adding water to the dry mixture at a fixed water/cement ratio. The cementitious paste is manually mixed to uniformity and it is then poured into a 9-cm diameter Petri dish filled to full height (about 1 cm). Each specimen is allowed to hydrate for 28 days at 23 °C and a relative humidity of >95%. After that period the specimens are exposed to 23 °C, 50% RH and allowed to equilibrate their humidity content with the ambient air. A blank sample without TiO_2 is also prepared.

Figure 8.8 shows a schematic diagram of the experimental set-up used for catalytic activity measurements. The system consists of an artificial atmosphere generator system, a photochemical reactor and an irradiation box.

The atmosphere generator system uses as VOC source a standard gas cylinder mixture containing BTEX at trace level in nitrogen. The standard cylinder effluent stream is mixed with a humidified, oxygen-enriched synthetic air stream. The humidification grade, oxygen level and flow of the latter are calculated to obtain 20% oxygen, 50% RH (at 23 °C) in the air delivered to the photocatalytic reactor.

The reaction chamber is a 2-l Duran 50 borosilicate glass (Schott) reaction vessel provided with four service ports and a teflonated o-ring sealed cover.

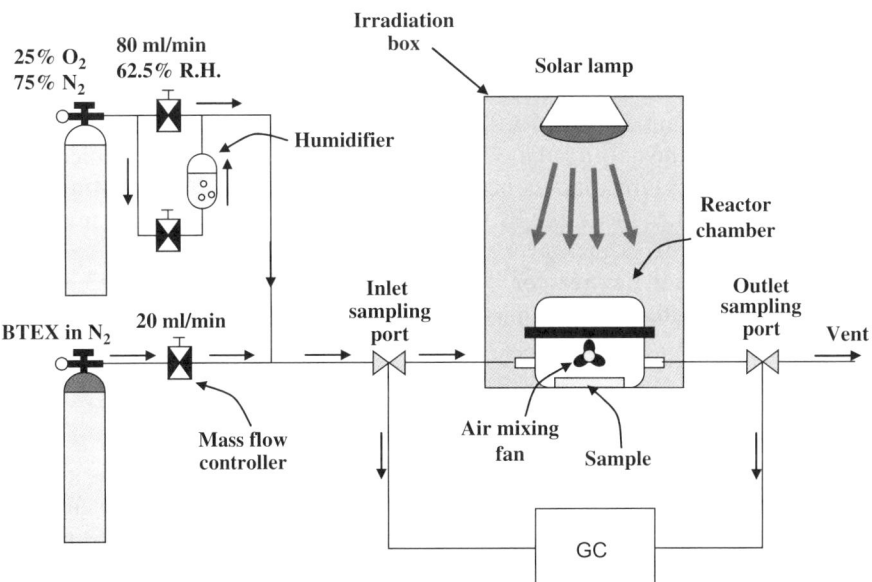

Figure 8.8 Schematic diagram of the experimental set-up used to measure the catalytic activity of cementitious materials.

Air mixing inside the chamber is ensured by a stainless-steel fan operated at 350 rpm. The independence of experimental BTEX concentration data from the fan speed is verified by performing a series of measurements operating at catalytic conditions under irradiation at different fan speeds.

The reaction chamber is contained inside an irradiation box provided with a 300 W solar lamp (Ultra Vitalux, Osram) placed 50 cm from the sample surface. The irradiation box shields the reaction chamber from possible interference by the ambient light, and two fans ensure reaction chamber cooling. The actual irradiance value at the sample level is controlled for each determination, because the lamp shows a considerable irradiation drift, particularly in the early life period. During irradiation, the reaction chamber warms up because of the strong visible and infrared components of the lamp radiation.

The BTEX concentration measurements are taken under steady-state conditions, so it is possible to acquire multiple samples separated in time for verification and statistical data processing. The catalytic activity determination proceeds as follows. The catalytic specimen is placed in the reaction chamber, which is then closed inside the exposure system in dark conditions. A purified air flow is established and maintained for 24 h, to allow possible pollutants to be desorbed from the specimen. A first air sample is taken to check the cleanliness of the system. The BTEX test flow is then established, connecting the chamber to the synthetic atmosphere generator. The system is allowed to equilibrate for at least 18 h and a series of six or more air samples are taken with a sampling interval of at least 1 h. These samples give the non-irradiated steady-state concentration for each VOC under study. The irradiation source is then activated, and the system is allowed to equilibrate for at least a further 18 h. A new series of at least six air samples are taken, giving the irradiated steady-state concentrations. If the analyses show that the concentration is not stabilized, the sampling is repeated after 1 day until the steady-state equilibrium is reached. After that the lamp is turned off and, if required, the system is allowed to re-equilibrate to the non-irradiated condition, a last sample is taken to confirm the previously measured dark concentration levels. The BTEX concentration of the supply air is then measured by taking a series of samples at the air inlet port at the end of the experiment.

This photocatalytic reactor system is characterized by the specific photocatalytic oxidation rate, measured with the following relationship:

$$r_{ox} = \frac{F(C_{in} - C_{irrad})}{A} \qquad (8.19)$$

where r_{ox} is the area specific oxidation rate (mmol m^{-2} h^{-1}), F is the air flow rate (m^3 h^{-1}), C_{in} is the reactant concentration in the supply air flow (mmol m^{-3}), C_{irrad} is the steady-state reactant concentrations in the reactor during irradiation and A the sample area exposed to the irradiation (m^2).

This method, proposed by Strini et al.,[45] allows determination in a reliable way of the surface catalytic activity in the photodestruction of organic micropollutants in air, at ppb concentration levels. The oxidation rates of cementitious samples are linearly dependent on the air concentration of the reactant and the catalytic activity is also linearly dependent on the irradiation in the measured ranges. However, the catalytic activity is not linearly dependent on the TiO_2 content in the samples. Possible causes of this are, among others, the formation of catalyst clusters in the cementitious paste at higher concentrations or segregation processes in the poured paste influenced by the different viscosity of the paste itself due to the presence of the catalyst.

8.3.3 Paving Blocks

The use of organic compounds for investigating the photocatalytic properties of cementitious materials has been questioned on the basis that during their degradation CO_2 may be produced as a consequence of mineralization. Considering the tendency of concrete to carbonation, part of this CO_2 may be transformed into solid $CaCO_3$ in the outer pore system. Consequently, part of the reaction products mineralize and a quantitative assessment is hampered as the amount of mineralized CO_2 consumed by carbonation is difficult to assess. From this fact it can be concluded that the application of dyes, VOCs and aromatic compounds is less suitable for the photocatalytic oxidation of concrete substrates under outdoor exposure.

NO_x is considered to be a suitable model pollutant for the assessment of photocatalytic efficiency of concrete products. NO_x, as also sulfur dioxide, not only poses a threat to human health and nature preservation but it is causing degradation to many inner-city buildings.[22] A large part of the NO_x in the atmosphere is emitted by car traffic and transport. Therefore, the application of a photocatalytic paving block (close to the source of emission) is a desirable solution. The literature[21,22] reports that the degradation of NO or more generally of NO_x, also referred to as the denitrogenization process, delivers a suitable model to assess the ability of surfaces for air purification. This denitrogenization process ultimately produces nitrate ions, which are flushed from the concrete surface as weak nitric acid. Quantitative analysis of the reaction products considering the initial pollutant concentration allows evaluation of the degradation ability.

On the basis of the above reported considerations, NO_x is a suitable model pollutant for testing the performance of photocatalytic paving blocks. A standard method has been proposed recently by the ISO, the ISO 22197-1:2007;[46] this test method utilizes NO_x for determining the air-purification performance shown by fine ceramics that contain a photocatalyst or have photocatalytic films on their surface. The method consists of measuring the photoactivity of a test piece continuously exposed to the model air pollutant under illumination with ultraviolet light. ISO 22197-1:2007 applies to different kinds of materials, such as construction materials in various shapes, to materials in honeycomb form and to plastic or paper materials, but only if they contain ceramic microcrystals and composites.

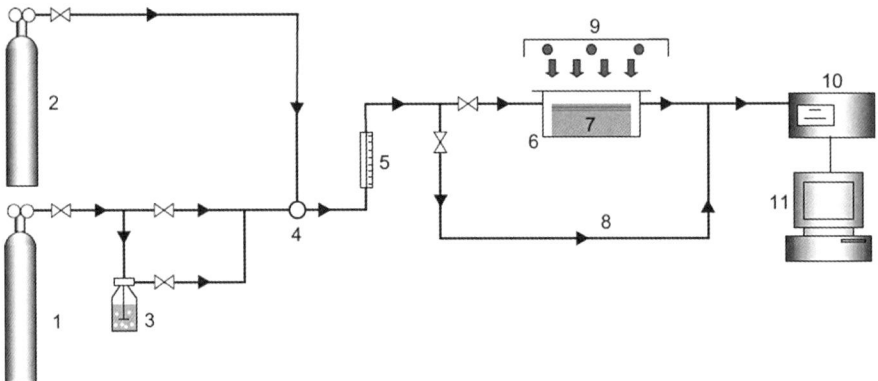

Figure 8.9 Schematic diagram of the test set-up: (1) synthetic air; (2) NO_x source; (3) gas-washing bottle; (4) temperature and relative humidity sensor; (5) flow controller; (6) reactor cell; (7) paving block; (8) side stream for bridged flow; (9) light source; (10) NO_x analyzer; and (11) computer. (Reproduced with permission from ref. 47.)

By considering that the standard ISO 22197-1:2007 also satisfies the need for photoactivity measurements on concrete specimens, Hüsken et al.[47] have proposed an appropriate test set-up as well as a reliable measuring procedure for determining the performance of photocatalytic active paving blocks. The used set-up is designed by following the recommendations given by ISO 22197-1:2007 standard and it is scaled according to the requirements on photocatalytic concrete products. The test set-up, which uses the UV-A induced degradation of NO, is schematically depicted in Figure 8.9 and it consists of:

- a reactor cell housing the paving block sample,
- a suitable UV-A light source,
- a NO_x analyzer,
- an appropriate gas supply.

The core of the experimental set-up is a gas reactor that allows a planar sample of size $100 \times 200\,mm^2$ to be embedded. Figure 8.10 gives a schematic representation of the reactor cell.

The reactor is made from materials that do not absorb the applied pollutant and do not degrade under the high irradiation of UV-A light. On top, the reactor is tightly closed with a glass plate made from borosilicate glass, which allows the UV-A radiation to pass through with almost no absorption. Within the reactor the planar surface of the specimen is fixed parallel to the covering glass, leaving a 3 mm gap for the gas to pass through. The active sample area used for the degradation process is higher than that suggested by the ISO standard. The area is enlarged from $50 \times 100\,mm$ to $100 \times 200\,mm$, to adjust for the standard paving block dimensions. By means of profiles and seals, the sample gas only passes the reactor through the gap between the sample surface

Standardization

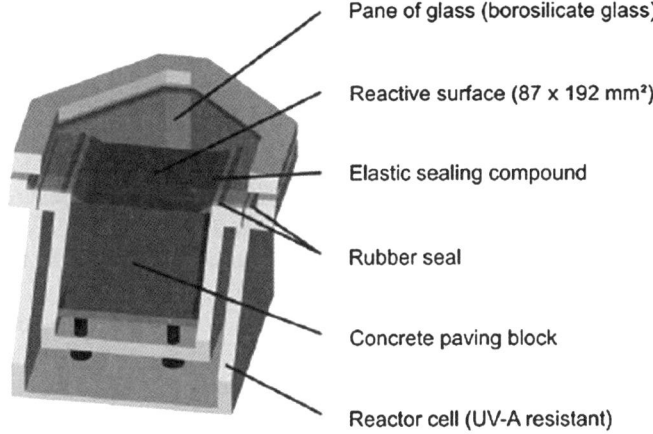

- Pane of glass (borosilicate glass)
- Reactive surface (87 x 192 mm²)
- Elastic sealing compound
- Rubber seal
- Concrete paving block
- Reactor cell (UV-A resistant)

Figure 8.10 Schematic diagram of the reactor cell used for the paving block test. (Reproduced with permission from ref. 47.)

and the glass cover. All structural parts inside the box are designed to enable laminar flow of the gas along the sample surface and to prevent turbulence.

The light source is chosen in order that its spectrum fits the cut-off wavelength of the applied powder. TiO_2 photocatalyst may be used as doped or undoped powder. For undoped TiO_2 the use of UV-A light is required as the cut-off wavelength of conventional TiO_2 is 388 nm;[12] fluorescent tubes are suitable lamps as they emit highly concentrated UV-A radiation in the range 300–400 nm with maximum intensity at about 345 nm. For doped powders, having a broader spectrum covering also the visible light spectrum, standard fluorescent lamps may be used as light source as they emit cool daylight in the 420–650 nm range with three prominent peaks at 460, 560 and 600 nm.

Warming of the reactor by the light source is prevented by the spatial separation of light source and reactor and a cooling of the lamps by means of fans. The irradiance of fluorescent tubes is adjusted by a dimmer to exactly $10\,W\,m^{-2}$ at the sample surface for the use of UV-A irradiation with the help of a calibrated UV-A radiometer. A lead time of about 15 min has to be considered for fluorescent tubes to obtain a stable UV-A radiation.

As previously reported, the model pollutant chosen for conducting the experiments on photocatalytic concrete products is NO. According to the requirements given in standard ISO 22197-1:2007,[46] the concentration of NO in the gas feeding the photoreactor is 1 ppmv. The model contaminant, contained in a nitrogen flow at a concentration of 50 ppmv, is mixed with synthetic air to obtain the desired concentration and flow rate. A high precision valve is used to adjust the concentration of the model pollutant to 1 ppmv NO. While adjusting the concentration, the photoreactor is bypassed to avoid pollution of the sample surface. During adjustment of the flow, the NO concentration is monitored by the NO analyzer, connected to the outlet of the gas supply unit. The overall gas flow rate is $16.67\,cm^3\,s^{-1}$ and the relative humidity is 50%.

For the sake of repeatability and accuracy, a defined measuring procedure using constant experimental conditions is also established. This procedure distinguishes a bridged and non-bridged flow and allows therefore for a much shorter period of adjusting the volumetric flow, pollutant concentration and humidity. When a stable pollutant concentration is achieved, the bridge is closed and the gas flows along the surface of the concrete sample. Initially, the pollutant concentration decreases since the air inside the photoreactor does not contain any NO and adsorption of NO at the sample surface takes place. After a certain time the surface is saturated and the NO concentrations at the photoreactor outlet and inlet show the same value.

For standard flow conditions a period of 5 min is needed to reach steady-state conditions. At this moment, the sample is exposed to irradiation. This leads to an immediate decrease in the NO concentration. This process of active degradation is conducted for 30 min. Finally, the light source is switched off and the NO concentration returns to the original inlet concentration.

References

1. http://www.iso.org/iso/home.htm
2. V. Augugliaro, L. Palmisano and M. Schiavello, Rate of photon absorption and turnover number: two parameters for the comparison of heterogeneous photocatalytic systems in a quantitative way, *Coord. Chem. Rev.*, 1993, **125**, 173–182.
3. M. Cabrera, O. Alfano and A. Cassano, Absorption and scattering coefficients of titanium dioxide particulate suspensions in water, *J. Phys. Chem.*, 1996, **100**, 20043–20050.
4. N. Serpone, G. Sauvé, R. Koch, H. Tahiri, P. Pichat, P. Piccinini, E. Pelizzetti and H. Hidaka, Standardization protocol of process efficiencies and activation parameters in heterogeneous photocatalysis: relative photonic efficiencies, *J. Photochem. Photobiol. A*, 1996, **94**, 191–203.
5. M. Schiavello, V. Augugliaro and L. Palmisano, An experimental method for the determination of the photon flow reflected and absorbed by aqueous dispersion containing polycrystalline solids in heterogeneous photocatalysis, *J. Catal.*, 1991, **127**, 332–341.
6. V. Augugliaro, L. Palmisano and M. Schiavello, Photon absorption by aqueous TiO_2 dispersion contained in a stirred photoreactor, *AIChE J.*, 1991, **37**, 1096–1100.
7. V. Augugliaro, V. Loddo, L. Palmisano and M. Schiavello, Performance of heterogeneous photocatalytic systems: influence of operational variables on photoactivity of aqueous suspension of TiO_2, *J. Catal.*, 1995, **153**, 32–40.
8. V. Augugliaro, V. Loddo, L. Palmisano and M. Schiavello, Heterogeneous photocatalytic systems: influence of some operational variables on actual photons absorbed by aqueous dispersions of TiO_2, *Sol. Energy Mater. Sol. Cells*, 1995, **38**, 411–419.

9. V. Augugliaro, V. Loddo, L. Palmisano and M. Schiavello, Bestimmung der quantenausbeute von heterogenen photokatalytischen systemen. bestimmung des absorbierten und reflektierten photonenflusses in wäßrigen suspensionen von polykristallinen titandioxid, in *Photochemie Konzepte, Methoden, Experimente*, ed. D. Wöhrle, M. W. Tausch and W. D. Stohrer, Wiley-VCH, Weinheim, 1998, pp. 459–465.
10. M. Schiavello, V. Augugliaro, V. Loddo, M. J. López-Muñoz and L. Palmisano, Quantum yield of heterogeneous photocatalytic systems: revisiting an experimental method for determining the absorbed photon flow, *Res. Chem. Intermed.*, 1999, **25**, 213–227.
11. S. Yurdakal, V. Loddo, B. Bayarri-Ferrer, G. Palmisano, V. Augugliaro, J. Giménez-Farreras and L. Palmisano, Optical properties of TiO_2 suspensions: Influence of pH and powder concentration on mean particle size, *Ind. Eng. Chem. Res.*, 2007, **46**, 7620–7626.
12. B. Serrano and H. de Lasa, Photocatalytic degradation of water organic pollutants. Kinetic modelling and energy efficiency, *Ind. Eng. Chem. Res.*, 1997, **36**, 4705–4711.
13. J. R. Bolton and S. Cater, Homogeneous photodegradation of pollutants on contaminated water: an introduction in *Aquatic and Surface Photochemistry*, ed. G. Hels, R. Zepp and D. Crosby, Lewis Publications, 1994, pp. 467–490.
14. J. R. Bolton, K. G. Bircher, W. Tumas and C. A. Tolman, Figures-of-merit for the technical development and application of advanced oxidation processes, *J. Adv. Oxid. Technol.*, 1996, **1**, 13–17.
15. J. R. Bolton, K. G. Bircher, W. Tumas and C. A. Tolman, Figures-of-merit for the technical development and application of advanced oxidation technologies for both electric- and solar-driven systems, *Pure Appl. Chem.*, 2001, **73**, 627–637.
16. H. de Lasa, B. Serrano and M. Salaices, *Photocatalytic Reaction Engineering*, Springer, 2005.
17. J.-M. Herrmann, Heterogeneous photocatalysis. Fundamentals and application to the removal of various types of aqueous pollutants, *Catal. Today*, 1999, **53**, 115–129.
18. K. Rajeshwar, C. R. Chenthamarakshan, S. Goeringer and M. Djukic, Titania-based heterogeneous photocatalysis – materials, mechanistic issues, and implications for environmental remediation, *Pure Appl. Chem.*, 2001, **73**, 1849–1860.
19. C. C. Liu, Y. H. Hsieh, P. F. Lai, C. H. Li and C. L. Kao, Photodegradation treatment of azo dye wastewater by UV/TiO_2 process, *Dyes and Pigments*, 2006, **68**, 191–195.
20. J. Zhao and X. D. Yang, Photocatalytic oxidation for indoor air purification: a literature review, *Building Environ.*, 2003, **38**, 645–654.
21. H. Q. Wang, Z. B. Wu, W. R. Zhao and B. H. Guan, Photocatalytic oxidation of nitrogen oxides using TiO_2 loading on woven glass fabric, *Chemosphere*, 2007, **66**, 185–190.

22. J. S. Dalton, P. A. Janes, N. G. Jones, J. A. Nicholson, K. R. Hallam and G. C. Allen, Photocatalytic oxidation of NO_x gases using TiO_2: a surface spectroscopic approach, *Environ. Pollut.*, 2002, **120**, 415–422.
23. W. A. Jacoby, D. M. Blake, R. D. Noble and C. A. Koval, Kinetics of the oxidation of trichloroethylene in air via heterogeneous photocatalysis, *J. Catal.*, 1995, **157**, 87–96.
24. J. Peral and D. F. Ollis, Heterogeneous photocatalytic oxidation of gas-phase organics for air purification: acetone, 1-butanol, butyraldehyde, formaldehyde, and m-xylene oxidation, *J. Catal.*, 1991, **136**, 554–565.
25. T. N. Obee and R. T. Brown, TiO_2 photocatalysis for indoor air applications: effects of humidity and trace contaminant levels on the oxidation rates of formaldehyde, toluene, and 1,3-butadiene, *Environ. Sci. Technol.*, 1995, **29**, 1223–1231.
26. A. Mills and M. McFarlane, Current and possible future methods of assessing the activities of photocatalyst films, *Catal. Today*, 2007, **129**, 22–28.
27. T. Sawunyama, L. Jiang, A. Fujishima and K. Hashimoto, Photodecomposition of a Langmuir-Blodgett film of stearic acid on TiO_2 film observed by in situ atomic force microscopy and FT-IR, *J. Phys. Chem. B*, 1997, **101**, 11000–11003.
28. T. Minabe, D. A. Tryk, P. Sawunyama, Y. Kikuchi, K. Hashimoto and A. Fujishima, TiO_2-mediated photodegradation of liquid and solid organic compounds, *J. Photochem. Photobiol. A*, 2000, **137**, 53–62.
29. R. Fretwell and P. Douglas, An active, robust and transparent nanocrystalline anatase TiO_2 thin film – preparation, characterization and the kinetics of photodegradation of model pollutants, *J. Photochem. Photobiol. A*, 2001, **143**, 229–240.
30. A. Mills and J. Wang, Simultaneous monitoring of the destruction of stearic acid and generation of carbon dioxide by self-cleaning semiconductor photocatalytic films, *J. Photochem. Photobiol. A*, 2006, **182**, 181–186.
31. R. W. Matthews, Photooxidative degradation of coloured organics in water using supported catalysts. TiO_2 on sand, *Water Res.*, 1991, **25**, 1169–1176.
32. P. Reeves, R. Ohlhausen, D. Sloan, K. Pamplin, T. Scoggins, C. Clark, B. Hutchinson and D. Green, Photocatalytic destruction of organic dyes in aqueous TiO_2 suspensions using concentrated simulated and natural solar energy, *Sol. Energy*, 1992, **48**, 413–420.
33. S. Lakshmi, R. Renganathan and S. Fujita, Study on TiO_2-mediated photocatalytic degradation of methylene blue, *J. Photochem. Photobiol. A*, 1995, **88**, 163–167.
34. B. Serrano and H. de Lasa, Photocatalytic degradation of water organic pollutants. Kinetic modelling and energy efficiency, *Ind. Eng. Chem. Res.*, 1997, **36**, 4705–4711.

35. T. Zhang, T. Oyama, A. Aoshima, H. Hidaka, J. Zhao and N. Serpone, Photooxidative N-demethylation of methylene blue in aqueous TiO_2 dispersions under UV irradiation, *J. Photochem. Photobiol. A*, 2001, **140**, 163–172.
36. A. Houas, H. Lachheb, M. Ksibi, E. Elaloui, C. Guillard and J. M. Herrmann, Photocatalytic degradation pathway of methylene blue in water, *Appl. Catal. B*, 2001, **31**, 145–157.
37. M. Wark, J. Tschirch, O. Bartels, D. Bahnemann and J. Rathoussky, Photocatalytic activity of hydrophobized mesoporous thin films of TiO_2, *Microporous Mesoporous Mater.*, 2005, **84**, 247–253.
38. D. Stephan, P. Wilhelm and M. Schmidt, Photocatalytic degradation of rhodamine B on building materials – influence of substrate and environment, in *Photocatalysis, Environment and Construction Materials*, ed. P. Baglioni and L. Cassar, RILEM, 2007, pp. 299–306, ISBN: 978-2-35158-056-1.
39. A. Mills and J. Wang, Photobleaching of methylene blue sensitised by TiO_2: an ambiguous system?, *J. Photochem. Photobiol. A*, 1999, **127**, 123–134.
40. A. Mills, J. Wang, S. K. Lee and M. Simonsen, An intelligent ink for photocatalytic films, *Chem. Commun.*, 2005, 2721–2723.
41. A. Mills, J. Wang and M. McGrady, Method of rapid assessment of photocatalytic activities of self-cleaning films, *J. Phys. Chem. B*, 2006, **110**, 18324–18331.
42. P. Evans, S. Mantke, A. Mills, A. Robinson and D. W. Sheel, A comparative study of three techniques for determining photocatalytic activity, *J. Photochem. Photobiol. A*, 2007, **188**, 387–391.
43. A. Kafizas, D. Adriaens, A. Mills and I. P. Parkin, Simple method for the rapid simultaneous screening of photocatalytic activity over multiple positions of self-cleaning films, *Phys. Chem. Chem. Phys.*, 2009, **11**, 8367–8375.
44. A. Mills and M. McGrady, A study of new photocatalyst indicator inks, *J. Photochem. Photobiol. A*, 2008, **193**, 228–236.
45. A. Strini, S. Cassese and L. Schiavi, Measurement of benzene, toluene, ethylbenzene and o-xylene gas phase photodegradation by titanium dioxide dispersed in cementitious materials using a mixed flow reactor, *Appl. Catal. B*, 2005, **61**, 90–97.
46. *ISO 22197-1:2007*, Fine ceramics (advanced ceramics, advanced technical ceramics) – Test method for air-purification performance of semiconducting photocatalytic materials – Part 1: Removal of nitric oxide, 2007.
47. G. Hüsken, M. Hunger and H. J. H. Brouwers, Experimental study of photocatalytic concrete products for air purification, *Building Environ.*, 2009, **44**, 2463–2474.

Subject Index

References to figures are given in *italic* type. References to tables are given in **bold** type.

A*STAR 156
acetaldehyde 195
acetone 244
adsorption 16–17
   chemisorption 16–17
   desorption equilibria 20
   isotherms 17–18
      Freundlich 17, 25–26
      Langmuir 17, 23–25
      Redlich-Peterson 26–28
   photoadsorption 17–18
   see also superhydrophilicity
Aero Super-Element Purifiers 221–222
aggregate 164
Ahlstrom 186, 195
Air Oasis purifiers 229–231, *230*
air pollution sources, vi
air purification 98, 116–117, 212–213
   Aero Super Element 221–222
   Air Oasis 229–231, *230*
   Airpura purifiers 226–228
   AirSteril 214, **215**
   Airwise 217–220, *224*
   Comefresh 224
   CREC 213
   Daikin 215–216
   Genesis 217
   ZAND-AIR 222–224
Airpura purifiers 226–228

AirSteril purifier 214, **215**
Airwise purifiers 217–220, *224*
ammonia 180
anatase 15, 43, *44*
antibacterial coatings 107–114, 118, 159–160, 192, 210, *212*
anticorrosive coatings 106–107
antireflective surfaces 102–105, 127–129
arc lamps 31
asphalt 151
atomic force microscopy (AFM) 108–109

bactericides *see* antibacterial coatings
band theory 1, 1–2
   band gap 11
   dopants 6–8
   electronic properties 2–5
   photocatalytic processes 11–16
   photoexcitation 10–11
   redox systems 8–10
   surface energy 5–6
benzene 252
Bioclean Cool-Lite ST 139–140
bitumen 180
bonding orbital 1
brookite 43, *44*, 51–52
   hydrolytic synthesis 50–51
BTEX 252–254

Subject Index 263

building materials *see* cementitious materials; glass
1,3-butadiene 244

car windshields 130
Cardinal Glass Industries 140
cathode arc deposition 75–76
cementitious materials
    glass as aggregate 164
    ix 145–156, **148–149**, 151–154
    spray coating 154
    standards 249–255
ceramic tiles 144, 156–162
    applications 157, *158*
    benefits 157
chemical bath deposition (CBD) 80–81
chemical potential 6
Chemical Reactor Engineering Center (CREC) 207, 213
chemical vapor deposition (CVD) 78–80
chemical vapor synthesis (CVS) 78
chemisorption 17
chloride process 45–46
chromium 13
churches 151
cigarette smoke 160, 178, 195
Cit de la Musique et des Beaux Arts 151
colloids 155–156
    see also sol-gels
Comefresh air purifiers 215–216, *229*
concrete *see* cementitious materials
conduction band 2
contact angle 101–102, *103*, 249, *251*
corrosion protection 106–107
cotton 178
cyanide degradation 200–206

Daikin air purifiers 215–216, *218*
de Broglie relation 2
defects 20
degradation 20–22
    standard tests 243–249
Deutsche Steinzeug Cremer 156–157

dichloroindophenol test 249, *250*
dip-coating 60–64, *61*
Dirac constant 3
direct band gap semiconductor 3
direct current magnetron sputtering 75
diuron 186, *188*
Dives in Misericordia 151–152
doping 6–8, *7*, 15–16, *16*, 124–126, 180

EcoLife homes 165–166
Ecopittura/Ecopaint 156
effective mass 3
efficiency parameters
    photochemical thermodynamic efficiency factor 240–241
    quantum yield 236–238
electric energy per mass (EEM) 241–242
electric energy per order (EEO) 242–243
electron beam vapor deposition 72
electron energy states 1
electronic band theory *see* band theory
electrophoresis 83–84
Eley-Rideal mechanism 21
energetic efficiency of degradation (EED) 243
energy bands *see* band theory
Erlus Lotus 162, *163*
Escherichia coli 107–108
ESSROC cement 154
ethylbenzene 252
Ever Fine Coat 193–195

Fermi-Dirac statistics 4
figures of merit *see* efficiency parameters
films *see* thin films
Fioravanti Hidra 130
flame pyrolysis 46
flat band potential 9
fluorescent lamps 31–32
    inbuilt photocatalyst 231–232

fogging 117–118
food packaging 113–114
formaldehyde 244
Freundlich isotherm 17, 25–26
Fujishima, Akira 156

Genesis air purifier 217, *223*
Gibbs free energy 5–6
glass
   as aggregate 164
   applications 117–118, 130, 131–140
   barrier films 120–121
   consumption rates 131
   impurities 119–121
   optical properties 122–123
   performance characterization 245–249
   photocatalytic activity 123–124
   polyethylene glycol 124
   soda-lime 118, **119**
   specifications 122–124
Global Engineering 156
gold nanoparticles 180

HEGE Solar 102–105
Hongsheng Ceramics 162
hydrolytic preparation
   nanostructures 49–57
   on polymers 170
hydrophilicity *see* superhydrophilicity
Hydrotect 117, 156
hydrothermal synthesis 49–57
hydroxyl radicals 29, 240

ilmenite 45
impurities 7
   glass 119–121
   see also doping
incandescent lamps 32
indirect band gap semiconductor 3–4
International Organization for Standardization (ISO) 235
iron 126–127
irradiation sources *see* radiation sources

ISO 22197–1:2007 255–256, 257
isotherms of adsorption 17
   Freundlich 25–26
   Langmuir 23–25
   Redlich-Peterson 26–28
Italcementi 147, 148, 152

Japan Synthetic Textile Inspection Institute Foundation 195

keratin 178, 180–182
kinetics 16–23
   adsorption 16–17
   Langmuir-Hinshelwood model 21–22
   light intensity 28–30
Kon Corporation 154
Kuraray Trading Company *194*, 195

lamps 31–35, 208
Landau-Levich equation 60–61
Langmuir isotherm 17, 23–25
Langmuir-Hinshelwood mechanism 21
laser pyrolysis 58
lasers 32–33
Legionella bacteria 210, *212*
light absorption 15–16
light transmittance 122–123
light-emitting diodes (LED) 33, *34*
lipid peroxidation 107–108
liquid-solid systems 22–23
low-pressure metal-organic chemical vapor deposition (LPMO CVD) 79–80
Luch air purifiers 220–221, *225*
lyocell fibers 183–184, **185**

magnetron sputtering 73–75, 172–173
malondialdehyde (MDA) 107
Matsushita Denso building *132*
mechanical properties
   lyocell fibers **185**
   polymers 174–175
Medicare Air Purifier 231
mercury arc lamps 31

metals 187–192
Methylene Blue 180, 244, 246–248
Meyerhofer equation 65
microspheres 54
microwave-assisted synthesis 58–59
Millennium Inorganic Chemical 156
Mitsubishi Materials Corporation 146–147, **148**
MOLZA Corporation 196
multilayer films 102, 128–130, 175–176

nanocrystals 46–47
nanoparticles 49–50
 deposition on metals 188–189
nanostructures 49–57
 fibers 182–183
 tubes 55–56
 wires 56–57
Neat glass 140, *141*
Nippon Paper 219
Nippon Soda Corporation *191*, 192
nitrogen oxides 144, 255
noble metals 13
Novosibirsk Institute of Catalysis 215
Noxer paving blocks *145*, 146

operating rooms 112
optical properties 15–16, 122–123, 127–128
organic compounds *see* volatile organic compounds
oxidation deposition techniques 81–83
Oxygena tiles 160–161, *161*

PanaHome 165–166
paper 184–187, *194*, 195, 195–196
Parylene 172
patents **148–149**
paving blocks 255–258
 see also cementitious materials
performance characterization, cements 251–255

pesticide degradation 199–200, *202*, *203*
phenol 48–49
Photo-CREC-Water reactors 207–209
photoadsorption 17–18
photocatalytic process 11–16
 advantages, vii
 kinetics 16–23
 light dependence 28–30
 photonic efficiency 237–238
 quantum yield 236–238
 superhydrophilicity and 100–102
photochemical thermodynamic efficiency factor (PTEF) 240–241
photoexcitation 10–11, 20
photon absorption rate 236–237
 determination 238–240
photonic efficiency 237–238
photoprotection 105–106
photovoltaics 102–105
physical vapor deposition 70–78, 119
Pilkington Activ 131–135, *132*, *133*, **136**
Pittsburgh Plate Glass Company 135–137
plasma deposition 75–76
plasma spray drying 68
polycarbonates 170–171, 172–173
poly(dimethylsiloxane-b-etherimide) 183
polyethylene 175
polyethylene oxide 183
polyhedral oligosilsesquioxane (POSS) 182–183
polymers
 antimicrobial 113–114
 applications 192–196
 inert 172
 mechanical properties 174–175
 nanofibers 182–183
 nanostructured coatings 177–178
 spray-coating 168–169
poly(methyl methacrylate) 170–171, 183

polymorphic forms 15–16, **42**, *44*
  see also anatase; brookite; rutile
polytetrafluoroethylene (PTFE) 192
poly(vinyl chloride) 175–176, *193*
powders
  preparation
    chloride process 45–46
    sulfate process 45
  preparation techniques 41, 43–49
  solvothermal preparation 57
  suspensions 12–13
preparation
  films 59–70
  laser pyrolysis 58
  microwave-assisted 58–59
  nanostructures 49–57
  powders 41, 43–49
  thin films 59–70
    wet coating 59–70
propanol *63*
pulsed laser deposition 76–78
pumice stone 164
purification devices *see* air
  purification; water purification
pyrolysis
  flame 46
  laser 58
  spray 69–70

radiation sources 30–35
  intensity 28–30
Raman spectroscopy *53*
reaction rates 28–30
Redlich-Peterson isotherm 26–28
redox systems 8–11
reflectance 127
refractive index 127
resazurin ink test 248–249
Rhodamine B 173–174, *175*
rutile 15, 43, *44*

Saint Gobain Glass 137–139
self-cleaning surfaces 101–102,
  117–118, 130
semiconductors
  band properties 2–11
  doping 6–8, *7*
  photoadsorption 19–20
  surface properties 4–6
silica *7*, 101, 120–121, 128
silicon nitride 121–122
silver 125–126
size-press treatment 184–185, *186*
slurry reactors 202–203
soda-lime glass 118, **119**
sodium ions 119–120
sol-gel methods 46–49, 57–58, 62,
  103
  on concrete 154–155
  on glass 119–120, 128–129
  on metals 187–188
  on polymers 177–178
  on textiles 180
solar energy production 102–105
solar radiation 34–35
SOLARDETOX 199
solvothermal synthesis 57
space-charge layer 19
spheroplasts 108
spin coating 63–67
spray coating
  concretes 154
  film quality 170
  polymers 168–169
spray pyrolysis 69–70
sputter deposition 72–73
stainless steel 69, 109, *111*, 189–190
standards
  cements 251–255
  film performance 244–249
  paving blocks 255–258
  photochemical thermodynamic
    efficiency factor
    (PTEF) 240–241
  quantum yield 236–238
  water purification 241–243
standards bodies 235
Staphylococcus aureus 109, *111*
stearic acid test 246
sterilization 109–110
sulfate process 45
Sunclean glass 135–137, *138*

Subject Index

superhydrophilicity 18–19, 98–100, *103*, 123–124
  films on metals 190–191
  Pilkington Activ *132–133*
  test 249
surface energy 5
surface tension 5–6
surfactants 57–58, 189–190

Taiyo Kogyo Corporation 192
tarpaulin *191*, 192
Tedlar 172
Temkin isotherm 17
textiles 178–184, *194*
  applications 195
  cotton 178–180
thermal oxidation 81–82
thermodynamic efficiency 240–241
thermoplastics *see* polycarbonates
thin films
  composite 67, 102
  deposition 59–70
    cathode arc 75–76
    chemical bath 80–81
    chemical vapor 79–81
    electrophoretic 83–84
    on glass 119–120
    optical properties and 127–128
    oxidative 81–83
    physical vapor 70–78, 119
    pulsed laser 76–78
    spray pyrolysis 69
    sputter 72–73
  disadvantages 42
  magnetron sputtering 73–75, 172–173
  nanoparticle deposition 188–189
  optical properties and 127–128
  performance
    characterisation 244–249
  plasma spray drying 68
  on polymers 172–173
tiles *see* ceramic tiles
TitanClean reflector 228

titanium isopropoxide 49–50
titanium tetrachloride 49–51
toluene 252
TOP-NANO-HEGE coating 103
Toto Ltd. 156
TPX 155–156
transmittance (optical) 122–123
trichloroethylene 244
tungsten-halogen lamp 32
TX Active cement 151–152, *152*

UBE group 209–210
ultraviolet irradiation 32, 33
urine 160

vacuum evaporation-deposition 70–72
valence band model 1–2
volatile organic compounds (VOC) 123–124, 243–244, 252–254, 255

water droplets 101–102, *103*, 249, *251*
  see also superhydrophilicity
water purification
  CREC reactors 207–209
  cyanide degradation 200–206
  fiber reactors 209–210
  pesticide degradation 199–200
  standards 241–243
wave number 3
wet coating technologies 59–60
  dip-coating 60–64
wet-end addition 184, 186
windshields 130

X-ray diffraction analysis 51–52
o-xylene 252

YKK Corporation 195, *196*

ZAND-AIR 222–224, *227*
zirconium dioxide 128, *129*